Gems and the New Science

synthesis

A series in the history of chemistry, broadly construed,
edited by Carin Berkowitz, Angela N. H. Creager, John E. Lesch,
Lawrence M. Principe, Alan Rocke, and E. C. Spary, in
partnership with the Science History Institute

Gems and the New Science

Matter and Value in the Scientific Revolution

MICHAEL BYCROFT

The University of Chicago Press
Chicago and London

The University of Chicago Press, Chicago 60637
The University of Chicago Press, Ltd., London
Published 2026
Printed in the United States of America

35 34 33 32 31 30 29 28 27 26   1 2 3 4 5

ISBN-13: 978-0-226-64460-8 (cloth)
ISBN-13: 978-0-226-82598-4 (ebook)
DOI: https://doi.org/10.7208/chicago/9780226825984.001.0001

Library of Congress Cataloging-in-Publication Data

Names: Bycroft, Michael author
Title: Gems and the new science : matter and value in the Scientific Revolution /
    Michael Bycroft.
Other titles: Synthesis (University of Chicago. Press)
Description: Chicago ; London : The University of Chicago Press, 2026. |
    Series: Synthesis | Includes bibliographical references and index.
Identifiers: LCCN 2025030741 | ISBN 9780226644608 cloth |
    ISBN 9780226825984 ebook
Subjects: LCSH: Gemology—History | Gems—Research—Europe—History |
    Gems—Classification—History | Science—Europe—History—16th century |
    Science—Europe—History—17th century | Science—Europe—History—
    18th century | Science—Europe—History—19th century
Classification: LCC QE392 .B95 2026
LC record available at https://lccn.loc.gov/2025030741

♾ This paper meets the requirements of ANSI/NISO Z39.48-1992 (Permanence of Paper).

Authorized Representative for EU General Product Safety Regulation (GPSR)
queries: **Easy Access System Europe**—Mustamäe tee 50, 10621 Tallinn, Estonia,
gpsr.requests@easproject.com
Any other queries: https://press.uchicago.edu/press/contact.html

Contents

Color illustrations follow page 166.

Abbreviations

A S   Archives of the Académie Royale des sciences, Institut de France (Paris)

A N   Archives Nationales (Paris)

D S B   *Complete Dictionary of Scientific Biography*, ed. Charles Coulston Gillispie, Frederic Lawrence Holmes, and Noretta Koertge (Detroit, MI: Charles Scribner's Sons, 2008), from Gale Virtual Reference Library.

H A S   *Histoire de l'Académie Royale des science*

M #   The #th of Charles-François de Cisternay Dufay's eight articles on electricity

M A S   *Mémoires de l'Académie Royale des sciences*

P T   *Philosophical Transactions of the Royal Society of London*

P V   *Procès-verbaux of the Académie Royale des sciences*

Works   *The Works of Robert Boyle*, ed. Michael Hunter and Edward Davis (London: Pickering and Chatto, 1999).

Note on Terminology

Terms for gemstones have varied widely across time, space, and languages. Where relevant for the argument of this book, I have made some effort to identify the modern-day equivalents of historical terms. For example, the stone that Albert the Great called "hyacinthus" bears a close resemblance to gem-grade corundum. For the most part, however, I have reproduced the terms used by the authors under study. Thus I write "adamas" when discussing Pliny the Elder, rather than listing the various modern names for the stones that Pliny included in that category. Where a convenient translation exists, terms in languages other than English are translated literally; otherwise, they are left untranslated. Thus the French "rubis" becomes "ruby," but the French "smaragdoprase" is not translated.

Introduction

Strictly speaking, gems do not exist. There is no category in systematic mineralogy that corresponds to the colloquial concept of a gem, gemstone, or precious stone. A reader of the *Strunz Mineralogical Tables*, a standard work of mineral classification, will search in vain for a class made up of diamond, ruby, sapphire, garnet, agate, tourmaline, and the handful of other minerals that are displayed in jewelers' shops and studied in gemology courses. Instead, we find a list of ten "primary groups," most of which are baffling to the nonspecialist: "sulfosalts," "halides," "vanadates," and so on. The minerals we know as gemstones are scattered through these groups in an equally bewildering manner. Diamond is classed as an "element," along with gold, tin, arsenic, and many other non-gems. Ruby is in a different group altogether, the "oxides." This red stone does not even qualify as a mineral species—it is merely the red variety of the species known as "corundum" (plate 1).[1] Garnet looks like ruby, but it is in the "silicate" group, a very long way from ruby. Hyacinth is also in the silicate group, but this has nothing to do with its color. "Hyacinth" is a colloquial name for zircon, a species that comes in a range colors, some of which are very different from the color of garnet (plate 5). The list goes on. "Gem" is not a scientific category, however significant it may be in commerce, industry, and popular culture.

It was not always so. Until about 1800, gems were a matter of fundamental importance in the study of nature in Europe. They were seen as a fully fledged class of mineral, as robust and respectable as simple oxides are for today's mineralogists. They were at the center of the dramatic changes in natural

---

1. Hugo Strunz and Ernest Nickel, *Strunz Mineralogical Tables: Chemical-Structural Mineral Classification System*, 9th ed. (Stuttgart: Schweizerbart, 2001 [1941]), 17–27, 52, 192, 540, 839.

knowledge that happened in the three centuries before 1800. They were the plaything of kings, queens, and aristocrats, precisely the sorts of people who patronized the sciences in early modern Europe. Gems were an obvious choice for naturalists—including natural historians, natural philosophers, experimental philosophers, chemists, and physicians—who were eager to show the utility of their new approaches to the study of nature. Many of the big names of European science were involved: Galileo Galilei invented a device for measuring the density of gems; Isaac Newton speculated about their composition; Antoine-Laurent Lavoisier turned diamonds into gas. Gems were rubbed, tasted, measured, split, dissolved, collected, classified, and delicately caressed by a range of lesser-known figures. Their findings were significant, not just for mineralogy but also for the sciences we now know as crystallography, chemistry, geology, experimental physics, and even cartography. To a surprising extent, the story of gems is the story of the new science.

One aim of this book is to tell this story. The other aim is to bring the story to bear on the history of science more broadly. Much research in this field in the last few decades may be summed up in the word "materialist." The aim has been to show that the material world mattered in past science. Material substances such as gold, material things such as fossils, material sites such as laboratories, and material practices such as mining and printing—all have been folded into the history of ideas about the natural world. This is a valuable development but an incomplete one. I hope to extend the materialist argument in two ways. One is to explore the relationship between different parts of the material world—between gold and gems, for example, or between collections and laboratories, or between mining and glassmaking. The name I shall use for this kind of exploration is "transmaterialism." The other novelty is to give as much attention to material evaluation as we have already done to material production. The practical role of science was (and is) not only to *make* things but also to determine the goodness of things. The point of the science of gold, for example, was not just to make better gold, or to make it more efficiently, but to distinguish between gold of different qualities and to distinguish real gold from fake gold.

These two themes—transmaterialism and material evaluation—help to account for the science of gems in early modern Europe. This book is not only about gems, however. It is also about gold, earths, drugs, mineral waters, glass, porcelain, and other material substances that fed into the history of gems. The book is also about the wider fields of study in which gems were embedded. Each chapter deals with one such field at a formative period in its history: natural history in the sixteenth century; technical writing and experimental philosophy in the seventeenth century; experimental physics, mineralogy, and chemistry in the eighteenth century; and applied science early in the nineteenth century. More ambitiously, evaluation and transmaterialism shed light on the event sometimes known as the Scientific Revolution. They illuminate five topics that are an awkward fit for the traditional narrative of the event: natural history, the eighteenth century, merchants and artisans, European expansion, and geographical mobility. This is not an argument for or against the existence of the Scientific Revolution. It is an argument for using an expanded form of materialism to make sense of the long-term evolution of science.

These are big claims. To make them more precise and more plausible, I shall now expand upon the key terms in the title of the book: "gems," "matter," "value," "the Scientific Revolution," and "new science," in that order.

Gems

What was the science of gems about? For Anselmus Boethius de Boodt, a physician at the Prague court of the Holy Roman Emperor Rudolf II, and the author of an influential book on the topic published in 1609, gemstones were small, rare, hard, and beautiful. This meant that diamond, opal, rock crystal, agate, and jasper were all gems—or *gemmae*, to use the Latin term that Boodt favored and that his French and English translators rendered as "pierres précieuses" and "precious stones." The boundaries of Boodt's category were clear in principle but fuzzy in practice. Marble, granite, and alabaster were obviously excluded, on the grounds of size, plainness, and softness, respectively. But toadstone was a borderline case, since it was small, rare, hard, and valuable, but not beautiful. Another complication is that Boodt picked out some gems as being particularly precious: diamond, ruby, pearl, and sapphire, in that order.[2] Boodt's list of gems was characteristic of the whole early modern period, the main variation being the gradual whittling down of the category, so that by the end of the eighteenth century the phrase "precious stones" was reserved for about ten of the most valuable gems.

---

2. Anselmus Boethius de Boodt, *Gemmarum et lapidum historia* (Hanover, 1609), "Ad lectorum" (gemmae defined implicitly), 2–3 (gemmae defined explicitly), 95 (the most precious stones), 134 (turquoise), 138 (lapis lazuli), 152 (toadstone), 153 (coral; setting as criterion), 161 (amber). Several stones are excluded from "gemmae" in the introduction and then described as "gemmae" in the main text, e.g., rock crystal on 2–3 and 108. "Pierres précieuses" in Anselmus Boethius de Boodt, *Le parfait joaillier ou Histoire des pierreries*, ed. Adrian Toll, trans. Jean Bachou (Lyon, 1644 [1609]), 4–5. "Precious stones" in the title of Thomas Nicols, *Lapidary; or, The History of Pretious Stones, with Cautions for the Undeceiving of All Those That Deal with Pretious Stones* (Cambridge, 1652).

Mathurin-Jacques Brisson, a French naturalist writing in 1787, gave a list of twelve stones he called "pierres précieuses": diamond, ruby, sapphire, girasol, Ceylon jargon, hyacinth, vermilion, garnet, emerald, chrysolite, aquamarine, and beryl.[3]

There is little in these definitions that would startle a modern gemologist. Diamond is still the king of gems in the West. Diamonds, rubies, emeralds, and a handful of other species are still distinguished from an array of cheaper stones, a distinction now marked in everyday language by the terms "precious stone" and "semiprecious stone." Gems are still defined by what can be found in a jeweler's shop. The strangeness of Boodt and Brisson does not lie in the content of these categories but in their ontological status. For Brisson, gems were every bit as real as copper, metals, and volcanic rocks, to mention three other categories in the "methodical table of the mineral kingdom" that prefaced his 1787 book. For him, there was no sense in which gems were artificial, conventional, or subjective. Nor was Brisson eccentric or outdated. He borrowed his classification of stones from Louis-Jean-Marie Daubenton, the main teacher of mineralogy at the leading institution of natural history in France at the time, the Jardin du Roi.[4] Chemists and crystallographers agreed that gems formed a natural kind, even if they disagreed about the definition of this kind.[5] This attitude changed only in the last decade of the century. In 1808, the term "pierres précieuses" was still used in commerce but had been "rejected in mineralogy as vague and misleading," in the words of an encyclopedia published in that year.[6] Before the 1790s, gems were as natural as gold is today. They were also intensely studied. Stones and metals were the most-studied category of mineral in the eighteenth century, to judge from the number of pages dedicated to them in books on mineralogy in the period. By the same measure, gems were the most-studied category of stone.[7] Gems had

3. Mathurin-Jacques Brisson, *Pesanteur spécifique des corps: Ouvrage utile à l'histoire naturelle, à la physique, aux arts et au commerce* (Paris, 1787), xx–xxi.

4. Ibid., vi, 59.

5. Examples are Torbern Bergman, *Opuscules chimiques et physiques*, trans. Louis-Bernard Guyton de Morveau and Claudine Picardet (Dijon, 1780–1785 [1779–1780]), vol. 2, 108–11; and Jean-Baptiste Louis de Romé de l'Isle, *Cristallographie, ou Description des formes propres à tous les corps du règne minéral* (Paris, 1783), vol. 2, 170.

6. Antoine-François Fourcroy, *Encyclopédie méthodique: Chymie, pharmacie et métallurgie*, vol. 5 (Paris, 1808), 618–19. For the recognition of precious stones as a category in earlier French encyclopedias, see Michael Bycroft, "Dossier critique de l'article PIERRES PRÉCIEUSES" (*Encyclopédie*, t. XII, p. 593b–595a), *Édition numérique collaborative et critique de l'Encyclopédie*, https://enccre.academie-sciences.fr/encyclopedie/article/v12-1449-47, accessed June 26, 2020.

7. See chap. 5 for details.

roughly the same position in the hierarchy of the earth sciences as fossils and crystals do today.

If the end of the early modern period marked the end of gems, the start of that period marked the start of minerals. The notion of the mineral kingdom has roots in the ancient world, but it was only in the sixteenth century that the concept became part of a coherent literary tradition. For most of the Middle Ages, writings on gems took the form of lapidaries, which were books about stones, not about minerals in general. In the Renaissance this genre merged with texts on metals, earths, and salts to produce a series of treatises that dealt with the whole mineral kingdom, and only the mineral kingdom. This is not to say that the lapidary disappeared overnight, or even that it disappeared at all. Indeed, one theme of this book is the remarkable persistence of the lapidary, both as a distinct literary genre and as a semi-autonomous component of natural histories of minerals. The most celebrated example of a new natural history of minerals, written by the German physician Georg Agricola and published in 1546, contained much that was copied verbatim from an ancient lapidary. The year 1808 saw the publication, not only of the skeptical encyclopedia article mentioned earlier, but also of a *Treatise on Precious Stones* that was an ancestor of the manuals on gems and gemology that remain popular today. In early modern Europe, an ancient tradition concerned with gems coexisted with a new tradition concerned with minerals.[8]

The peculiar identity of gems—not quite real, not quite mineralogy—explains the indifference of historians of science to these attractive objects. Minerals, fossils, and crystals are all the subject of scholarly monographs that cover the whole early modern period in a systematic, interpretative fashion.[9] There are no such works on gems. Relevant material can be found in a range of other places: histories of the earth sciences; histories of physics and chemistry; histories of jewelry and the gem trade; encyclopedic histories of

8. On Agricola, the shift from lapidaries to natural history, and Cyprien-Prosper Brard's *Traité des pierres précieuses, des porphyres, granits, marbres, albâtres, et autres roches, propres à recevoir le poli et à orner les monumens publics et les édifices particuliers* (Paris, 1808), see chaps. 1 and 7, this volume.

9. Hélène Metzger, *La genèse de la science des cristaux* (Paris: Albert Blanchard, 1918); Frank Adams, *The Birth and Development of the Geological Sciences* (Baltimore, MD: Williams & Wilkins, 1938); John G. Burke, *Origins of the Science of Crystals* (Berkeley: University of California Press, 1966); Martin J. Rudwick, *The Meaning of Fossils: Episodes in the History of Palaeontology*, 2nd ed. (Chicago: University of Chicago Press, 1985 [1972]); Rachel Laudan, *From Mineralogy to Geology: The Foundations of a Science, 1650–1830* (Chicago: University of Chicago Press, 1987); David Oldroyd, *Sciences of the Earth: Studies in the History of Mineralogy and Geology* (Brookfield, VT: Ashgate, 1998).

one or other species of gem; gemology textbooks; and the remarkable bibliography of gemology compiled by the cutter, collector, and naval pilot John Sinkankas.[10] A few pioneering historians have published studies on this or that aspect of early gem science. I am especially indebted to work by Derek J. Content,[11] Sven Dupré,[12] François Farges,[13] Joshua Hillman,[14] Stephen Irish,[15]

10. On electricity, luminescence, and chemistry, I have found these surveys especially useful: John L. Heilbron, *Electricity in the 17th and 18th Centuries: A Study of Early Modern Physics* (Berkeley: University of California Press, 1979); Edmund Harvey, *A History of Luminescence from the Earliest Times Until 1900* (Philadelphia: American Philosophical Society, 1957); and James R. Partington, *A History of Chemistry* (London: Macmillan, 1960–1964), vols. 1–3. Encyclopedic histories include George Frederick Kunz and Charles Hugh Stevenson, *The Book of the Pearl: Its History, Art, Science and Industry* (Courier Dover Publications, 2002 [1908]); and Robert A. Donkin, *Beyond Price: Pearls and Pearl-Fishing, Origins to the Age of Discoveries* (Philadelphia: American Philosophical Society, 1998). There is a vast literature on the history of diamonds, mostly on trade, mining, and jewelry. Two foundational works are Godehard Lenzen, *The History of Diamond Production and the Diamond Trade* (London: Barrie & Jenkins, 1970); and Herbert Tillander, *Diamond Cuts in Historic Jewelry: 1381–1910* (London: Art Books, 1995). Recent contributions include Marcia Pointon, *Rocks, Ice and Dirty Stones: Histories of Diamonds* (London: Reaktion Books, 2017); Jack Ogden, *Diamonds: An Early History of the King of Gems* (New Haven, CT: Yale University Press, 2018); Tijl Vanneste, *Blood, Sweat and Earth: The Struggle for Control over the World's Diamonds Throughout History* (London: Reaktion Books, 2021). For more on the historiography of early modern gems, see the works cited in Michael Bycroft and Sven Dupré, "Introduction," in *Gems in the Early Modern World: Materials, Knowledge and Global Trade, 1450–1800*, ed. Michael Bycroft and Sven Dupré (London: Palgrave Macmillan, 2019), 1–32. See also John Sinkankas, *Gemology: An Annotated Bibliography* (Metuchen, NJ: Scarecrow Press, 1993). Equally compendious works on minerals in general, not just gems, are Wendell E. Wilson, *The History of Mineral Collecting, 1530–1799* (Tucson, AZ: The Mineralogical Record, 1994); Curtis P. Schuh, *Mineralogy and Crystallography: An Annotated Biobibliography of Books Published 1469 Through 1919* (Tucson, AZ: pub. by author, 2007); and idem, *Mineralogy and Crystallography: On the History of These Sciences from Beginnings Through 1919* (Tucson, AZ: pub. by author, 2007). On the modern history of gemology, see Peter G. Read, *Gemmology*, 3rd ed. (London: Elsevier, 2005 [1991]), 1–9.

11. Derek J. Content, *Ruby, Sapphire & Spinel: An Archaeological, Textural, and Cultural Study* (Turnhout: Brepols, 2016), vol. 1, chaps. 1 and 2.

12. Sven Dupré, "The Art of Glassmaking and the Nature of Stones: The Role of Imitation in Anselm de Boodt's Classification of Stones," in *Steinformen: Materialität, Qualität, Imitation*, ed. Isabella Augart, Maurice Saß, and Iris Wenderholm (Berlin: De Gruyter, 2019), 207–20.

13. François Farges, *Gems* (Paris: Flammarion, Muséum national d'histoire naturelle, van Cleef et Arpels, 2020); François Farges and Johan Kjellman, "Bicentenaire du décès de René-Just Haüy: Les dernières découvertes au Muséum national d'histoire naturelle," *Le règne minéral* 165 (2022): 7–42.

14. Joshua Hillman, "Invisible Labour in the Woodwardian Collection," *Museum & Society* 22, no. 2–3 (2024): 43–58, 50–53.

15. Stephen T. Irish, "The Corundum Stone and Crystallographic Chemistry," *Ambix* 64, no. 4 (2017): 301–25.

Arash Khazeni,[16] Christine Lehman,[17] Annibale Mottana,[18] and Claire Sabel.[19] But there are no synthetic works on gems as a category of scientific study in early modern Europe.

The demise of the category around 1800 helps to explain this gap. Gems never became the subject of a discipline in science faculties in Western universities. As a result, they never became the subject of discipline-oriented histories of science in the West. Gemology is a flourishing field today, but there are far fewer gemology departments in universities than there are chemistry departments and earth science departments.[20] It is therefore not surprising that we have many book-length histories of chemistry and the earth sciences but no book-length histories of the science of gems. The notion that gems are a prescientific category may also explain why there is a large literature on medieval lapidaries and their Renaissance successors. These topics are usually

16. Arash Khazeni, *Sky Blue Stone: The Turquoise Trade in World History* (Berkeley: University of California Press, 2014).

17. Christine Lehman, "What Is the 'True' Nature of Diamond?," *Nuncius* 31 (2016): 361–407.

18. Annibale Mottana, "Galileo as Gemmologist: The First Attempt in Europe at Scientifically Testing Gemstones," *Journal of Gemmology* 34, no. 1 (2014): 24–31; idem, "Italian Gemology During the Renaissance: A Step Towards Modern Mineralogy," in *The Origins of Geology in Italy*, ed. Gian Battista Vai and W. G. E. Caldwell (Boulder, CO: Geological Society of America, 2006), 1–21.

19. Claire Conklin Sabel, "Rare Earth: Gemstones, Geohistory, and Commercial Geography, c. 1600–1750" (PhD diss., University of Pennsylvania, 2024), ProQuest (31485274), with author's permission. The publication of this wide-ranging thesis is eagerly awaited. In the meantime, see idem, "The Impact of European Trade with Southeast Asia on the Mineralogical Studies of Robert Boyle," in Bycroft and Dupré, *Gems in the Early Modern World*, 87–116; idem, "'Glass Worke': Precious Minerals and the Archives of the Early Modern Early Sciences," in *New Earth Histories: Geo-Cosmologies and the Making of the Modern World*, ed. Alison Bashford, Emily M. Kern, and Adam Bobbette (Chicago: University of Chicago Press, 2023), 145–62.

20. In the UK, the only university-based gemology qualification of which I am aware is the BSc in Gemmology and Jewellery Studies at Birmingham City University. On the history of gemological institutes, see Read, *Gemmology*, 2–3; and François Farges and Olivier Segura, *Pierres précieuses: Guide visual* (Paris: Dunod, 2023), 66. The interwar period was key for the formation of the British, German, and US gemological institutes, to judge from the short histories on their respective websites: https://gem-a.com/about/history, https://www.dgemg.com/en/we-about-us.html, https://www.gia.edu/gia-about. In the UK, the National Association of Goldsmiths began issuing a gemological qualification in 1908, though the Gemmological Association was founded in 1931. The French association was created in 1962, though the Laboratoire français de gemmologie had been around since 1929: https://ingemmologie.com/linstitut/lhistoire-de-ling/. Websites accessed Oct. 20, 2021. For another angle, see James Evans, "A History of Gemmology, 1912–1972," *Gemmology Bulletin* 1 (2024): 1–14.

treated under such headings as "the lore of gems" and "magical jewels," suggesting that they were ancestors of science rather than the thing itself.[21]

Looking beyond Europe, there are substantial studies of "the Arab roots of gemology," to borrow a phrase from Samar Najm Abul Huda.[22] The measurements of gem density by Abū Rayḥān al-Bīrūnī (973–ca. 1050) and Abū'l-Fatḥ al-Khāzini (fl. 1115–1130) are well known to historians of science.[23] Aḥmad al Tīfāshī's "Best Thoughts on the Best of Stones," written in about 1240, is available to the Anglophone world thanks to Huda's translation and notes. These works were part of a long tradition of Arabic and Persian texts on the natural history of precious stones, a tradition that stretched from at least the eighth century (AD) to the nineteenth century.[24] European scholars have known about this tradition, and admired the measurements of al-Bīrūnī and al-Khāzini, since the middle of the nineteenth century.[25] But this has rarely led to an interest in the corresponding developments in early modern Europe. It may be that the perception of gems as inherently unscientific, combined with the notion that science is inherently Western, has discouraged the idea that there was such a thing as gem science in Europe. In any case, it is only very recently that scholars have begun to join up the Asian and European histories of gem science. Sabel's studies of gem knowledge in the Indian Ocean world are one example; Arash Khazeni's monograph on the natural history of turquoise across Eurasia is another.[26] The geographical and cultural breadth of these

21. Examples are Joan Evans, *Magical Jewels of the Middle Ages and the Renaissance, Particularly in England* (Oxford: Clarendon Press, 1922); and Lynn Thorndike, *A History of Magic and Experimental Science*, vol. 6 (New York: Macmillan, 1941), chap. 39 ("lore of gems").

22. Aḥmad ibn Yūsuf al Tīfāshī, *Arab Roots of Gemology: Ahmad ibn Yusuf al Tifaschi's Best Thoughts on the Best of Stones*, ed. and trans. Samar Najm Abul Huda (London: Scarecrow Press, 1998).

23. The standard English translations are Abū Rayḥān al-Bīrūnī, *Al-Beruni's Book on Mineralogy: The Book Most Comprehensive in Knowledge on Precious Stones*, ed. and trans. Hakim Mohammad Said (Islamabad: Pakistan Hijara Council, 1989 [ca. 1050]); and Nikolaï Vladimirovich Khanykov, "Analysis and Extracts of كتاب ميزان الحكمة: Book of the Balance of Wisdom, an Arabic Work on the Water-Balance, Written by 'Al-Khâzinî in the Twelfth Century," *Journal of the American Oriental Society* 6 (1858): 1–128. The latter translation is incomplete. For details, see the articles on al-Bīrūnī and al-Khāzini in the *Dictionary of Scientific Biography* and the *Encyclopaedia of the History of Science, Technology, and Medicine in Non-Western Cultures*, 3rd ed., ed. Helaine Selin (Dordrecht: Springer, 2016).

24. For overviews, see Khazeni, *Sky Blue Stone*, 22–26; Manfred Ullmann, *Die Natur- und Geheimwissenschaften im Islam* (Leiden: E. J. Brill, 1972), 114–44.

25. Jean-Jacques Clément-Mullet, *Recherches sur l'histoire naturelle et la physique chez les Arabes* (Paris, 1858); Khanykov, "Book of the Balance of Wisdom."

26. Sabel, "Impact of European Trade"; idem, "Rare Earth"; Khazeni, *Sky Blue Stone*, esp. chap. 4; see also Content, *Ruby, Sapphire & Spinel*, vol. 1, chaps. 1 and 7.

studies is remarkable, and I do not claim to match it in this book. However, I do hope that a thorough study of the European case will help to correct the traditional picture of an "Eastern" science that centers gems and a "Western" science that leaves them out.

Gems were central to early European science, but they were also in a state of flux. Much of this flux can be summed up in a phrase: the rise and fall of the transparency-color-locality (TCL) scheme. Transparency, color, and locality had always been used to some extent for classifying stones, but the scheme was formalized by Boodt and other Renaissance naturalists.[27] Boodt grouped all gems into three classes: transparent, semitransparent, and opaque. The transparent ones were grouped by color: red, blue, green, and yellow. Each species was subdivided by locality. For example, Boodt's *granatus* was a transparent, red stone that was divided into occidental and oriental varieties. Now fast-forward to 1801, when the French mineralogist and gem connoisseur René-Just Haüy published his *Treatise on Mineralogy*. Haüy's classification was based on chemical composition, crystal form, and a range of physical properties of which hardness, density, and double refraction (now known as birefringence) were the most important. For example, Haüy's *grenat* was one of the large class of "nonacidic earthy substances." Grenat differed from other species in this group by having crystals based on the rhomboidal dodecahedron and by having a density 3.6 to 4.1 times that of water. Haüy did list the colors of grenat, but he called these "accidental" properties, in contrast to "essential" properties such as density and crystal form. Locality and transparency, like color, were no longer what distinguished one kind of gem from another. In one sense, Boodt and Haüy were writing about the same stone, now known as "garnet." In another sense, granatus and grenat were radically different entities.[28] They stood for two different social and intellectual worlds. For Boodt, gems were part of nature; for Haüy they were not. For Boodt, the study of their nature and the study of their value were one and the same; for Haüy, these were two different projects, mineralogy and the applied science of gems.

27. For the sake of simplicity, I am silent in this book about changes in the meaning of individual gem properties, although a full history of gem classification would take these changes into account. Haüy's *rouge* did not have the same meaning as Boodt's *ruber*, for example. The same goes for hardness, locality, transparency, etc.

28. Pliny the Elder, *Natural History*, vol. 10, trans. David E. Eichholz (Cambridge, MA: Harvard University Press, 1989 [ca. AD 80]), bk. 37, 248–51 (sarda); Boodt, *Gemmarum*, 75; René-Just Haüy, *Traité de minéralogie*, vol. 2 (Paris, 1801), 403–6 (nonacidic earthy substances), 540 (essential properties), 544–49 (varieties), 549–51 (color, transparency, locality).

How did this happen? How did the TCL scheme become the crystal-chemistry-physics scheme? These are questions about scientific change, and to understand scientific change we need to think carefully about matter and value.

Matter

The obvious thing to say about the material world is that it is inseparable from the mental world. Take the discovery that diamonds are combustible, an event usually traced to Paris in the years around 1770. This was not only an event in the history of ideas but also in the history of experiment, instruments, and the crafts. The discovery was made by manipulating matter, not just by thinking things through. Diamonds were heated to high temperatures. This was done with elaborate instruments such as a large lens that focused the rays of the sun. Jewelers were involved, supplying diamonds and some of the equipment used to heat them. And the most dramatic experiments, with the burning lens, were done in the gardens of the Louvre, an ideal site for spectacular science. The discovery of the combustion of diamonds was a meeting of mind and matter.

But there is another way of looking at the same discovery. Diamonds were not the only materials involved in the discovery—porcelain and metallic ores also played a role. Jewelers were not the only artisans—porcelain makers were equally important. There were multiple instruments used to heat the diamonds, some drawn from the world of jewelry and others from the world of porcelain making; the decisive experiments combined the two worlds. There was a convergence of sites as well. Gem collections were as important to the discovery as the burning lens in the gardens of the Louvre. One might say, anachronistically but usefully, that the experiments combined aspects of the museum and of the laboratory. There was certainly a meeting of mind and matter in these experiments, but the meeting of matters was equally important. Indeed, it was the meeting of matters that made the meeting of mind and matter possible.

"Materialist" is a name for histories that focus on the material world. "Transmaterialist" is therefore a natural name for histories that focus on the relationship between different material worlds. The term has the advantage of suggesting an analogy to "transnational" histories, those that look at the movement of people and ideas across national borders. These movements change the borders; they also change the things that move across the borders.[29] Analogously, the transmaterialist sees the history of science as a series

---

29. A recent summary is Jürgen Osterhammel, "Global History," in *Debating New Approaches to History*, ed. Marek Tamm and Peter Burke (London: Bloomsbury, 2018), 28.

of consequential exchanges between different parts of the material world. The word "transmaterialist" has the further advantage of having the same breadth as its root word. A full account of materialism in the history of science would include material substances,[30] material things,[31] material sites,[32] the materiality of the natural environment,[33] and political and economic life in contradistinction to intellectual life.[34] In addition, there is an older tradition of understanding past science in terms of its diverse phenomena or subject matters, which include everything from atoms to asteroids.[35] I intend "transmaterialist" to cover all these different ways of thinking about the materiality of science. I shall be especially concerned with the interaction between gems and other material substances, between gem cutters and other artisans, between gem collections and other kinds of hardware, and between the various phenomena associated with gems, such as color, density, and magnetism.

My approach builds on the last two decades of research on the material culture of early modern science.[36] This literature has played up the role of the material world in scientific change. This has drawn attention to many

30. This is the sense of "materials" in the title of Ursula Klein and Emma Spary, eds., *Materials and Expertise in Early Modern Europe: Between Market and Laboratory* (Chicago: University of Chicago Press, 2010); and Lissa Roberts and Simon Werrett, eds., *Compound Histories: Materials, Governance and Production, 1760–1840* (Leiden: Brill, 2018).

31. These are the main subject of Anita Guerrini, "The Material Turn in the History of Life Science," *Literature Compass* 13, no. 7 (2016): 469–80.

32. Diarmid A. Finnegan, "The Spatial Turn: Geographical Approaches in the History of Science," *Journal of the History of Biology* 41 (2008): 374–78.

33. Mark D. Hersey and Jeremy Vetter, "Shared Ground: Between Environmental History and the History of Science," *History of Science* 57, no. 4 (2019): 403–40, esp. 427.

34. Harun Küçük, "Early Modern Ottoman Science: A New Materialist Framework," *Journal of Early Modern History* 21 (2017): 407–19, esp. 408 ("economic activity, political action, and science"). Note also the political and economic crises in Gianamar Giovannetti-Singh and Rory Kent, "Crises and the History of Science: A Materialist Rehabilitation," *British Journal for the History of Science* 9 (2024): 39–57.

35. For an attempt to revive this tradition, see Michael Bycroft, "A Neo-Positivist Theory of Scientific Change," *British Journal for the History of Science* 9 (2024): 129–48.

36. The following are some landmark works in this movement. Pamela Smith and Paula Findlen, eds., *Merchants and Marvels: Commerce and the Representation of Nature in Early Modern Europe* (Hoboken, NJ: Routledge, 2001); Lissa Roberts, Simon Schaffer, and Peter Dear, eds., *The Mindful Hand: Inquiry and Invention from the Late Renaissance to Early Industrialisation* (Amsterdam: Koninklijke Nederlandse Akademie van Wetenschappen, 2007); Klein and Spary, *Materials and Expertise*; Paula Findlen, ed., *Early Modern Things: Objects and Their Histories, 1500–1800* (London: Routledge, 2013); Pamela H. Smith, Amy R. W. Meyers, and Harold J. Cook, eds., *Ways of Making and Knowing: The Material Culture of Empirical Knowledge* (Ann Arbor, MI: University of Michigan Press, 2014); Roberts and Werrett, *Compound Histories*; Pamela Smith, *From Lived Experience to the Written Word: Reconstructing Practical Knowledge in the Early Modern*

topics that were once slighted by historians of science: the nuts and bolts of experimental research; the material properties of plants and animals; the role of useful materials such as coal and coffee; the role of commerce, consumption, and global trade; and the specific sites where science was done, from ships to kitchens to cathedrals. There is already an element of transmaterialism in this literature, though not under that name. Examples include Dániel Margócsy's observation that animals were taxonomized after plants; Pamela Smith's remarks on the sensitivity of artisans to differences in their materials; Deborah Harkness's study of the "complementary trades" involved in science in Elizabethan London; and studies by Paula Findlen and Mary Terrall of early modern overlaps between experiments and collections.[37]

These and other examples have shaped my own thinking about the history of gem science. But they can be taken much further than they have been so far. In the history of natural history, the three kingdoms (plants, animals, minerals) had very different histories. Even within one kingdom, such as the minerals, there is a range of chronologies and causal factors depending on whether one considers metals, stones, earths, salts, or other classes of mineral. In the history of chemistry, there is much to be gained by studying the relationship between different chemical crafts—between mining and medicine, for example, or between gold making and glassmaking. In the history of physics, there is a vast literature on instruments and a much smaller literature on the objects to which those instruments were applied. Each of these fields has benefited from the surge of interest in the material world of science; yet they would all benefit from more attention to the relationship between different material worlds. This is not to deny the importance of the mental side of science, its concepts, methods, and theories. After all, the present study is organized around one such theory, the TCL scheme. The point is that we are in a better position to make sense of this scheme when we study the meeting of matters, not just the meeting of mind and matter.

World (Chicago: University of Chicago Press, 2022). A useful survey is Paula Findlen, "Early Modern Things: Objects in Motion, 1500–1800," in Findlen, *Early Modern Things*, 3–28.

   37. Dániel Margócsy, *Commercial Visions: Science, Trade, and Visual Culture in the Dutch Golden Age* (Chicago: University of Chicago Press, 2014), chap. 2, esp. 33–38, 65–70; Pamela Smith, "Making as Knowing: Craft as Natural Philosophy," in Smith et al., *Ways of Making and Knowing*, 33–34; Deborah Harkness, *The Jewel House: Elizabethan London and the Scientific Revolution* (New Haven, CT: Yale University Press, 2007), 30, and chaps. 1, 2, 4, and 6; Mary Terrall, *Catching Nature in the Act: Réaumur and the Practice of Natural History in the Eighteenth Century* (Chicago: University of Chicago Press, 2014), esp. 48–50, 89–90; Paula Findlen, *Possessing Nature: Museums, Collecting, and Scientific Culture in Early Modern Italy* (Berkeley: University of California Press, 1994), chap. 5.

Value

What drove the interaction between different material worlds? A common answer is production. Gems were certainly "produced" in early modern Europe,
whether by being cut, polished, pierced, mounted, or otherwise enhanced or
fabricated.[38] But this is not a book about material production, which may be
defined as the effort to make good things. It is a book about material evaluation, which may be defined as the effort to determine how good things are. The
detection of counterfeit gems was one form of evaluation, but most acts of gem
evaluation did not involve fake or disguised gems. The art of gem appraisal was
difficult enough even before dishonesty entered the equation. Some parts of the
art could be reduced to rule, such as the square rule for the prices of diamond,
which stated that the price of diamonds increases in proportion to the square
of their weight. But this was never more than a rule of thumb, one that applied
only to perfect stones of medium size. Small, large, and imperfect stones called
for expert judgment, as did any stones that were not diamonds. Finally, there
was the question of which species the stone belonged to. Was it a diamond, or
a transparent sapphire? A ruby, or an unusually brilliant garnet? A topaz, or a
piece of yellow rock crystal? These were all pressing questions in early modern
Europe, when gems were highly prized, artfully imitated, and widely traded.
They were also new questions, because gems were worked in new ways, found
in new places, and available to new classes of consumer.[39]

38. A useful summary of such practices is Boodt, *Gemmarum*, 31–33. On cutting and polishing, see Paul Grodzinski, "The History of Diamond Polishing," *Industrial Diamond Review* 1,
special supplement (1953): 1–13; Lenzen, *History of Diamond Production*, 68–81; Tillander, *Diamond Cuts*; Ogden, *Diamonds*, chaps. 5–10, chap. 14. On early modern gem fabrication, useful
starting points are Marjolijn Bol, "The Emerald and the Eye: On Sight and Light in the Artisan's
Workshop and the Scholar's Study," in *Perspective as Practice: Renaissance Cultures of Optics*,
ed. Sven Dupré (Turnhout: Brepols, 2019), 77–101; and Federica Boldrini, "All That Glitters Is Not
Gold: False Jewellery and Its Juridical Regulation in Italy Between the Late Middle Ages and the
Early Modern Period," in *Faking It! The Performance of Forgery in Late Medieval and Early Modern Culture*, ed. Philip Lavender and Matilda Amundsen Bergström (Leiden: Brill, 2023), 52–74.

39. Most past research on the history of gem evaluation is by professional geologists or
gemologists. The standout example is Ogden, *Diamonds*, chap. 11. See also the two pioneering
articles by Mottana, "Galileo as Gemmologist" and "Italian Gemology During the Renaissance";
and the richly researched account of practical gem evaluation in early modern Asia in Sabel,
"Rare Earth," 187–93, 197–98, 216–21. There is also scattered data on gem evaluation in works on
particular gems, such as Kris Lane, *Colour of Paradise: The Emerald in the Age of Gunpowder
Empires* (New Haven, NJ: Yale University Press, 2002); Gedalia Yogev, *Diamonds and Coral:
Anglo-Dutch Jews and Eighteenth-Century Trade* (Leicester: Leicester University Press, 1978);
Tijl Vanneste, *Global Trade and Commercial Networks: Eighteenth-Century Diamond Merchants*
(London: Pickering and Chatto, 2011); idem, *Blood, Sweat and Earth*; Khazeni, *Sky Blue Stone*;

Gem appraisal shaped the sciences in multiple ways, most obviously as a motive for invention and discovery. Naturalists came up with new techniques that could (at least in theory) do a better job of evaluating gems than the traditional techniques. Most of these proposals did not work in practice, but they often led to new theories or methods that were significant for natural history or natural philosophy. Naturalists also used *existing* modes of gem appraisal—precise balances, sensitive grinding machines, visual inspection, naming conventions, and so on—to advance their own inquiries on topics that went well beyond gem appraisal. The world of craft and commerce also supplied something less tangible than specific techniques: the habit of doing the same test on many samples and comparing the results. Artisanal evaluation involved "rapid enquiry into a large volume of similar things," to quote the economic historians Patrick Wallis and Catherine Wright; gem appraisal was no exception to this rule.[40]

Evaluation also supplied a set of concepts that helped to explain what natural knowledge was about and how it should be acquired. The Latin "virtus," the English "virtue," the French "connoissance" and the French "qualité" were especially important. Each term had a wider meaning in the early modern period than it does today; each also had a closer association with natural science than it does today. The virtues (*virtutes*) of gems helped to define the species (*genera*) to which they belonged. To study the "quality" of a gem was to study its "nature" as well. "Connoissance" appeared in the titles of mineralogical treatises and academic articles as well as in the titles of manuals for working jewelers.[41] Material evaluation was not just a matter of techniques but also of personnel, methodologies, and metaphysical and epistemological categories.

The time is ripe for a synthetic study of these topics. Historians of science are familiar with the idea that "making and knowing" go together in material life, in Pamela Smith's evocative phrase.[42] We now need to see that judging

Donkin, *Beyond Price*; Content, *Ruby, Sapphire & Spinel*; Francesca Trivellato, *The Familiarity of Strangers: The Sephardic Diaspora, Livorno, and Cross-Cultural Trade in the Early Modern Period* (New Haven, CT: Yale University Press, 2009), chaps. 9 and 10; and Molly Warsh, *American Baroque: Pearls and the Nature of Empire, 1492–1700* (Chapel Hill, NC: University of North Carolina Press, 2018), esp. 227–30, 97–101.

40. Patrick Wallis and Catherine Wright, "Evidence, Artisan Experience, and Authority in Early Modern England," in Smith et al., *Ways of Making and Knowing*, 139.

41. On the scope of "connoissance," see Michael Bycroft and Alexander Wragge-Morley, "Science and Connoisseurship in the European Enlightenment," *History of Science* 60, no. 4 (2022): 450–54.

42. See The Making and Knowing Project: Intersections of Craft Making and Scientific Knowing, https://www.makingandknowing.org/, accessed May 26, 2023. Note also Pamela Smith,

and knowing go together. It is true that material evaluation is often studied under the heading of material production. Some of the best recent studies of evaluation have appeared in edited collections that are nominally dedicated to production.[43] But the fact remains that "making" and "production," not "judging" and "evaluation," are the terms that are usually used to connect early modern science to the world of art, craft, and industry.[44] Material evaluation has certainly been studied, but under a range of different headings and by a range of historians working on different materials. Medical historians write about drug testing;[45] economic historians about quality control;[46] historians

---

The Body of the Artisan: Art and Experience in the Scientific Revolution (Chicago: University of Chicago Press, 2004); and Smith et al., *Ways of Making and Knowing.*

43. Examples are Wallis and Wright, "Evidence, Artisan Experience, and Authority"; and Lissa Roberts and Joppe van Driel, "The Case of Coal," in Roberts and Werrett, *Compound Histories*, 57–84.

44. Note the titles of these recent edited collections: Roberts and Werrett, *Compound Histories: Materials, Governance and Production, 1760–1840*; Richard J. Oosterhoff, José Ramón Marcaida, and Alexander Marr, eds., *Ingenuity in the Making: Matter and Technique in Early Modern Europe* (Pittsburgh, PA: University of Pittsburgh Press, 2021); Smith et al., *Ways of Making and Knowing.* There are sections with "production" or "making" in the title, but none with "evaluation" or similar in the title, in each of these edited collections: Findlen, *Early Modern Things*; Klein and Spary, *Materials and Expertise*; Roberts and Werrett, *Compound Histories.* There's a tendency to run together the terms "production" and "making" with terms like "technical" or "artisanal" or "material" or "manual" or "manipulation." Examples are Findlen, *Early Modern Things*, 15; Lissa Roberts and Simon Schaffer, "Preface," in Roberts et al., *Mindful Hand*, xiii. There is a similar slippage in the wider culture, as in the titles of the Institute of Making at University College London, and the Nantes Maker Campus in France. See https://www.institute ofmaking.org.uk/ and https://nantesmakercampus.fr/, accessed Aug. 1, 2022.

45. Elaine Leong and Alisha Rankin, "Testing Drugs and Trying Cures: Experiment and Medicine in Medieval and Early Modern Europe," *Bulletin of the History of Medicine* 91, no. 2 (2017): 157–82, and subsequent articles. Notable works on medical testing since published are Elaine Leong, *Recipes and Everyday Knowledge: Medicine, Science and the Household in Early Modern England* (Chicago: University of Chicago Press, 2018), chap. 4; and Alisha Rankin, *Wonder Drugs, Experiment, and the Battle for Authority in Renaissance Science* (Chicago: University of Chicago Press, 2021). Testing of exotic drugs is a theme of Antonio Barrera-Osorio, *Experiencing Nature: The Spanish American Empire and the Early Scientific Revolution* (Austin, TX: University of Texas Press, 2006); and Samir Boumediene, *La colonisation du savoir: Une histoire des plantes médicinales du Nouveau monde, 1492–1750* (Vaulx-en-Velin: Les Éditions des Mondes à faire, 2019). Diagnostics often meant evaluating materials, especially urine: Michael Stolberg, *Uroscopy in Early Modern Europe* (Farnham: Ashgate, 2015).

46. William Ashworth, *Customs and Excise: Trade, Production, and Consumption in England, 1640–1845* (Oxford: Oxford University Press, 2003); idem, "Quality and the Roots of Manufacturing 'Expertise' in Eighteenth-Century Britain," *Osiris* 25, no. 1 (2010): 231–54; Philippe Minard, *La fortune du colbertisme: Etat et industrie dans la France des lumières* (Paris: Fayard,

of physics and chemistry about precision measurement.[47] Meanwhile, scientists have documented the history of materials science and analytic chemistry.[48] Other historians study antiquarianism,[49] colonial bioprospecting,[50] the assaying of gold,[51] the assaying of metallic ores,[52] and taste and connoisseur-

---

1998). There are a few early modern examples in Alessandro Stanziani, *La qualité des produits en France: XVIII–XXe siècles* (Paris: Belin, 2003).

47. A classic survey is Witold Kula, *Measures and Men*, trans. R. Szreter (Princeton, NJ: Princeton University Press, 1986 [1970]). Contributions from historians of early modern science include Simon Schaffer, "Golden Means: Assay Instruments and the Geography of Precision in the Guinea Trade," in *Instruments, Travel and Science: Itineraries of Precision from the Seventeenth to the Twentieth Century*, ed. Marie Noëlle Bourguet, Christian Licoppe, and H. Otto Sibum (London: Routledge, 2003), 20–50; idem, "Measuring Virtue: Eudiometry, Enlightenment, and Pneumatic Medicine," in *The Medical Enlightenment of the Eighteenth Century*, ed. Andrew Cunningham and Roger French (Cambridge, UK: Cambridge University Press, 1990), 281–318; idem, "Ceremonies of Measurement: Rethinking the World History of Science," *Annales* 70 (2015): 335–60; William Ashworth, "'Between the Trader and the Public': British Alcohol Standards and the Proof of Good Governance," *Technology and Culture* 42, no. 1 (2001): 27–50; and James Sumner, *Brewing Science, Technology and Print, 1700–1880* (Pittsburgh, PA: University of Pittsburgh Press, 2016), esp. chap. 4.

48. Examples are Stephen Timoshenko, *History of Strength of Materials: With a Brief Account of the History of Theory of Elasticity and Theory of Structures* (London: McGraw-Hill, 1953 [1930]); and Ferenc Szabadváry, *History of Analytical Chemistry*, International Series of Monographs in Analytical Chemistry (Oxford: Pergamon Press, 1966 [1960]).

49. Anna Marie Roos, "Taking Newton on Tour: The Scientific Travels of Martin Folkes, 1733–1735," *British Journal for the History of Science* 50, no. 4 (2017): 601; Alessio Mattana and Giacomo Savani, "Introduction: The Antique and the Natural: Exploring the Eighteenth-Century Textual Network," *Journal for Eighteenth-Century Studies* 43, no. 4 (2020): 423–32, and subsequent articles.

50. I borrow the phrase from Londa L. Schiebinger, *Plants and Empire: Colonial Bioprospecting in the Atlantic World* (Cambridge, MA: Harvard University Press, 2004). For a survey of Enlightenment bioprospecting, see James Poskett, *Horizons: The Global Origins of Modern Science* (London: Penguin, 2022), chap. 4.

51. Stephen Stigler, "Eight Centuries of Sampling Inspection: The Trial of the Pyx," *Journal of the American Statistical Association* 72, no. 359 (1977): 493–500; Lawrence M. Principe and William R. Newman, *Alchemy Tried in the Fire: Starkey, Boyle, and the Fate of Helmontian Chymistry* (Chicago: University of Chicago Press, 2002), 38–49; William R. Newman, "Alchemy, Assaying and Experiment," in *Instruments and Experimentation in the History of Chemistry*, ed. Frederic L. Holmes and Trevor H. Levere (Cambridge, MA: MIT Press, 2000), 35–54; Schaffer, "Golden Means"; Jasmine Kilburn-Toppin, "'A Place of Great Trust to Be Supplied by Men of Skill and Integrity': Assayers and Knowledge Cultures in Late Sixteenth- and Seventeenth-Century London," *British Journal for the History of Science* 52, no. 2 (2019): 197–223; Sebastian Felten, *Money in the Dutch Republic: Everyday Practice and Circuits of Exchange* (Cambridge, UK: Cambridge University Press, 2022), chap. 4.

52. Anon., *Bergwerk- und Probierbüchlein: A Translation from the German of the Bergbüchlein, a Sixteenth-Century Book on Mining Geology, and of the Probierbüchlein, a Sixteenth-Century Work on Assaying*, ed. Cyril Stanley Smith, trans. Anneliese Grünhaldt Sisco (New

ship.[53] These are all forms of material evaluation, but they are studied in a fragmented way that obscures the breadth and coherence of the practice. It is as if we had studied steam engines, potters, and the factory system, without noticing that they are all forms of material production.

Some historians of early modern knowledge have begun to connect the dots. In a remarkable article published several decades ago, Carlo Ginzburg drew a series of historical connections between medical diagnosis, the detection of criminals, the study of physiognomy, and connoisseurship in the fine arts—in other words, between the evaluation of people and the evaluation of paintings. According to Ginzburg, these are all forms of "conjectural knowledge," in the sense that they make use of apparently insignificant details such as the shape of a brushstroke or the grooves on a person's fingertips.[54] There are a range of more recent proposals. Ursula Klein and Emma Spary note that naturalists staked a claim to useful knowledge by making "prescriptive statements about the authenticity, rarity, and value of individual materials."[55] Steven Shapin's notion of "the sciences of subjectivity" captures the idea that a great deal of present-day science is involved in quantifying the goodness of wine, food, painting, and much else. Alexander Wragge-Morley has shown that aesthetic judgments were part of the practice of early British science,

---

York: American Institute of Mining and Metallurgical Engineers, 1949 [ca. 1500 (*Bergbüchlein*); 1578 (*Probierbüchlein*)]); Cyril Stanley Smith and R. J. Forbes, "Metallurgy and Assaying," in *A History of Technology*, vol. 3, ed. Charles J. Singer (Oxford: Clarendon Press, 1954–1978), 27–71; Harkness, *Jewel House*, 174–79; Hjalmar Fors, "Elements in the Melting Pot: Merging Chemistry, Assaying and Natural History, c. 1730–1760," *Osiris* 29 (2014): 230–44; idem, *The Limits of Matter: Chemistry, Mining and Enlightenment* (Chicago: University of Chicago Press, 2015); Charlotte A. Abney Salomon, "The Pocket Laboratory: The Blowpipe in Eighteenth-Century Swedish Chemistry," *Ambix* 66, no. 1 (2019): 1–22.

53. Harold Cook, *Matters of Exchange: Commerce, Medicine, and Science in the Dutch Golden Age* (New Haven, CT: Yale University Press, 2007), 6–20; Emma Spary, "Scientific Symmetries," *History of Science* 42 (2004): 1–46; Bettina Dietz and Thomas Nutz, "Collections Curieuses: The Aesthetics of Curiosity and Elite Lifestyle in Eighteenth-Century Paris," *Eighteenth-Century Life* 29, no. 3 (2005): 44–75; Jonathan Simon, "Taste, Order and Aesthetics in Eighteenth-Century Mineral Collections," in *From Private to Public: Natural Collections and Museums*, ed. Marco Beretta (Sagamore Beach, MA: Science History Publications, 2005), 97–112; Daniela Bleichmar, "Learning to Look: Visual Expertise Across Art and Science in Eighteenth-Century France," *Eighteenth-Century Studies* 46, no. 1 (2012): 85–111; Sarah Easterby-Smith, *Cultivating Commerce: Cultures of Botany in Britain and France, 1760–1815* (Cambridge, UK: Cambridge University Press, 2017).

54. Carlo Ginzburg, "Morelli, Freud and Sherlock Holmes: Clues and Scientific Method," trans. Anna David, *History Workshop Journal* 9 (1980): 5–36.

55. Klein and Spary, "Introduction: Why Materials?," in Klein and Spary, *Materials and Expertise*, 21–22.

especially in the context of natural theology and the Royal Society of London.[56] On a larger scale, John Pickstone has argued that all materials, from soil to paper, were studied in the way that connoisseurs study wine and paintings today—by classifying them according to their qualities and their places of origin, and by the direct use of the senses.[57] Most recently, in a study of the French chemist Jean Hellot, Lisa Coulardot has shown that a great deal of eighteenth-century chemistry hinged on the notion of the *essai*, a term that may be translated as "assay" or "trial."[58]

Each of these notions joins up some forms of material evaluation while leaving others out. Ginzburg's "conjectural knowledge" excluded physics, chemistry, and what he called "Galilean science." Klein and Spary focused on chemistry and natural history, at the expense of experimental physics and mixed mathematics, in their discussion of "systems of valuation." The *essais* studied by Coulardot are centered on the same two disciplines, chemistry and natural history. Shapin's "sciences of subjectivity" include many things, but they are weighted toward culinary and aesthetic judgments and the twentieth-century science of consumer preference. Wragge-Morley's "aesthetic science" focuses on the aesthetic dimension of judgment, whereas there was a medical and commercial dimension as well. Pickstone overlooks the elaborate instrumentation that was used to evaluate some materials in preindustrial Europe. In addition, Pickstone implies that early modern "connoisseurship" was replaced by an "analysis of production" around 1800. This suggests, in my view misleadingly, that material evaluation is marginal in modern economies.

We therefore have multiple accounts of material evaluation in early modern Europe that each focus on a different aspect of the topic. The way to unify these accounts is to define evaluation by its ends rather than by its means. To evaluate a material is to determine how good it is, irrespective of how the determination is done. Indeed, one reason I use the phrase "material evaluation" is that it is neutral with respect to methods in a way that the alternatives are not. The word "connoisseurship" suggests a sensual, qualitative, tacit form of knowledge as opposed to an instrumental, quantitative, explicit form of knowledge. Conversely, "testing" suggests a more deliberate and elaborate process than the "tweaking" that historians of early modern medicine have

56. Alexander Wragge-Morley, *Aesthetic Science: Representing Nature in the Royal Society of London, 1650–1720* (Chicago: University of Chicago Press, 2020).

57. John Pickstone, "Thinking over Wine and Blood: Craft-Products, Foucault, and the Reconstruction of Enlightenment Knowledges," *Social Analysis: The International Journal of Social and Cultural Practice* 41, no. 1 (1997): 97–105.

58. Lisa Coulardot, "Les lumières au banc d'essai: Science, économie et environnement autour de Jean Hellot (1685–1766)" (PhD diss., European University Institute, 2024).

found in apothecary shops and private households—tweaking that neverthe-less aimed to determine the goodness of ingredients and recipes.[59] "Standard-ization" suggests a level of uniformity and centralization that was rarely pres-ent in early modern evaluation. The phrase "material evaluation," by contrast, suggests no particular way of knowing beyond that which aims to determine the value of material things.

The phrase has other advantages as well. It is continuous with current usage.[60] It links the history of science to the emerging field of the sociology of valuation.[61] It distinguishes the evaluation of materials from the evaluation of people and ideas, large topics in their own right. The phrase also indi-cates the economic and historical importance of the practice. It implies that material evaluation is analogous to, but distinct from, material production. I sometimes use "value" as a shorthand for material evaluation, as in the title of this book, but I prefer "material evaluation" since it draws attention to the techniques used to determine the value of things. There are many studies of value that are not studies of material evaluation. Indeed, material evalu-ation is absent in much of what economists and anthropologists call "value theory."[62] Adam Smith may have used diamonds to explain the distinction between "use value" and "exchange value," but he had nothing to say about the instruments and gestures that jewelers used to judge diamonds.[63]

Smith was right in one respect, however: gems are an ideal subject matter for exploring material evaluation in the broad sense defined here. Gem ap-praisal was an unusually versatile practice, one that cut across institutions,

59. Valentina Pugliano, "Pharmacy, Testing, and the Language of Truth in Renaissance Italy," *Bulletin of the History of Medicine* 91, no. 2 (2017): 232–73, esp. 238; Leong, *Recipes and Everyday Knowledge*, chap. 4.

60. Examples are Schaffer, "Golden Means," 39 ("devices to evaluate gold"); Wallis and Wright, "Evidence, Artisan Experience, and Authority," 147 ("artisanal evaluation"); Klein and Spary, "Introduction," 21 ("systems of valuation"); Harkness, *Jewel House*, xvii ("urban ways of knowing and evaluating nature"); Ursula Klein and Wolfgang Lefèvre, *Materials in Eighteenth-Century Science: A Historical Ontology* (Cambridge, MA: MIT Press, 2007), 208 ("evaluative gaze"); Coulardot, "Lumières au banc d'essai," passim (*évaluation*).

61. Michael Hutter and David Stark, "Pragmatist Perspectives on Valuation: An Introduc-tion," in *Moments of Valuation: Exploring Sites of Dissonance*, ed. Ariane Berthoin Antal, Michael Hutter, and David Stark (Oxford: Oxford University Press, 2015), 1–12, and references therein.

62. This remark is based on a study of David Graeber, *Toward an Anthropological Theory of Value: The False Coin of Our Own Dreams* (New York: Palgrave, 2001); Mariana Mazzucato, *The Value of Everything: Making and Taking in the Global Economy* (London: Penguin, 2018); and David Harvey, "Value in Motion," *New Left Review* 126 (Nov./Dec. 2020): 99–126.

63. Adam Smith, *An Inquiry into the Nature and Causes of the Wealth of Nations* (India-napolis, IN: Liberty Fund and Oxford University Press, 1981 [1776]), vol. 1, bk. 1, chap. 4, para. 13.

disciplines, and techniques that are usually separated in existing studies. It was linked to assaying and mining, since gems were often grouped with gold and silver in the study of precious minerals. Yet it also had affinities to medical testing, since the ancient tradition of lapidary medicine persisted well into the seventeenth century. It was also tied to connoisseurship, in the modern sense of the term—like paintings and horses, gems were collected by the social elite and admired for their beauty and provenance. In this context, gem appraisal was usually a visual affair. Among jewelers and gem cutters, however, it was done with the hand and the tongue as well as with the eye. At the same time, gems were the subject of some of the most precise measurements done in early modern Europe. In terms of institutions, gem appraisal became part of the apparatus of European states in the eighteenth century, mainly through the scientific and technical academies sponsored by these states. But most gem appraisal continued to be done by members of craft guilds, especially guilds of goldsmiths and gem cutters. Finally, appraisal was bound up with the transformations in the trade and production of gems that shaped early modern Europe.[64] The process of diamond faceting, introduced around 1400, helped to judge gems by their hardness. The advent of specialized gem-cutting guilds in the sixteenth century raised the question of who—goldsmiths or gem cutters—were best equipped to judge their wares. Encounters with gems in new places, such as emeralds in the New World and diamonds in the Old, were a challenge for old forms of appraisal. European rulers staked a claim for the value of their own gems in their interactions with each other and with rulers outside Europe. All these developments were tied up with the science of gems. No substance can cover everything, of course. But gems, more than most materials, allow us to study material evaluation in the round.

The Scientific Revolution

Transmaterialism and material evaluation are the main themes of the story of the science of gems. But a story needs a plot as well as themes, and here the historian of early European science runs into a problem. On the one hand, we are urged to think in terms of the "big picture," to locate our case studies in a wider chronological and geographical framework.[65] On the other hand, there

64. On these transformations, see the works cited in note 10; the summaries in Bycroft and Dupré, "Introduction," 5–10; and Vanneste, *Global Trade and Commercial Networks*, chap. 2.

65. James A. Secord, "Introduction," *British Journal for the History of Science* 26, no. 4 (1993): 387–89; Robert Kohler, "A Generalist's Vision," *Isis* 96, no. 2 (2005): 224–29.

is no agreement on what this framework is. In particular, there is no agreement on whether something called "the Scientific Revolution" is part of the framework. The *Oxford Companion to the History of Modern Science*, published in 2003, says that it is.[66] The fourth volume of the *Cambridge History of Science*, published in 2006, says that it is not.[67] Since then, the tendency has been to dismiss the idea of the Scientific Revolution as a product of the Cold War, or perhaps of the Progressive Era.[68] But the idea still has vigorous advocates.[69] Arguments against it are not clear-cut. The idea is often dismissed on the grounds that science itself did not exist before 1800.[70] But some scholars have found the word "science" being used in early modern Europe in roughly the same way as it is used today.[71] The Scientific Revolution is sometimes seen as an outdated narrative of inevitable progress. Yet the label is sometimes used by scholars who reject narratives of inevitable progress.[72] Others suggest that the Scientific Revolution is a Eurocentric notion that has no place in today's global and anti-colonial academy.[73] Yet the most ambitious global his-

---

66. John L. Heilbron, "History of Science," in *The Oxford Companion to the History of Modern Science* (Oxford: Oxford University Press, 2003), https://doi.org/10.1093/acref/9780195112290.001.0001.

67. Lorraine Daston and Katharine Park, "Introduction," in *Cambridge History of Science*, vol. 3: *Early Modern Science*, ed. Lorraine Daston and Katharine Park (Cambridge, UK: Cambridge University Press, 2006), 12–16; cf. Stéphane van Damme, "Un ancien régime des sciences et des savoirs," in *Histoire des sciences et des savoirs*, vol. 1: *De la Renaissance aux Lumières*, ed. Stéphane van Damme (Paris: Seuil, 2015), 19–20.

68. J. B. Shank, "After the Scientific Revolution: Thinking Globally About the Histories of the Modern Sciences," *Journal of Early Modern History* 21, no. 5 (2017): 377–93; James A. Secord, "Inventing the Scientific Revolution," *Isis* 114, no. 1 (2023): 50–76, esp. 76 ("dislodge"); idem, "Against Revolutions," *BJHS Themes* 9 (2024): 17–37.

69. David Wootton, *The Invention of Science: A New History of the Scientific Revolution* (London: Penguin, 2015), 1, 41. Earlier literature is surveyed in Floris H. Cohen, *The Scientific Revolution: A Historiographical Inquiry* (Chicago: University of Chicago Press, 1994); and idem, "Postscript 2012 to *The Scientific Revolution: A Historiographical Inquiry*," July 2012, https://hfloriscohen.wordpress.com/wp-content/uploads/2020/06/postscript-chinese-srhi.pdf. See also John Henry, "The Scientific Revolution: Five Books About It," *Isis* 107, no. 4 (2016): 809–17.

70. For example, Andrew Cunningham and Perry Williams, "De-centring the 'Big Picture': 'The Origins of Modern Science' and the Modern Origins of Science," *British Journal for the History of Science* 26, no. 4 (1993): 407–32.

71. Harkness, *Jewel House*, xv–xviii; Wootton, *Invention of Science*, passim.

72. E.g., Pamela Smith and Benjamin Schmidt, "Knowledge and Its Making," in *Making Knowledge in Early Modern Europe: Practices, Objects, and Texts, 1400–1800*, ed. Pamela Smith and Benjamin Schmidt (Chicago: University of Chicago Press, 2007), 8 ("no inevitable march"), 14 ("the Scientific Revolution").

73. Shank's "After the Scientific Revolution" and Secord's "Inventing the Scientific Revolution" are written in this spirit.

tory of science published to date has two chapters entitled "Scientific Revolution, c. 1450–1700."[74]

It is tempting to retreat to neutral ground and refer to "science in early modern Europe" rather than "the Scientific Revolution." But this comes with its own problems. Textbooks on the general history of early modern Europe tell us that there *was* a Scientific Revolution, along with a Renaissance, a Reformation, and an Enlightenment.[75] In any case, the category "early modern Europe" is not as natural as it might appear. Scholars do not agree on whether the eighteenth century is part of it. The *Cambridge Companion to Early Modern Science* excludes that century, whereas the equivalent French textbook includes it. The fourteenth century is also contentious, with some including it in a capacious "late medieval / early modern" category.[76] "Early modern Europe" is as much a battleground as "the Scientific Revolution."

One way out of this maze is to focus on events rather than labels. There is a core set of changes that happened in Europe from roughly 1500 to 1800 and that are recognized by leading specialists, whether or not they endorse the phrase "the Scientific Revolution." Lorraine Daston put it like this in a recent survey: "It is undisputed that during this period [1500 to 1750] every aspect of natural knowledge in Europe was radically reconfigured."[77] The changes to which Daston refers may be summed up as the collapse of Aristotelean natural philosophy.[78] A worldview based on four causes, four elements, and a sharp distinction between the terrestrial and celestial realms, was superseded by a very different worldview. This was accompanied by changes in theories, methods, disciplines and institutions. New theories, such as the circulation of blood and the composite nature of white light, were radically different from the theories they replaced. The bookish methods of medieval natural philos-

74. Poskett, *Horizons*, pt. 1.

75. Beat Kümin, *The European World 1500–1800: An Introduction to Early Modern History*, 4th ed. (London: Taylor & Francis, 2022 [2009]), chap. 28; Merry E. Wiesner-Hanks, *Early Modern Europe, 1450–1789*, 3rd ed. (Cambridge, UK: Cambridge University Press, 2023 [2006]), chap. 10.

76. The eighteenth century is covered in van Damme, *Histoire des sciences et des savoirs*, vol. 1, but not in Daston and Park, *Cambridge History of Science*, vol. 3. The fourteenth century is in neither of these works, but it is in the "late medieval / early modern" category in Pamela H. Smith, "Nodes of Convergence, Material Complexes, and Entangled Itineraries," in *Entangled Itineraries: Materials, Practices, and Knowledges Across Eurasia*, ed. Pamela H. Smith (Pittsburgh, PA: University of Pittsburgh Press, 2019), 16–19.

77. Lorraine Daston, "Philosophie de la nature et philosophie naturelle (1500–1750)," in *Histoire des sciences et des savoirs*, vol. 1, 183, cf. 177 ("profondément transformée"). The translation is mine.

78. The following sketch is meant to capture the main points in Daston, "Philosophie de la nature," and Heilbron, "History of Science." Daston rejects the phrase "the Scientific Revolution" in her piece; Heilbron endorses it in his.

ophers were supplemented by contrived experiments, elaborate instruments, large collections of plants and minerals, and other techniques for reading the book of nature. Mixed mathematics, natural history, and the mechanical arts became part of natural philosophy rather than separate territories on the map of knowledge. The new philosophy was not only carried out in universities, the bastion of the old philosophy, but also in many other places, notably royal courts and scientific academies. These changes were multiple, haphazard, contested, and spread out in time and space—but so were the changes that historians still call "the Reformation" and "the Industrial Revolution." I shall therefore continue to use the phrase "the Scientific Revolution" in what follows. Readers who object to this phrase may wish to mentally substitute "the new science," "the so-called Scientific Revolution," or similar. The important thing is the events themselves, not the name we give to them.

Matter and Value in the Scientific Revolution

Once we identify these events, we can ask general questions about them. How exactly does the eighteenth century fit into the Scientific Revolution? What about natural history? What difference did European expansion make to the intellectual life of the sciences? What role did merchants and artisans play in these events? These are familiar questions, but we get fresh answers when we rethink matter and value in the ways described earlier in this chapter. The answers sketched in this section are not entirely new, as the footnotes will show. But they are new enough that they lack systematic exposition. They lack names, for example. The names I shall use are "two-stage classification," "physics as stamp collecting," "spatial species," "material specialties," and "hand-hand coordination." I shall take these terms in turn.

"Two-stage classification" is a name for a way of thinking about science in the eighteenth century. That century is omitted in most book-length accounts of the Scientific Revolution; in an odd asymmetry, the sixteenth century is included in nearly all such accounts.[79] The eighteenth century is a problem for historians of science, as Geoffrey Cantor pointed out forty years ago.[80]

79. Eighteenth-century science is virtually absent in Wootton, *Invention of Science*; Floris H. Cohen, *How Modern Science Came into the World: Four Civilizations, One 17th-Century Breakthrough* (Amsterdam: Amsterdam University Press, 2010); and in textbooks on the Scientific Revolution. An exception is David Knight, *Voyaging in Strange Seas: The Great Revolution in Science* (New Haven, CT: Yale University Press, 2015), chap. 12 and passim.

80. Geoffrey Cantor, "The Eighteenth Century Problem," *History of Science* 20 (1982): 44–63. For details see Michael Bycroft, "Introduction: Science Beyond the Enlightenment," *Journal of Early Modern Studies* 12, no. 1 (2023): 9–31; and other articles in the same special issue.

The solution offered in this book is a transmaterialist one. The naturalists of the eighteenth century took methods that they had inherited from the seventeenth century and adapted them to new phenomena. There was an ancient tradition of gem classification based on color, transparency, and locality. This tradition was consolidated in the seventeenth century (the first stage). In the same century, the experimental philosophy brought new properties of gems to the fore, especially their density, crystal form, and double refraction. But these properties were not integrated into the ancient tradition of classification until the end of the eighteenth century (the second stage). This second stage was no mere extension of the first. It was accompanied by new methods, most notably the increased use of numbers to express the properties of gems and of precise instruments to measure those properties. Some readers may recognize this as a version of a pattern sketched out by Thomas Kuhn in a 1976 article on the history of quantification in the physical sciences.[81] Kuhn noted that some phenomena (such as planetary orbits) were quantified in the seventeenth century whereas other phenomena (such as electricity) had to wait another century for the same treatment. Kuhn's scheme has been widely adopted since then, but in a way that separates the two stages, with some historians looking at the first stage and others at the second.[82] I propose to reconnect the two stages and to extend them to the history of classification.

Classification in the early modern period is usually associated with natural history, another topic that has often been an awkward fit for the Scientific Revolution. The historiography of natural history has blossomed since the 1990s.[83] But this literature has often played up the distinctiveness of natural history as compared to astronomy and experimental physics, disciplines that have long been placed at the core of the Scientific Revolution.[84] Accordingly,

81. Thomas Kuhn, "Mathematical Versus Experimental Traditions in the Development of Physical Science," *Journal of Interdisciplinary History* 7, no. 1 (1976): 1–31.

82. On the first stage, see Cohen, *Scientific Revolution*, 126–34; idem, *How Modern Science Came into the World*, 368; idem, "Postscript," 19. On the second stage, see Tore Frängsmyr, John L. Heilbron, and Robin E. Rider, eds., *The Quantifying Spirit in the Eighteenth Century* (Berkeley: University of California Press, 1990); John L. Heilbron, *Weighing Imponderables and Other Quantitative Science Around 1800* (Berkeley: University of California Press, 1993).

83. For reviews, see Nicholas Jardine, James Secord, and Emma Spary, eds., *Cultures of Natural History* (Cambridge, UK: Cambridge University Press, 1996); Paula Findlen, "Natural History," in Daston and Park, *Cambridge History of Science*, vol. 3, 435–68; Helen Curry, Nicholas Jardine, James Secord, and Emma Spary, eds., *Worlds of Natural History* (Cambridge, UK: Cambridge University Press, 2018).

84. Here are some examples of sharp contrasts being drawn between natural history and experimental philosophy. The first term in each pair represents natural history. Particulars/

natural history seems to show the disunity of early modern science and the fu-
tility of reducing the period to a single revolutionary change.[85] When natural
history *is* integrated into the bigger narrative, it is usually analyzed in the way
the English philosopher Francis Bacon analyzed it early in the seventeenth
century—as a source of new particulars that helped to reform the universal
theories of natural philosophers.[86] Yet the tradition of "experimental natural
history" that Bacon inaugurated was only loosely related to the practices of
classification and collecting often associated with natural history.[87] The upshot is the Scientific Revolution seems to confirm the modern prejudice that
physics has little to do with natural history—that "all science is either physics
or stamp collecting," as the physicist Ernest Rutherford is supposed to have
said.[88] This does not do justice to the role of natural history in the science of
gems. What we find is a continuous and multilayered interaction between
the ancestors of mineralogy and the ancestors of experimental physics. Collections met instruments; botanical gardens met cabinets of experimental
philosophy; the commodities of merchants met the tools of artisans; and so
on. The material culture of natural history met the material culture of experimental philosophy and that of its eighteenth-century successor, *la physique
expérimentale*. Rutherford notwithstanding, we may say that early modern
physics was akin to stamp collecting.[89]

---

universals in Alix Cooper, *Inventing the Indigenous: Local Knowledge and Natural History in
Early Modern Europe* (Cambridge, UK: Cambridge University Press, 2007), 13. Description/ex-
planation in John Pickstone, "Ways of Knowing: Towards a Historical Sociology of Science,
Technology and Medicine," *British Journal for the History of Science*, 26, no. 4 (1993): 439. Incre-
mental/revolutionary change in Findlen, "Natural History," 437. Economy/spectacle in Simon
Schaffer, "Natural Philosophy and Public Spectacle in the Eighteenth Century," *History of Science*
21, no. 1 (1983): 22. Heterogeneous/homogeneous in Emma Spary, *Utopia's Garden: French Natural History from Old Regime to Revolution* (Chicago: University of Chicago Press, 2000), 6, 246.
Materials/instruments in Klein and Spary, "Introduction," 11–16. Classification/quantification in
John Lesch, "Systematics and the Quantifying Spirit," in Frängsmyr et al., *Quantifying Spirit*, 88.

85. Daston and Park, "Introduction," 13–14; Findlen, "Natural History," 437.

86. Daston and Park, "Introduction," 4, 11, 14; Pickstone, "Ways of Knowing," 439; Richard
Yeo, "Classifying the Sciences," in *The Cambridge History of Science*, vol. 4: *Eighteenth Century
Science*, ed. Roy Porter (Cambridge, UK: Cambridge University Press, 2003), 241–66; Cohen,
How Modern Science Came into the World, 113–41.

87. "Experimental natural history" in Peter Anstey and Alberto Vanzo, *Experimental Philosophy and the Origins of Empiricism* (Cambridge, UK: Cambridge University Press, 2023), chap. 3.

88. On the origin and fate of Rutherford's remark, see Kristin Johnson, "Natural History as
Stamp Collecting: A Brief History," *Archives of Natural History* 34, no. 2 (2007): 244–58.

89. This claim synthesizes and extends several areas of existing research. Animal electricity: James Delbourgo, *A Most Amazing Scene of Wonders: Electricity and Enlightenment in Early
America* (Cambridge, MA: Harvard University Press, 2006), chap. 5; Marco Piccolino, *The*

Any discussion of early modern natural history must make room for European expansion, a third aspect of the Scientific Revolution that is now getting the attention it deserves. It has been estimated that western Europeans controlled 5 percent of the surface of the earth in 1400, compared to 35 percent in 1800.[90] What difference did this make to science in western Europe? Various answers have been given, from the "species explosion" in botany to the invention of the idea of discovery to the appropriation of the expertise of conquered peoples.[91] Gems suggest another answer: Classification was a geographical exercise, in the sense that gem species were defined in terms of where they came from. The nature of a gem was related to its native land, just as people were defined in terms of their place of origin. This was a longstanding feature of gem classification in Latin natural history, going back at least to Pliny the Elder in the first century AD. The notion that the best gems came from outside Europe was equally persistent. As a result, European schemes for classifying gems were sensitive to changes in European conceptions of global geography. Terms such as "oriental ruby" and "Bohemian garnet" were not merely ornamental, commercial, or symbolic, and they were not spurned by the Enlightenment. They were at the heart of gem classification in European natural history until at least 1800. They persisted because they described the qualities of gems as well as their natures. To call a pearl "oriental" was not only to say that it came from the East but also to say that it was better than an "occidental" pearl. These evaluative terms drove the search for proxies for geography—simple tests that could help to assign labels such

Taming of the Ray: Electric Fish Research in the Enlightenment, from John Walsh to Alessandro Volta (Florence: Olschki, 2003). Instruments and travel: Bourguet et al., Instruments, Travel and Science. Experiments and collections: Findlen, Possessing Nature, chap. 5; Terrall, Catching Nature in the Act; Anita Guerrini, The Courtiers' Anatomists: Animals and Humans in Louis XIV's Paris (Chicago: University of Chicago Press, 2015); Marco Beretta, "Collected, Analysed, Displayed: Lavoisier and Minerals," in Beretta, From Private to Public, 113–40. Wonders and experiments: Lorraine Daston, "The Factual Sensibility," Isis 79, no. 3 (1988): 452–67; Lorraine Daston and Katharine Park, Wonders and the Order of Nature, 1150–1750 (New York: Zone Books, 1998), chaps. 6 and 7.

90. James Belich, The World the Plague Made: The Black Death and the Rise of Europe (Princeton, NJ: Princeton University Press, 2022), 3.

91. Surveys of this topic include Cohen, Scientific Revolution, 354–57; Knight, Voyaging in Strange Seas, chap. 11; Poskett, Horizons, chaps. 1–4; Paula Findlen, ed., Empires of Knowledge: Scientific Networks in the Early Modern World (London: Routledge, 2018). The invention of discovery is the theme of Wootton, Invention of Science, chap. 3. "Species explosion" in Justin E. H. Smith and James Delbourgo, "In Kind: Species of Exchange in Early Modern Science," Annals of Science 70, no. 3 (2013): 300.

as "oriental" and "occidental" but without the hassle of tracing specimens to their source. These proxies included the electrical, optical, and mechanical properties of gems, and eventually their chemical properties as well. The recognition that species were spatial helps us to link the qualities of gems to their natures and the material culture of natural history to that of physics and chemistry.[92]

Of course, the geographical distribution of gems was not merely a matter of perception. Some places really did specialize in particular gems. The diamond mines of the Deccan plateau are obvious examples. Nishapur, now in northern Iran, had a virtual monopoly on the global supply of turquoise in the early modern period. The same goes for Europe itself, and for materials other than gems. The metal mines of Sweden and Saxony help to explain why chemical and mineralogical treatises from these places were weighted toward metals. French chemistry in the seventeenth century was weighted toward plants, perhaps because of its close ties to medicine, where plants rather than minerals were the key substances. The history of gem science may be seen as the history of the interaction between these regional specialties. In a sense, this is old news. Historians of science, and historians more generally, have been preoccupied with geographical mobility for some time.[93] But this mobility takes on new significance when we take seriously the differences between materials. Corundum came in a range of vivid and varied colors, all of them found in the same mines in Ceylon—a challenge for the classical idea that gem species are defined by their color. Diamonds had the peculiarity of being colorless and extremely hard, which meant that they lent themselves to faceting, which in turn drew attention to their crystal form. Metal ores were valued for what they contained rather than what they looked like, which suggested a particular view of chemical composition. In short, different materials suggested different ideas about gems. As a result, changes in the science of gems usually involved the interaction between material expertise from different places. Mobility mattered because matters mattered.

92. Here I build on works such as: Cooper, *Inventing the Indigenous*; Samir Boumediene and Valentina Pugliano, "La route des succédanés: Les remèdes exotiques, l'innovation médicale et le marché des substituts au XVIe siècle," *Revue d'histoire moderne et contemporaine* 3, no. 66.3 (2019): 24–54; Schaffer, "Golden Means"; Emma Spary, *Eating the Enlightenment: Food and the Sciences in Paris* (Chicago: University of Chicago Press, 2012), chap. 2; Margócsy, *Commercial Visions*, chap. 2; Benjamin Schmidt, *Inventing Exoticism: Geography, Globalism, and Europe's Early Modern World* (Philadelphia: University of Pennsylvania Press, 2015).

93. For an overview, see James Poskett and Gianamar Giovannetti-Singh, "Global History of Science," in *Debating Contemporary Approaches to the History of Science*, ed. Lukas M. Verburgt (London: Bloomsbury, 2024), 22–25.

Finally, what role did merchants and artisans play in all this? The transmaterialist answer is that they interacted with each other. This may sound obvious, but it is rarely brought to the fore in a historiography that focuses on the relationship between scholars on the one hand, and merchants and artisans on the other, rather than on the relationship between different sorts of merchant, between different sorts of artisan, and between merchant groups and artisan groups. In the history of gem science, the interaction between gem cutters and goldsmiths was just as important as the interaction between either of those groups and university-trained scholars. Assayers, porcelain makers, glassmakers, and apothecaries also intermingled, with significant results. This does not mean that scholars were unimportant. On the contrary, they were important precisely because they mediated between different groups of practitioners. Much has been written about the integration of "mind" and "hand" in early science. But hand-hand coordination was just as important as hand-mind coordination.

Does this strengthen the case for the reality of the Scientific Revolution? No. Does it weaken the case? No. Does it show that transmaterialism and material evaluation help to make sense of the overall development of science in Europe between 1500 and 1800? Yes.

Seven New Sciences

To sum up: we have two themes, material evaluation and transmaterialism. We have a plot, the Scientific Revolution. And we have five subplots: two-stage classification, physics as stamp collecting, spatial species, material specialties, and hand-hand coordination. The story will be told in seven chapters, each corresponding to one of the new sciences that emerged in early modern Europe. There is background material on the ancient and medieval periods, and a concluding chapter that takes the story into the nineteenth century, but the bulk of the action takes place in the period 1500 to 1800. Each chapter is dedicated to a new science at a key moment in its formation, from descriptive natural history in the sixteenth century to applied science at the start of the nineteenth century. In each chapter, I shall argue that gems were central to the new science in question and that their role in that science can be understood in terms of material evaluation and the meeting of matters. Some chapters focus more on matter, others more on value, but together they cover both themes evenly. The seven new sciences are covered roughly in chronological order, as follows:

Natural history is often studied on the model of botany, a field that was transformed in the sixteenth century by the arrival of many new plant species in Europe. But there was no comparable proliferation of gem species. The real novelty for gems was not the species but the places from which these species came. This helps to explain the rise of the distinction between "oriental" and "occidental" gems. The distinction between hard and soft stones was equally important. The convergence of these distinctions was a major event in the history of gem classification, one that makes sense when we pay attention to the relationship between different material worlds—between plants and gems, and between European expansion and the codification of the mechanical arts.

Technical writing is the next new science to consider. As the trade and production of gems expanded in Paris, so did the production of texts about them. Paris under Louis XIV became a "trading zone" for gem knowledge, in Pamela Long's sense of the term. But the knowledge in question was not only about material production and it was not only traded between scholars and practitioners. The evaluation of gems was the key question in a new generation of maps, travel narratives, inventories and craft manuals. And interactions between different practitioners—between goldsmiths, gem cutters, merchants, painters, and mapmakers—were as much a part of these texts as interactions between practitioners and scholars.

Experimental philosophy came of age in England in the same period. A key text was Robert Boyle's *Essay About the Origine and Virtues of Gems*, published in 1672. This was part of wider project on gems that was embedded in the global gem trade, the mining industry in the British Isles, the culture of collecting, and Boyle's own domestic arrangements. These were the sources of the hundreds of gem specimens that Boyle examined as part of his corpuscular account of the origin and nature of gems. Boyle's experimental philosophy of gems is a striking example of the role of collections in two fields—the history of physics and the history of experiment—that are still dominated by instruments. At the heart of this philosophy lay the practical project of determining the "goodness" of gems.

Experimental physics, or la physique expérimentale, can be traced to the work of two French gem connoisseurs, René-Antoine Ferchault de Réaumur and Charles-François de Cisternay Dufay. Dufay used his gem collection to study light and electricity, formulating the basic laws of electrostatics in the process. His systematic approach to experimentation was new, but not in the way that historians have supposed. Rather than inventing an entirely new method, he adapted a method that members of the Parisian Academy of Sciences had already applied to plants, animals, metals and mineral waters. The academicians had built large collections of objects, done similar tests on the entire collection, and compared the results, often with a view to evaluating the objects in question. Their experiments were driven by materials as much as by instruments. Dufay's experimental physics was the result of the migration of this method to new materials.

Mineralogy was a new science in the middle of the eighteenth century, a science dedicated to the classification of the mineral kingdom. It was in mineralogical treatises that the findings of experimental philosophy and experimental physics were integrated into the ancient tradition of dividing gems into kinds. Density, double refraction, and crystal form emerged as the new principles of gem classification. Gem classification became quantitative in the process, but it did not become any less evaluative. On the contrary, the aim of many quantifiers was to make judgments about gems more precise. This project brought together systematic natural history and quantitative physics, two Enlightenment worlds that are usually studied separately.

Chemistry was a third factor in the reclassification of gems, alongside physics and crystallography. The decomposition of gems—separating them into their chemical components—was an achievement of the last third of the eighteenth century. This was part of a wider trend toward understanding substances in terms of their composition, a trend that included, but was not limited to, the Chemical Revolution. This trend was driven by chemical crafts such as mining, pharmacy, glassmaking, and porcelain making. It was also driven by the interaction between these crafts, as the example of gems shows. The end result was not only the decomposition of gem specimens but also the dissolution of gems as a general category.

Gemology has its origins in the collapse of this category. A "science of precious stones applied to the arts" emerged in France in the first three decades of the nineteenth century. This field was defined by a new generation of books on gems, and by the tests, collections, and careers that accompanied the books. This was a paradoxical field, one that aimed to use the value-free science of mineralogy to make better evaluations of gems. The solution to the paradox is not to deny that mineralogy was value-free but to deny that it made evaluations on its own. Many hands were required to bring the findings of the new mineralogy to bear on practical questions, whether in the cabinet, the workshop, the newspaper, or the courtroom.

The conclusion begins with a survey of the history of gem classification, one that connects the dots between the episodes considered in the foregoing chapters, using garnet as a test case. This survey reminds us why we ought to be transmaterialists as well as materialists and why we should consider material evaluation as well as material production. Can these lessons be applied to the present as well as the past, to the modern period as well as the early

modern period, and to the world outside Europe as well as to Europe? I end
the book by sketching some answers to these questions.

Alternative Histories of Gem Science

As this overview indicates, the following chapters range over three centuries,
multiple scientific disciplines, and many types of material evaluation. Still,
this book is far from complete. It is weighted toward men, elite institutions,
western Europe, and (in the latter chapters) France. It contains more intellectual history, and less political and economic history, than one would normally find in a book written under the banner of "materialism." I shall not
bore the reader by explaining or justifying these choices here. Like any book,
this one is a reflection of the expertise and enthusiasms of the author at the
time of writing. A different set of preoccupations would result in a different
set of primary sources and a different set of insights.

Women, empire and natural kinds are three topics that seem especially
well-suited to the study of gem knowledge and that are given short shrift in
this book. Women not only wore gems in early modern Europe, but also cut,
traded and polished them. European empires were not only a source of gem
specimens, and of ideas about the global distribution of gems, topics that *are*
covered here. They were also the site of conflicts over the ownership of gems,
gem mines, and the knowledge associated with them—topics that are *not*
covered here. Gender and empire may seem a long way from natural kinds,
a philosophers' term for categories that reflect the structure of the natural
world rather than the interests of human beings. And yet, gems loom large in
the Anglophone literature on natural kinds, with jade, emerald and diamond
being especially common examples in this literature. The history of gem classification might be told as a form of "applied metaphysics," to use Lorraine
Daston's term—as an effort by naturalists to work out the relationship between particulars and universals.[94]

Readers looking for a thorough account of these themes in the following
pages will be disappointed. That said, there are a few glimpses of how they
could be developed. Women evaluated gems as much as men, sometimes in

---

94. Colonial commerce is a major theme of Sabel's thesis on gem knowledge in early mod-
ern Eurasia, "Rare Earth." For a survey of gems as natural kinds, see Joseph LaPorte, *Natural
Kinds and Conceptual Change* (Cambridge, UK: Cambridge University Press, 2004), 94–100. On
jade, see also Ian Hacking, "The Contingencies of Ambiguity," *Analysis* 67, no. 4 (2007): 269–77.
"Applied metaphysics" in Lorraine Daston, "Type Specimens and Scientific Memory," *Critical
Inquiry* 31, no. 1 (2004): 157.

ways that fed into male-dominated scientific institutions such as the Royal Society of London and the Paris Academy of Sciences. Katherine Jones, also known as Lady Ranelagh and as Robert Boyle's sister, is an example discussed in chapter 3. More broadly, the science of gems was often done in private households, spaces that were more open to women in early modern Europe than the lecture halls of universities or the meeting rooms of scientific societies. Gem science may be seen as a form of thrifty science, to borrow Simon Werrett's phrase—a by-product of the material culture of the early modern household.[95]

Similar points hold for empire. Much gem science done in Europe relied on the evaluative expertise of people outside Europe. Much of this expertise became available because of imperial projects that were driven in part by the European quest for gems in Asia and the Americas. This expertise included some of the central techniques of European gem appraisal, such as the hydrostatic balance and the square rule for pricing diamonds. There is a tantalizing possibility that the first European effort to quantify gem color, at the Jardin du Roi in Paris, was inspired by the South Asian practice of judging gems in a dark room.[96] There is every reason to think that gender and empire were constitutive of gem science, just as they were constitutive of other branches of early modern science.

Finally, regarding natural kinds, there is much more to say about the historical relationship between natures and qualities. The natures and qualities of gems were treated as two sides of the same coin until about 1800. Even after 1800, the division between the two was a division of labor rather than something more fundamental, as argued in chapter 7. Perhaps we need a philosophy of natural qualities to go with the philosophy of natural kinds. What such a philosophy would look like is beyond the scope of this book. My hope is that the following chapters will stimulate further research on these and other possibilities raised by the history of gem science. One possibility, of course, is that "the Scientific Revolution" and "the collapse of Aristotelean natural philosophy" are the wrong concepts for understanding gem knowledge in Europe in the centuries before 1800. Only time will tell whether this possibility is favored by the evidence.

<hr />

95. Simon Werrett, *Thrifty Science: Making the Most of Materials in the History of Experiment* (Chicago: University of Chicago Press, 2019).

96. See chap. 5, note 44. For more on gem evaluation in early modern Asia, see Sabel, "Rare Earth," 187–93, 197–98, 216–21. More broadly, the gem expertise of South Asian and Southeast Asian people is a major theme of Sabel's thesis.

Gem Classification and Renaissance Natural History

The number of plant species known in Europe increased by an order of magnitude in the sixteenth century, from about eight hundred species to about six thousand.[1] Not so for gems. In the same period, the number of gems known in Europe rose from a modest thirty species to an equally modest forty species. The difference is all the more striking given the significance of the "species explosion" in the history of botany.[2] The new plant species led to new classification schemes, ones that were sufficiently capacious, flexible and precise to contain the tide of novelty. The strangeness of the new species—twenty times more new species were introduced into Europe in the sixteenth century than in the previous 2,000 years[3]—amplified their impact.[4] This account, so compelling in the case of plants, makes a mystery of the changes in the study of gems in the same period. The ancient way of classifying gems was transformed by a new concern for the hardness of gems and for the distinction between "oriental" and "occidental" gems. If new species did not drive this change, what did?

I shall give a transmaterialist answer to this question. There are two parts to the answer. First, the natural history of gems did not follow the same

---

1. Brian Ogilvie, *The Science of Describing: Natural History in Renaissance Europe* (Chicago: University of Chicago Press, 2006), 139 and 208 (new species).

2. Smith and Delbourgo, "In Kind," 300.

3. Alan G. Morton, *History of Botanical Science: An Account of the Development of Botany from Ancient Times to the Present Day* (London: Academic Press, 1981), 118.

4. The quantity and strangeness of new plant species is emphasized in: Ogilvie, *Science of Describing*, 208 (quantity), chap. 5 (quantity and strangeness); Daston and Park, *Wonders and the Order of Nature*, esp. chaps. 6 and 7 (strangeness); Findlen, "Natural History," 448–54 (strangeness and quantity).

trajectory as the natural history of plants. There was an influx of foreign gems in sixteenth-century Europe, as there was of plants, but the key in the history of gems was not the abundance or strangeness of the new specimens but their new origins. The new gems were of the same kind as the old gems, but they came from new places. This mattered for gem classification because these objects had always been classified according to where they came from. When European perceptions of the global distribution of gems changed, so did European classifications of gems. This explains the gradual rise in the distinction between "oriental" and "occidental" gems in natural history in the sixteenth century, a period when the same distinction became salient as a result of European expansion in the Americas and the Indian Ocean.

The second part of the answer is this: geographical expansion was not the only factor in the reclassification of gemstones. The growth of the lapidary arts in Europe, especially in Florence and Prague, was also significant. Gem cutters were sensitive to the hardness of stones, which they equated with brilliance. As cutters came into contact with university-trained naturalists, the latter began to pay more attention to hardness. Of course, historians have long argued for the significance of European expansion and the codification of the crafts in Renaissance science.[5] But these two material drivers of science have usually been studied separately, with some historians focusing on the "coercive empiricism" of the crafts and others on the "possessive empiricism" of merchants and imperialists.[6] The former are associated with experimental sciences such as physics and chemistry; the latter with sciences of collecting and surveying, such as botany and cartography. Yet both types of empiricism were important in the natural history of gems. They combined to produce the notion of a hard oriental stone, along with its mirror image, the soft occidental stone.[7]

This may give the impression that the Renaissance was a clean break with the past as far as the natural history of gems was concerned. Certainly, the

5. There is a large literature on both topics. For recent surveys, see Pamela Long, *Artisan/Practitioners and the Rise of the New Science, 1400–1600* (Corvallis, OR: Oregon State University Press, 2011); and Poskett, *Horizons,* chap. 1.

6. "Coercive empiricism" is Floris Cohen's term: *How Modern Science Came into the World,* 113–34. "Possessive empiricism" is my adaptation of the title of Findlen, *Possessing Nature.* The two sorts of empiricism have sometimes been studied together, especially with regards to mining and pharmacy. Examples are Barrera-Osorio, *Experiencing Nature,* chap. 3; Boumediene, *Colonisation du savoir;* Valentina Pugliano, "Natural History in the Apothecary's Shop," in Curry et al., *Worlds of Natural History,* 44–60; and Boumediene and Pugliano, "La route des succédanés."

7. This chapter expands on Michael Bycroft, "Anselmus Boethius de Boodt and the Emergence of the Oriental/Occidental Distinction in European Mineralogy," in Bycroft and Dupré, *Gems in the Early Modern World,* 149–72.

distinction just mentioned was much more widespread in 1600 than it had been a century earlier. But the distinction emerged within an older framework. The basic practice of grouping gems by color and locality was ancient. It was disseminated by Pliny the Elder in a hugely influential chapter of his *Natural History*, written in the first century AD. This practice persisted deep into the eighteenth century, despite the growing importance of the distinction between hard and soft stones. Medieval lapidaries should not be discounted either, despite their preoccupation with the medical, moral, and spiritual virtues of stones. These virtues were a classificatory device; they helped to define gem species. They are easy to dismiss as unscientific because they were evaluative. But they were no less evaluative than the judgments about the price and brilliance of gems that underlay the Renaissance novelties already mentioned.

This chapter is therefore a call to reconsider some widespread assumptions about the history of natural history. One assumption is that there was a sharp contrast between "magical" and "scientific" approaches to gems—in fact they shared an interest in classification and evaluation.[8] Another is that classification only really began in the eighteenth century—in fact it was an ancient, medieval, and Renaissance practice.[9] A third tendency is to focus on the formal trappings of classification, such as tables and nomenclatures—in fact there was no correlation between the use of such devices and the uptake

8. This is the main organizing principle of Joan Evans, *Magical Jewels of the Middle Ages and the Renaissance, Particularly in England* (Oxford: Clarendon Press, 1922). A compressed version of the narrative is the introduction to Joan Evans and Paul Studer, *Anglo-Norman Lapidaries* (Paris: Edouard Champion, 1924), ix–xx. See also the short discussion in George Sarton, *Introduction to the History of Science*, vol. 1 (Malabar, FL: Robert E. Krieger, 1975 [1927]), 764–65. Adams, *Birth and Development of the Geological Sciences*, 143–61, is less erudite and less interpretative than Evans's book but written in the same spirit; see especially pp. 143, 149, 159, 169. The same goes for Thorndike, *History of Magic and Experimental Science*, vol. 6, chap. 36. My account builds on the "secular" view of medieval lapidaries in Brigitte Buettner, *The Mineral and the Visual: Precious Stones in Medieval Secular Culture* (University Park, PA: Penn State University Press, 2022), esp. part II.

9. Key sources for this assumption are Foucault, *Les mots et les choses: Une archéologie des sciences humaines* (Paris: Gallimard, 1966), chap. 5; and William B. Ashworth, "Natural History and the Emblematic World View," in *Reappraisals of the Scientific Revolution*, ed. David C. Lindberg and Robert S. Westman (Cambridge, UK: Cambridge University Press, 1990), 303–32. The wide acceptance of Foucault's account in the twenty-first century is noted in Margócsy, *Commercial Visions*, 44–45. My approach is indebted to works such as Andrea Guasparri, "Explicit Nomenclature and Classification in Pliny's Natural History XXXII," *Studies in the History and Philosophy of Science Part A* 44, no. 3 (2013): 347–53; Ogilvie, *Science of Describing*, chap. 5; and Sachiko Kusukawa, *Picturing the Book of Nature: Image, Text, and Argument in Sixteenth-Century Human Anatomy and Medical Botany* (Chicago: University of Chicago Press, 2011), chaps. 5 and 6.

of the notion of a hard, oriental gem. Informal classification was as innovative as the formal kind.

From Luxuries to Virtues

Perhaps the most important of these informal classifications of gems, at least for naturalists in Renaissance Europe, was the one in book 37 of Pliny's *Natural History*.[10] Pliny drew on many earlier authors, including Aristotle and his student Theophrastus, but his account of gems was longer and more detailed than either of its Greek precursors.[11] The bulk of Pliny's book concerned twenty-six species that he called "the principal gemstones" (*principales gemmae*). These included what would now be called precious stones, such as diamond and emerald, as well as what are now sometimes called semiprecious stones, such as amethyst and chalcedony. To these we can add amber and rock crystal, which Pliny described early in the book but which were not, he maintained, recognized as gemstones. Pearls are another borderline case, since Pliny recognized them as gemstones but described them in a different book in the same work. At the end of book 37 he also listed a large number of stones that were not the "principal" gemstones and that he described more briefly than the principal ones; some of these were what he called "plebian gems" (*plebeiae gemmae*) as opposed to "noble gems" (*nobiles gemmae*).[12] Most of these stones are now impossible to identify. Those that can be identified (such as agate, jet, ambergris, and geodes) suggest that this was a miscellaneous collection of minerals that had little in common except the possession of some notable property, such as a striking shape, color, or pattern. It makes sense, then, to think of Pliny's gems as being his twenty-six principal gemstones plus amber, rock crystal, and pearl.[13]

10. This section builds on Sydney H. Ball, *A Roman Book on Precious Stones* (Los Angeles, CA: Gemological Institute of America, 1950), chap. 3.

11. On Aristotle's scheme, see Adams, *Birth and Development of the Geological Sciences*, 78–84. On Theophrastus, see *Theophrastus on Stones: Introduction, Greek Text, English Translation, and Commentary*, ed. Earle Radcliffe Caley and John F. C. Richards (Columbus, OH: Ohio State University Press, 1956 [ca. 300 BC]).

12. Pliny, *Natural History*, vol. 10, 276, 324.

13. Pliny, *Natural History*, vol. 10, bk. 37, chaps. 9–11 (amber and rock crystal), p. 205 (not "acknowledged" as gemstones), p. 213 (pearls), chaps. 15–53 (the twenty-six principal gemstones), p. 277 ("principal gemstones"), chaps. 54–74 (miscellaneous stones), p. 197 ("common" stones), and pp. 283 (jet and ambergris), 285 (sardonyx), and 289 (geodes). I have omitted coral from my account of Pliny's gems, since he did not mention coral in bk. 37, though he did describe it in detail in bk. 32, chap. 11.

These twenty-nine stones were neatly classified by Pliny. They were organized by rank, color, and geographical origin. Rank determined the overall order in which the stones appeared in the chapters. As Pliny announced at the start of the work, he was going to deal with "gemstones that are acknowledged as such, beginning with the finest." He made it clear that diamond was the most valuable gem, followed by pearl, the green stones known as *smaragdus*, and opal; he also described the stones in that order.[14] He was equally explicit about color, writing near the end of the work that he had "discussed the principal gemstones, classifying them according to their color." Accordingly, he led the reader through the red, green, purple and white stones, in that order. Statements such as "We must describe the properties of all the other fiery red gemstones," and "we shall assign to another category purple stones," marked the transition from one color-category to the next.[15] Geographical origin was the basis for the division of the species into varieties. Consider the *carbunculi*, the first of the fiery red gemstones. These came in several varieties, according to Pliny, the main ones being the Indian, Garamantic, Ethiopian, and Alabandian. Pliny went on to distinguish between Indian sardonyx and Arabian sardonyx, Egyptian amethyst and Galatian amethyst, Ethiopian topaz and Arabian topaz, and so on for most of his twenty-nine species. Only a few species—eight out of twenty-nine—were not divided into varieties based on their place of origin.[16] In sum, there was a consistent division of gems into what later naturalists would have called classes, species, and varieties, all of which Pliny called *genera*. Species such as *adamas* and smaragdus each corresponded to one chapter of the book; varieties were covered within each chapter; and classes were groups of chapters indicated by sentences such as the ones quoted.

Pliny's classification makes sense in light of his conception of gemstones as luxuries—as expensive objects prized mainly for the pleasure they gave to the viewer. It was "the variety, the colors, the texture, and the elegance of

14. Ibid., 297 (diamond), 213 (pearl, smaragdus), 229 (opal), 205 ("the finest")

15. Ibid., 277 ("according to their color"), 239 ("fiery red gemstones"), chaps. 32–39 (green stones, and some blue ones, though not labeled as such), 263 ("purple stones"), 129 and 131 ("white stones," or "bright colorless stones").

16. Localities in ibid., 239 (carbunculi), 233 (sardonyx), 263 (amethyst), 267 (topaz). Not divided by locality: lyncurium, opal, anthracites, malachite, lapis lazuli, hyacinthus, leucochrysi, iris. My account expands on remarks in Patrick R. Crowley, "Factitious Gems and the Matter of Facts in Pliny's Natural History," in *The Nature of Art: Pliny the Elder on Materials*, ed. Anna Anguissola and Andreas Grüner (Turnhout: Brepols, 2020), 248–49; Michaeli Maria Elisa, "Agate: Fortunes and Misfortunes," in *Pliny on Materials*, ed. Anna Anguissola and Andreas Grüner (Turnhout: Brepols, 2020), 288, 295; and Géraldine Vœlke-Viscardi, "Les gemmes dans l'*Histoire Naturelle* de Pline l'Ancien: Discours et modes de fonctionnement de l'univers," *Museum Helveticum* 58, no. 2 (2001): 113–18, 120–22.

gems" that appealed to Pliny and his Roman readers.[17] This justified the inclusion of amber and rock crystal in the chapter, as well as the vessels made of
agate or fluor-spar and known as *murrina*. Though these were not gemstones
(*gemmae*), they were luxuries (*deliciae*), the Latin word suggesting pleasure
and delight as well as great expense.[18] Pliny was wary of luxury consumption,
but he made a partial exception for gems. His own descriptions show that
he valued them primarily for the pleasure they gave to the eye. Pearl ranked
highly in his scheme because of its luster, roundness, size and smoothness;
emerald came next because "no color has a more pleasing appearance."[19]
Color helped to distinguish between classes, as we have seen. Color, along
other visual properties, was also a way of a distinguishing varieties. Among
the carbunculi, for example, the best stones had a "brilliant, colorless reful-
gence, so that when placed on a surface [they] enhance the luster of other
stones that are clouded at the edges."[20] Color and brilliance were the basis
for the distinction between "male" and "female" varieties, a distinction Pliny
took over from Theophrastus and applied to several species.[21] Visual properties were much more important to Pliny than hardness, which featured in the
description of three species whereas color or brilliance featured in descriptions of all of them.[22] Medical and moral properties were even more marginal
than hardness, with Pliny dismissing the idea that—for example—amethyst
could shield a wearer from spells, hail, locusts, and drunkenness.[23]

   The irony is that it was precisely such properties that filled the lapidaries of the Middle Ages. The lapidary was a literary genre that drew heavily,
though indirectly, on Pliny's *Natural History* and Theophrastus's *On Stones*.
Lapidaries had the same structure as their ancient predecessors: a list of species of gems accompanied by short descriptions of each species. The novelty
was the inclusion in these descriptions of many properties that Pliny and
Theophrastus omitted or slighted. Modern historians have several labels for
these properties: lore, magic, allegory, medicine, and Christian symbolism.
The first label is a reminder that many of the properties in question had been
ejected from mainstream natural history by the eighteenth century. The other

17. Pliny, *Natural History*, vol. 10, bk. 37, 165.
18. Ibid., chap. 8, p. 186 (deliciae).
19. Ibid., 213.
20. Ibid., 241.
21. Male/female distinction in 239, 239 n. g (carbunculi), 245 (sandastros), 252 (sard), 263
(cyanus). Cf. Caley and Richards, *Theophrastus on Stones*, 124–26. Note also Pliny's distinction
between stones valued by women only and those valued by men and women, 231–33.
22. Pliny, *Natural History*, vol. 10, bk. 37, 209 (adamas), 247 (sandastros), 253 (peridot).
23. Ibid., 265–67.

labels are a reminder that we should not write off these beliefs as mere ignorance or credulity. They were grounded in the formidable intellectual apparatus of pagan philosophy and Christian theology. Aristotle's cosmos, with its nested, crystalline spheres, lent itself to the idea that the stars and planets can endow bodies on earth with remarkable powers.[24] Aristotle's distinction between matter and form helped to account for phenomena that could not be explained in terms of the elements (earth, air, fire, and water) and the elementary qualities (heat, cold, dryness, and wetness). The ability of sapphire to cure abscesses was due, not to these elements or qualities, but to something else that philosophers came to call the "specific substantial form" of sapphire.[25] This substantial form was invisible, something that chimed with the philosophy of Plato and the theology of the Church Fathers, which shared the idea that visible things are a window onto invisible ones.[26] These theories were backed up by discussions of specific gems in authoritative texts. Ancient Greek physicians, Arabic astronomers, Latin encyclopedists, and medieval moralists, all supplied material for the authors of lapidaries, as did the Christian Bible.[27] Gems packed a symbolic punch out of all proportion to their size.

What is the best label for this nest of associations? "Virtue" is a good candidate. The word is widely used in the secondary literature on medieval lapidaries, but without much attention to its scope and significance, or even to the fact that it was used in the lapidaries themselves.[28] The Latin *virtus* is usually translated as "power," but it also denoted excellence, virility, and moral probity. The word was favored by Marbode of Rennes and Albert the Great, the authors of the two most important lapidaries written in Europe between

24. Brian Copenhaver, *Magic in Western Culture: From Antiquity to the Enlightenment* (Cambridge, UK: Cambridge University Press, 2015), 36.

25. Ibid., 115–20. Albert the Great, *Albertus Magnus' Book of Minerals*, trans. Dorothy Wyckoff (Oxford: Clarendon Press, 1967 [1569]), 65.

26. Peter Harrison, *The Bible, Protestantism, and the Rise of Natural Science* (Cambridge, UK: Cambridge University Press, 2001), chap. 1.

27. For details, see Evans, *Magical Jewels*, chaps. 1–5. See also the introductions to critical editions, such as Dorothy Wyckoff, "Introduction," in Albert, *Book of Minerals*, xiii–xlii; and John M. Riddle, "Preface," in Marbode of Rennes, *Marbode of Rennes' (1035–1123) De lapidibus*, trans. C. W. King (Wiesbaden: Steiner, 1977 [ca. 1096]), ix–xii.

28. For example, Evans used the word "virtue" frequently but was much less explicit about its meaning than she was about the word "magic." In practice, she tended to use the words "virtue," "magical," "magical virtue" and "magical property" interchangeably. Examples are Evans, *Magical Jewels*, "Preface," 14, 16, 17, and 20; and Joan Evans and Mary Sergeantson, *English Mediaeval Lapidaries* (London: Oxford University Press, 1933), xi. Partial exceptions to this rule are Riddle, "Preface," x; Buettner, *The Mineral and the Visual*, 3 and passim.

the first century and the sixteenth century.[29] Marbode began his lapidary *On Stones* by referring to "the native virtue by each stone possessed."[30] Albert gave his lapidary the title "On the names and virtues of stones." He also gave a detailed account of what he called "The causes of the virtues of stones."[31] This account was reproduced in a Latin lapidary published in 1502; the eighteenth-century translation of this lapidary rendered the Latin "virtus" as "virtue."[32] Meanwhile, Marbode's lapidary was translated into French and English, with the terms "vertues" and "vertus" appearing frequently in these translations and in other vernacular lapidaries from the twelfth century to the end of the fifteenth.[33] The title of Robert Boyle's *Essay About the Origine and Virtues of Gems*, published in 1672, is an echo of the medieval usage.[34]

The word could refer to many things in medieval lapidaries. Albert, for example, drew no distinction between the medical, moral, magical, and practical virtues of stones. "Curing abscesses, expelling poison, reconciling the hearts of men, bringing victory, and the like," was his summary of the sorts of virtues he had in mind.[35] He later added two properties that we now think of as physical rather than moral: the power of the magnet to attract iron and the (alleged) power of the diamond to restrict this power in the magnet.[36] Albert distinguished between stones that operate through their substantial form and those that derive their powers from their resemblance to a celestial body, such as a stone carved with the likeness of the constellation Aquarius.[37] The latter stones, he explained, have a virtus that is derived from the virtus of the heavens and that has much in common with the virtus of substantial forms, such as the tendency to decline in strength over time.[38] The notion of a virtus was equally broad in Marbode's lapidary, where moral and medical virtues

29. On Albert's life and work, see Wyckoff, "Introduction," xiii–xxx. And on Marbode, see Riddle, "Introduction," in *Marbode of Rennes' (1035–1123) De lapidibus*, 1–27.

30. Marbode, *De lapidibus*, trans. King, 34 ("Quin sua sit gemmis divinitas insita virtus").

31. Albert the Great, *Liber mineralium* (Cologne, 1569), xv ("de caussis virtutum lapidu pretiosorum"), 117 ("de lapidibus nominatis et eorum virtutibus"). The first print edition of Albert's work appeared in 1476: Sinkankas, *Gemology*, vol. 1, 11.

32. Camillo Leonardi, *Speculum lapidum* (Venice, 1502), xiii–xvii; idem, *The Mirror of Stones* (London, 1750 [1502]), 48–59.

33. Examples of this usage are in Studer and Evans, *Anglo-Norman Lapidaries*, 29, 22–23, 97, 118, 139, 265; and Evans and Sergeantson, *English Mediaeval Lapidaries*, 17–18, 38, 51, 58, 63.

34. On this work by Boyle, see chap. 3, this volume.

35. Albert, *Book of Minerals*, 55–56.

36. Ibid., 56.

37. Ibid., 128, 134–37, 140–45.

38. Key instances of virtus are in Albert, *Liber mineralium*, 199 ("moveri a virtutibus caelestibus"), 203–4 ("sicut virtutes naturales perdurant in quodam tempore & non ultra, ita est etiam de

were blended with explicitly religious ones. The link between the virtues of people and those of stones was even more explicit in a shorter lapidary, one based on gemstones mentioned in the Bible. Here Marbode associated each gem with a set of moral virtues, with chrysolite denoting wisdom and charity, chalcedony representing those who do good works in secret, and so on.[39]

It may seem odd to speak of classification with regards to medieval lapidaries, with their far-fetched symbolism and their rather haphazard listing of species. But the virtues of stones were bound up with their division into species and varieties, and with the very idea of a gem. Take the division into species. The virtues of gems were due to their specific substantial forms, as Albert explained in his *Book of Minerals*. This meant, for example, that the ability of sapphire to cure abscesses was part of the identity of sapphire. A stone could be large or small, hot or cold, and still be a sapphire; but if it was unable to cure abscesses, there was a question as to whether it was a sapphire or some other species. To use Albert's analogy, intelligence is part of the specific substantial form of humans; if a human lacks this property there is a question as to whether he or she is a human at all.[40] This meant that the virtues of gems could be used as tests of their authenticity, as in the test for garnet shown in figure 1.1.[41] The varieties of gems, no less than their species, acquired a virtuous interpretation. One kind of agate is an antidote against poison, Marbode maintained; another appeases thirst; a third gives off fumes when strewn on an altar.[42] The very idea of a gem was revised in these texts, with Albert and Marbode dealing indiscriminately with the greater and lesser gemstones described by Pliny. Magnetite appeared alongside amethyst in Marbode's lapidary, bloodstone alongside rock crystal.[43] After all, magnetite could scarcely be omitted from a list of virtuous stones if it had the power to save marriages, expose adultery, heal burns, and confer eloquence, as Marbode claimed. These texts are called "lapidaries" for a reason: they are about stones in general (*lapides*) rather than gems in particular (*gemmae*).[44]

This framework could lead to surprisingly modern results. For one thing, the virtues of stones included properties that were later studied under the

virtutibus imaginum"), 218 ("virtutis ligaturarum et suspensionum"), 224 ("lapides virtutibus suarum formarum specierum operantur"). In each instance, "virtus" is translated as "power" by Wyckoff.

39. Marbode, *De lapidibus*, 125–29.

40. Albert, *Book of Minerals*, 65–67.

41. Ibid., 95. Albert called the stone "gerachidem." Of this test, Wyckoff writes: "Possibly this is, in part, a distorted account of the use of arsenic to kill insects" (95).

42. Marbode, *De lapidibus*, 38.

43. Ibid., 53 (amethyst), 57 (magnetite). Cf. Buettner, *The Mineral and the Visual*, 5–6.

44. Albert used "lapis" in the title of his lapidary, and in the body of his discussion of the power of gems, though he used "lapis preciosus" in the title of the latter: *Liber mineralium*, 90, 117.

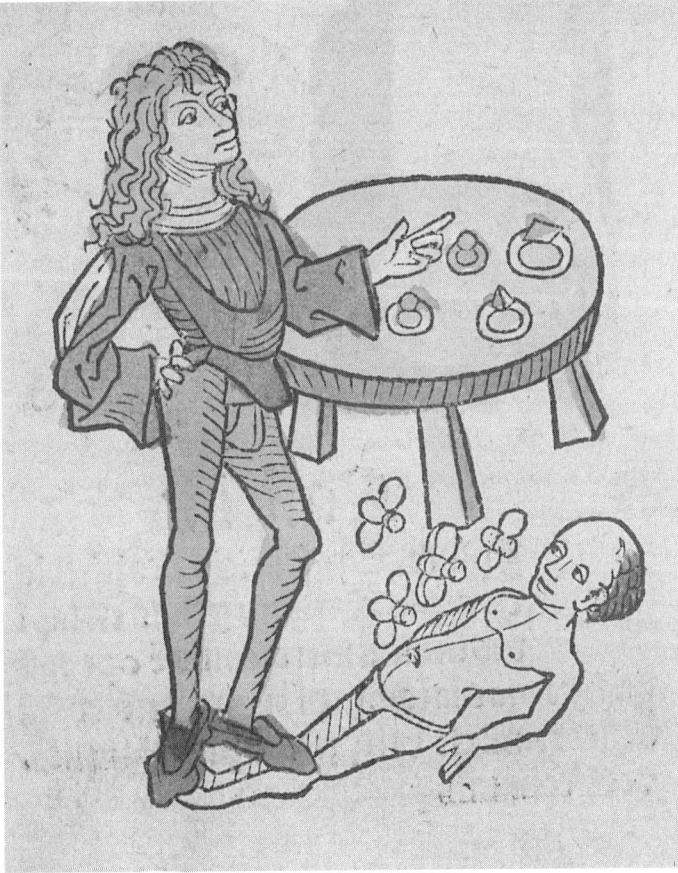

FIGURE 1.1. Albert the Great's test for garnet
One of the alleged "virtues" of garnets was to ward off wasps and flies. Since this was part of the definition of the stone, it was seen as a reliable test of the stone's authenticity. To judge from the behavior of the
insects in the image, the stones on the table are not real garnets. The test was described in Albert, *Book
of Minerals* (c. 1260), though this image is from *Ortus Sanitatis* (Mainz, 1491), "De lapidibus," chap. 58.
Courtesy of the Wellcome Collection.

heading of experimental physics, such as the ability to glow in the dark and
to attract pieces of straw when rubbed. Albert ascribed the latter property
to jet, chalcedony, and a stone that may have been tourmaline, as well as to
amber. Albert went so far as to say that nearly all precious stones possessed
this virtue, a discovery that is usually traced to an article published in 1733.[45]

45. Marbode, *De lapidibus*, 56 (jet), 62 (amber), and 125 (chalcedony). Albert, *Book of Minerals*, 102 (lyncurium, possibly tourmaline), 102 (all precious stones), and 121 (amber). On the 1733
article and its author (Dufay), see chap. 4, this volume.

Marbode had no trouble incorporating the effect into his biblical lapidary, comparing the power of chalcedony to draw chaff to the power of good Christians to draw sinners by the force of their example.[46] Medieval authors also reshuffled the species and varieties of stones in interesting ways.[47] They brought new species to the fore, such as garnet (*granatus*) and turquoise (*turchois*). Pliny had probably encountered both stones, but he did not use their modern names or treat them as distinct species, or even as distinct varieties. On some occasions, medieval authors moved away from Pliny's focus on color, preferring other properties such as hardness and locality instead. *Hyacinthus* is a striking example, with Albert using the term for a range of stones that were very hard and that came in different colors, whereas Pliny had used the term for a pale variety of amethyst. Finally, the old association between gems and the East began to crystallize into a taxonomic principle, with individual species being divided into a superior "oriental" variety and an inferior "occidental" one.[48] Albert applied this principle to sapphire. Another thirteenth-century writer did the same for amethyst, this time using the hardness of the oriental variety as a mark of its quality. There were scattered references to *perles d'Ecosse* and *saphir de Puy* in late medieval jewelry inventories.[49] The principle was slow to spread to other species. As late as 1502, Camillo Leonardi only applied it to three kinds of gem, namely pearl, carnelian, and topaz.[50] This was about to change, however, under the combined pressure of trade and tools.

46. Marbode, *De lapidibus*, 125–26.

47. Here I rely on the modern editors of Pliny and Albert, Eichholz and Wyckoff, who compare historical descriptions of stones with the species of modern mineralogy. Albert, *Book of Minerals*, 96 (granatus), 97 (hyacinthus), 123 (turchois); Pliny, *Natural History*, bk. 37, 252 n. d (callaina), 218 n. c (Cyprian smaragdus), 238 n. c and 239 n. h (carbunculi includes garnet), 147 n. a and 248 nn. b and c (lychnis includes garnet). As Wyckoff notes, Albert's hyacinthus is surprisingly similar to the modern category of gem-grade corundum. I do not mean to imply that Albert and Marbode had firsthand knowledge of all the gem species they named. That they did not is the moral of Urban T. Holmes, "Mediaeval Gem Stones," *Speculum* 9, no. 2 (1934): 195–204. Several earlier authors, writing in Latin, Sanskrit, and Arabic, had hinted at the similarities between red, blue, and yellow corundum: Content, *Ruby, Sapphire & Spinel*, vol. 1, 17, 28, 42–43, 143, 154.

48. On this association, see Bycroft, "Oriental/Occidental Distinction," 153–55; Buettner, *The Mineral and the Visual*, chap. 7; Paul Freedman, *Out of the East: Spices and the Medieval Imagination* (New Haven, CT: Yale University Press, 2008), 89–103; Marjolijn Bol, "Gems in the Water of Paradise: The Iconography and Reception of Heavenly Stones in the Ghent Altarpiece," in *Van Eyck Studies: Papers Presented at the Eighteenth Symposium for the Study of Understanding and Technology in Painting*, ed. Christina Currie, Bart Fransen, Valentine Henderiks, Cyriel Stroo, and Dominique Vanwijnsberghe (Leuven: Peeters Publishers, 2017), 34–48.

49. Holmes, "Mediaeval Gem Stones," 197, 198, 199.

50. Leonardi, *Speculum lapidum*, f. 27r (carnelian), 36v–37r (pearl), 43v–44r (topaz).

Trade and the Orient

In the year Leonardi's book appeared in Venice, the Portuguese explorer Vasco da Gama made his second voyage to the Indian Ocean. On his first voyage, completed in 1499, he had become the first European to round the Cape of Good Hope by sea, thereby linking the Atlantic and Indian Oceans. In the course of the second voyage, he installed a new official in Cannanore, a town on the Malabar coast of India known as Kāhānānūr in Persian sources. This official traveled with his nephew, a young man by the name of Duarte Barbosa. Over the next decade, Barbosa traveled widely in Asia while working as a clerk and translator in the service of the king of Portugal. He built up a detailed knowledge of the gem trade in the Indian Ocean, from the pearl fisheries in the Persian Gulf to the ruby mines in mainland Southeast Asia. When he returned to Lisbon around 1516, Barbosa worked these observations into a narrative that was eventually published in Italian in 1563.[51] In all this, Barbosa was a man of his time. Many other Europeans commented on the gem trade in the Americas and the Indian Ocean in the wake of the Iberian voyages of conquest and exploration. Narratives by the Portuguese physician Garcia da Orta, the Spanish missionary José de Acosta, and the Dutch merchant Jan Huyghen van Linschoten, were the principal sources for European naturalists writing about gems in the century, but they were not alone.[52] These

51. Mansel Longworth Dames, "Introduction," in Duarte Barbosa, *The Book of Duarte Barbosa, an Account of the Countries Bordering on the Indian Ocean and Their Inhabitants*, ed. Mansel Longworth Dames (Hakluyt Society: Farnham, 1918 [1563]), xxxiii (publication), xxxiv–xxxvi (Barbosa's life), xlviii (return to Lisbon); Barbosa, *Book of Duarte Barbosa*, 82 (Persian pearls), cf. 142–44 and 156–57 (Gujarat carnelian), 202–3 (Bisnager diamonds); Duarte Barbosa, *A Description of the Coasts of East Africa and Malabar*, ed. and trans. Henry E. J. Stanley (Hakluyt Society: Farnham, 2010 [1563]), 208 (Southeast Asian rubies). Barbosa's work was first published in Giovanni Battista Ramusio, *Navigationi et viaggi* (Venice, 1563). Identification of "Cannanore" in Irfan Habib, *An Atlas of the Mughal Empire: Political and Economic Maps with Detailed Notes, Bibliography and Index* (Oxford: Oxford University Press, 1982).

52. Garcia da Orta, *Colloquies on the Simples and Drugs of India*, trans. Clements Markham (London: Henry Sotheran, 1913 [1563]), 296–301 (pearls), 342–52 (diamonds), 353–61 (other precious stones); José de Acosta, *Natural and Moral History of the Indies*, ed. Jane E. Mangan, trans. Frances M. López-Morillas (Durham, NC: Duke University Press, 2002 [1590]), 193–95 (emeralds), 195–96 (pearls); Jan Huyghen van Linschoten, *Voyage of John Huyghen van Linschoten to the East Indies*, vol. 2, ed. Arthur C. Burnell (Cambridge, UK: Cambridge University Press, 1885 [1596]), 133–36 (pearls), 136–38 (diamonds), 139–42 (colored gems), 141–57 (pricing gems), 157–58 (pearls). Comparable narratives are Ludovico de Varthema, *The Travels of Ludovico di Varthema*, ed. George Percy Badger, trans. John Winter Jones (Cambridge, UK: Cambridge University Press, 1863 [1510]), 220 (Pegu rubies); Richard Hakluyt, *The Principal Navigations Voyages Traffiques and Discoveries of the English Nation*, vol. 5 (Cambridge, UK: Cambridge University

writers were stimulated by the stories of Asian gems they found in medieval travel narratives such as those by Marco Polo and John Mandeville, stories they enlarged and corrected.[53]

In the new narratives, gems were treated not as luxuries or as nuggets of virtue but as items of trade. The narratives were filled with data on the price of gems, whether in the form of tables of gem prices or formulas for deriving the price of a gem from its weight.[54] The price of a gem depended on its quality as well as its size, and the quality of a gem depended on where it came from. Barbosa's narrative illustrates the point. His description of gems was divided into sections that were each headed with the name of one kind of stone: diamond, sapphire, topaz, turquoise, and so on.[55] Some of these sections were united by a common term, such as the four sections that all belonged to "the class of rubies," as Barbosa put it. Most sections were divided into varieties, and these varieties were distinguished mainly on commercial and geographical grounds. Barbosa explained that there are two kinds of diamonds in circulation in India, those from the "old mine" in the "Decan kingdom" (really a group of sultanates on the Deccan plateau) and those from the "new mine" in the "kingdom of Narsynga" (really the Vijayanagara empire). The latter were "not so good" as the former, going for two-thirds the price. They also had a different commercial trajectory, with the Vijayanagara diamonds being cut and polished in the empire and the Deccan diamonds being exported in their raw form.[56] Barbosa treated colored gems in the same way. His discussion of rubies is especially interesting because it anticipates the modern distinction between red corundum and other kinds of red gem. The best rubies, Barbosa explained, come from the island of Ceylon and the "country of Peygu" (the Burmese Taungoo Empire, with its capital at modern-day Bago). Another

Press, 1903–1905 [1589–1600]), 463–512 (Fitch voyage), esp. 506–7 and 510 (gem trade); Jacques de Coutre, *The Memoirs and Memorials of Jacques de Coutre: Security, Trade and Society in 16th- and 17th-Century Southeast Asia*, ed. Peter Borschberg, trans. Roopanjali Roy (Singapore: NUS Press, 2014 [1620s]). The context for these and other gem-encrusted travelogues is surveyed in Sabel, "Rare Earth," chap. 2.

53. Bruce Lenman, "England, the International Gem Trade and the Growth of Geographical Knowledge from Columbus to James I," in *Renaissance Culture in Context: Theory and Practice*, ed. Jean R. Brink and William F. Gentrup (Brookfield, VT: Ashgate, 1993), 86–99.

54. Price data in Barbosa, *East Africa and Malabar*, 209–10, 212, 214–15, 217. A prose account with similar data is in Linschoten, *Voyage to the East Indies*, vol. 2, 141–57; note the general statements about price and quality (146–47). Formulas in Lenzen, *History of Diamond Production*, 94–96; Tillander, *Diamond Cuts*, 219; Ogden, *Diamonds*, 203–12.

55. Barbosa, *East Africa and Malabar*, 208–18.

56. Ibid., 213. Names of places and polities from Habib, *Atlas of the Mughal Empire*; Sabel, "Rare Earth," 73–77.

kind of ruby, also from Pegu, cost half as much. A third kind cost the same
as the ones from Pegu, but came from a "kingdom of Balaxayo" located near
Pegu and Bengal (possibly the Mruak-U Kingdom).[57] To judge from Barbosa's
descriptions, these three varieties correspond closely to the modern distinction between red corundum, red spinel, and rose spinel.[58] Price and locality
helped to make fine distinctions between different grades of gem.

The oriental/occidental distinction was a large-scale application of this
way of thinking about gems. An "oriental" gem was understood to come from
a particular place—in modern terms, Asia or the Middle East. But it was also
understood to be a gem of unusually high quality and price. The term was
widely used in sixteenth-century books on the gem trade. It was most often
applied to pearls, as had been done since at least the fourteenth century.[59]
Orta, Linschoten, and the English adventurer Ralph Fitch all used the term
"oriental" to refer to pearls originating from Asia, especially those from the
Persian Gulf. According to Orta, they were so named because "they come
from the East, the straight of Ormuz being to the east with respect to our
Europe." Yet it is clear from other narratives that the term had an evaluative
meaning as well as a geographical one. For instance, Fitch wrote that pearls
from the Coromandel coast of India are "not so orient" as pearls from the
Persian Gulf, a statement that makes little geographical sense. Fitch meant
that the Persian pearls were better than Coromandel ones, not that Persia
was further to the east than the Coromandel coast.[60] In the case of emeralds,
too, there was more to the term "oriental" than geography. When Europeans
came across emeralds in South America, they associated them with the green
stones that Pliny had called smaragdus and that were thought to originate in
Asia. The question arose as to the relative merits of the old "oriental" gems
and the new "Peruvian" ones. Linschoten was undecided on this question,
but other writers assumed that new emeralds were an inferior version of the
old ones, just as the people of the New World were seen as inferior versions

57. Barbosa, *East Africa and Malabar*, 208–13. On the identity of "Balaxayo," see also ibid.,
213n1; Sabel, "Rare Earth," 87–88; Jean-Baptiste Tavernier, *Travels in India*, edited by Valentine
Ball (London: Macmillan, 1889 [1676]), vol. 1, 382n1.

58. Identification with modern species based on Jiří Kouřimský, *Encyclopédie des minéraux*
(Paris: Gründ, 1985), 97, 104–5. Earlier efforts to distinguish red corundum from spinel are noted
in Holmes, "Mediaeval Gem Stones," 198–99; Content, *Ruby, Sapphire & Spinel*, vol. 1, 14n24, 18,
46, 75, 146–47, 149–50, 154.

59. Pierre Leroy, *Statuts et privilèges du corps des marchands Orfèvres-Joyailliers de la ville de
Paris* (Paris, 1734), 133.

60. Orta, *Colloquies on the Simples and Drugs*, 297; Hakluyt, *Principal Navigations*, vol. 5, 501
("not so orient"), cf. 468 and 501; Linschoten, *Voyage to the East Indies*, vol. 2, 133, 157.

of the people of the Old World.[61] This language seems to have spread to other species in the second half of the sixteenth century. In a 1572 book, the Spanish goldsmith Juan de Arfe y Villafañe gave tables of prices for *diamant oriental, rubi oriental, esmeralda oriental,* and *espinela oriental.* Linschoten added sapphire and topaz to this list in his *Itinerario.*[62] By 1600, European traders recognized an "oriental" variety of each of the gem species they considered most precious.

Why did this happen? Part of the explanation is the emergence of the oriental/occidental distinction in the wider culture of sixteenth-century Europe, following Christopher Columbus's encounter with the "Occidental Indies" and the subsequent partition of the globe between Spain and Portugal in the Treaty of Tordesillas of 1494. But traders also had their own reasons for reorienting the geography of gems. They had a new picture of the global distribution of gems, one that was hard to reconcile with the old distinction between gems from India and gems from everywhere else. Barbosa noted that the finest rubies occur as far east as Pegu, an observation repeated by later authors.[63] There was a similar story for pearls, which Albert the Great had located in "India" but which sixteenth-century travelers associated with the Persian Gulf, whose pearls they considered superior to those from the southern tip of India.[64] The far west also turned out to be a source of high-quality gems, as illustrated by the emerald mines that Acosta located in "Peru" and the "New Kingdom of Granada."[65] These stones may have been inferior to "occidental" emeralds, but they were emeralds nonetheless. Finally, the geographical gray area of northern Africa faded away as a source of gems. Pliny had located the mines of several major species, including carbunculi, in the region of northeast Africa that he called "Ethiopia." Most of these species were probably not mined in this region, even in Pliny's time, although they did pass through on their way to the Roman Empire.[66] This confusion was cleared up in the sixteenth century, when European travelers stopped

---

61. Linschoten, *Voyage to the East Indies,* vol. 2, 140, 154; Lane, *Colour of Paradise,* 101, 241–44.

62. Juan de Arfe y Villafañe, *Quilatador de la plata, oro, y piedras* (Valladolid, 1572), 50v; Linschoten, *Voyage to the East Indies,* vol. 2, 133.

63. Barbosa, *East Africa and Malabar,* 208; Orta, *Colloquies on the Simples and Drugs,* 361; Linschoten, *Voyage to the East Indies,* vol. 2, 140.

64. Albert, *Liber mineralium,* 105; Barbosa, *Book of Duarte Barbosa,* 82; Linschoten, *Voyage to the East Indies,* vol. 2, 140.

65. Acosta, *Natural and Moral History,* 194.

66. Pliny, *Natural History,* bk. 37, 208 n. c (adamas), 238 n. e (carbunculi), 215 n. f and 218 n. b (smaragdus), 267 n. d (chrysolitus). In Pliny's time there were indeed productive emerald mines in northeast Africa, near the town of Qift in modern Egypt.

reporting locations in Africa for diamonds, emeralds, and carbunculi. These stones were now thought to come from places that lay somewhere east of the Red Sea or somewhere west of Cyprus. A gap between eastern and western stones emerged just as India lost its monopoly on good eastern stones, at least in the eyes of Europeans. The oriental/occidental distinction began to make gemological as well as geopolitical sense.

The local distribution of gems mattered as much as their global distribution. Simply put, gems of different colors were often found in the same place. This was unknown to Pliny and to Albert, who had much to say about the regions or islands on which gems occurred but very little to say about the locations of the mines in which they were found. By contrast, the proximity of gems of different colors was familiar to miners and merchants who worked the ruby mines in Pegu and Ceylon. According to Barbosa, a single handful of gravel from Ceylon could contain stones of several different colors, which the skilled cutters on the island were able to differentiate.[67] The same cutters knew that stones of different colors could have similar properties. Rubies were red, sapphires were blue, and topazes were yellow, but they were all hard and dense and felt cold to the touch, which suggested that they were in fact the same kind of stone. Barbosa almost certainly obtained this information from local people in Ceylon or on the Malabar coast, whom he cited explicitly for similar information in other parts of his narrative.[68] Barbosa also reported the existence of transparent rubies, sapphires, and topazes, and of specimens that combined two or more of these colors.[69] In South Asia, bicolored rubies were common enough to have a name, *nilacandi*, in a local language.[70] In a language used in Pegu, sapphires were referred to as "blue rubies" and topazes as "yellow rubies."[71] European travelers theorized that rubies were white to begin with and changed color as they ripened, like fruit, in the tropical sun.[72] Data such as these were later used—for example, by Robert Boyle in 1672 and by René-Just Haüy in 1801—to argue that the classification of gems by color

67. Hugo Miguel Crespo, "The Plundering of the Ceylonese Royal Treasure, 1551–1553: Its Character, Cost, and Dispersal," in Bycroft and Dupré, *Gems in the Early Modern World*, 46.

68. Barbosa, *East Africa and Malabar*, 208, 211–12, 216.

69. Ibid., 213.

70. Orta, *Colloquies on the Simples and Drugs*, 357; Linschoten, *Voyage to the East Indies*, vol. 2, 139–40.

71. Tavernier, *Travels*, vol. 2, 101. See also the summary of historical names for spinel and corundum, in Content, *Ruby, Sapphire & Spinel*, vol. 1, chap. 7.

72. Orta, *Colloquies on the Simples and Drugs*, 357; Acosta, *Natural and Moral History*, 194–95.

was a mistake.[73] In hindsight, we can say that color-based classification was one of the casualties of European engagement with the wider world (plate 1).

Tools and Hardness

In the sixteenth century, however, this anti-color argument was not explicit. Barbosa may have written that ruby, sapphire, and topaz were the same kind of stone; but he still defined these stones by their color, and he dealt with them in three different sections in his narrative. Boyle and Haüy were able to dispense with color because they had other criteria to replace it, such as hardness. It is significant, then, that cutters began to draw attention to the hardness of gems at the same time that traders hinted at the variability of their color. Hardness had always mattered to people who worked stones, whether by drilling, engraving, or polishing them. The link between hardness and classification had already been made by Theophrastus, who distinguished between stones that were soft enough to be cut with iron tools and those that could only be cut with tools made of other stones.[74] But hardness became more prominent with the expansion of the lapidary arts in Europe in the fifteenth and sixteenth centuries. The faceting of diamonds was practiced in Europe from around 1400, with powdered diamond being used as an abrasive to smooth the natural faces of diamonds and, eventually, to create new faces.[75] Specialized communities of gem cutters were established in Venice, Antwerp, London, and Paris by 1600.[76] The ancient arts of gem engraving and

73. For example, in Robert Boyle, *An Essay About the Origine and Virtue of Gems* (1672), in *The Works of Robert Boyle*, ed. Michael Hunter and Edward Davis (London: Pickering and Chatto, 1999), 22 (hereafter *Gems*); Haüy, *Traité de minéralogie*, vol. 2, 488–89; Romé de l'Isle, *Cristallographie*, vol. 2, 220–21. Haüy and Romé de l'Isle read about nilacandi in Pierre de Rosnel, *Mercure Indien, ou Le trésor des Indes* (Paris, 1667), vol. 2, 13–14, whose account closely resembles those cited in note 70 as well as Boodt, *Gemmarum*, 72. On the bicolored sapphire in plate 1, see Romé de l'Isle, *Cristallographie*, vol. 2, 219; Bernard Morel, *Les joyaux de la couronne de France* (Paris: A. Michel Fonds Mercator, 1988), 218–19; François Farges and Olivier Segura, *Pierres précieuses: Guide visual* (Paris: Dunod, 2023), 170; and Farges and Kjellman, "Bicentenaire du décès de René-Just Haüy," 22.

74. *Theophrastus on Stones*, 53–54; cf. Marjolijn Bol, "Polito et Claro: The Art and Knowledge of Polishing, 1100–1500," in Bycroft and Dupré, *Gems in the Early Modern World*, 226.

75. Ogden, *Diamonds*, chaps. 5–10.

76. Karin Hofmeester, "Shifting Trajectories of Diamond Processing: From India to Europe and Back, from the Fifteenth Century to the Twentieth," *Journal of Global History* 8, no. 1 (2013): 34–39; René de Lespinasse, *Les métiers et corporations de la ville de Paris*, vol. 2 (Paris, 1892),

gem carving were revived and remastered, especially in Florence.[77] Meanwhile, the first printed descriptions of these arts appeared as part of a wider
trend toward the publication of craft secrets in Renaissance courts.

These trends came together in the work of Benvenuto Cellini, the Florentine sculptor and goldsmith who published a description of his two crafts in
1569. Cellini began with the claim that there are only four precious stones,
each corresponding to one of the Aristotelean elements: ruby is made of fire,
sapphire of air, emerald of earth, and diamond of water.[78] Having stated these
associations, however, Cellini ignored them in the rest of his treatise, where
he played up hardness at the expense of color. Topaz and sapphire may have
different colors, he wrote, but they have the same hardness and therefore belong to the same species (*una medesima spezie*).[79] He made a similar statement about the oriental ruby and the ballas ruby, saying that they had the
same hardness and therefore belonged to the same species, despite their difference in color and their even more striking difference in price.[80] Conversely,
he noted the wide variation in the colors of stones that belonged to the same
species. Hardness was the only constant in diamonds: "however great the variety of these colors [of diamond] is, the wondrous hardness of the stone is
similar in all cases."[81] Cellini had seen white rubies as well as red ones, and he
maintained that these were white by nature (*bianchi naturali*) rather than as
a result of being heated by humans.[82] He also referred to white topaz, white

82. The Antwerp guild of diamond- and ruby-cutters was formed in 1582, the Paris equivalent
in 1584.

77. Louis de Jaucourt, "Pierre gravée," in *Encyclopédie, ou dictionnaire raisonné des sciences,
des arts et des métiers*, vol. 12, ed. Denis Diderot and Jean le Rond d'Alembert (Paris, 1765), 587–
88; Suzanne Butters, "'Una pietra eppure non una pietra': Pietre dure e le botteghe medicee nella
Firenze del Cinquecento," in *Arti fiorentine: La grande storia dell'Artigianato*, vol. 3: *Il cinquecento*
(Florence: Giunti Gruppo, 2000), 133–85.

78. Benvenuto Cellini, *The Treatises of Benvenuto Cellini on Goldsmithing and Sculpture*,
trans. Charles R. Ashbee (New York: Dover Publications, 1967 [1569]), 22. The original work,
in Italian, was published in 1569, as clarified in Suzanne Butters, *The Triumph of Vulcan: Sculptors' Tools, Porphyry, and the Prince in Ducal Florence* (Florence: Olschki, 1996), 152n22. Cellini's
place in the codification of the arts is noted in Paolo Rossi, "'Parrem uno, e pur saremo dua':
The Genesis and Fate of Benvenuto Cellini's Trattati," in *Benvenuto Cellini: Sculptor, Goldsmith,
Writer*, ed. Margaret Gallucci and Paolo Rossi (Cambridge, UK: Cambridge University Press,
2004), 171–200.

79. Cellini, *Goldsmithing and Sculpture*, 23, 41; "una medesima spezie" in Benvenuto Cellini,
I trattati dell'oreficeria e della scultura, ed. Carlo Milanesi (Florence, 1857 [1569]), 66.

80. Cellini, *Goldsmithing and Sculpture*, 23.

81. Ibid., 31, 32.

82. Ibid., 42; "bianchi naturali" in Cellini, *Oreficeria e scultura*, 67.

sapphire, and white amethyst.[83] No wonder that Robert Boyle, reading Cellini's book a century later, reached this conclusion: "Italian Jewellers did not look upon the Tinctures [i.e., the colors] of Gems as anything near so essential to them, as they are commonly reputed."[84]

Cellini's emphasis on hardness makes sense in light of his work as a jeweler, goldsmith, and sculptor. As a jeweler, he appraised gems for his patrons, which meant seeing through the ruses of gem sellers. To distinguish between diamonds and white sapphires, he advised, remember that the former are much harder than the latter. Likewise, rubies were much harder than the colored rock crystal that was sometimes used to imitate them.[85] Although he did not polish diamonds himself, Cellini gave the first published description of the method that is still used today. In doing so he specified the materials involved: a steel wheel coated in oil and diamond dust. The wheel, he wrote, must be made of "the finest steel excellently tempered."[86] Here he touched on a topic that was central to his own practice, especially his invention of a pair of steel die for striking coins and medals. He made these die himself, a process that required careful regulation of the hardness of steel.[87] Like other Florentine sculptors, he made many of his own tools and was all too familiar with their short lifetimes when used on hard materials such as bronze and marble.[88] Tools for use on harder stones, such as porphyry and granite, tended to be heavier and take longer to make than those used on softer species.[89] Tools made of low-quality steel tended to shatter, sending shards of metal into the eyes of the user, as Cellini knew from his own experience.[90] Tool quality was a major concern in the circle of sculptors who worked at the court of the Duke of Tuscany, Cosimo I, in the third quarter of the sixteenth century. Cosimo himself has been credited with one of the major technical breakthroughs of the period, a method of tempering steel so that it was hard enough to work porphyry. The discovery was significant for a dynasty (the Medicis) that had long used the hardness of gemstones as a symbol of justice and durability.[91] The carver who perfected the technique, a certain Tadda,

83. Cellini, *Goldsmithing and Sculpture*, 40.
84. Boyle, *Gems*, 21.
85. Cellini, *Goldsmithing and Sculpture*, 27, 41, cf. 26.
86. Ibid., 31–32; Grodzinski, "Diamond Polishing," 2.
87. Butters, *Triumph of Vulcan*, 198–99.
88. Ibid., 197–98.
89. Ibid., 195, 202.
90. Ibid., 191.
91. Ibid., 94, 108.

worked for Cellini.[92] In sum, hardness was a theme of Cellini's career that cut across a range of materials, from ruby to marble to steel. It makes sense that he saw this property, and not color, as a reliable test of the identity of gems.

Oriental Hardness

Trade and tools both raised problems for color. Trade drew attention to geography, and tools to hardness. But these emerging themes had not yet merged at the end of the sixteenth century. Cellini had little to say about the geographical distribution of gems, whether local or global, in his treatise on jewelry. Barbosa had much less to say than Cellini about their hardness or about the tools for working them. Merchants and artisans both wrote about gems, but rarely together. They were brought together by someone who was neither a merchant nor an artisan but a university-trained physician. Anselm Boethius de Boodt was well-placed to achieve this task during his time as physician to the Holy Roman Emperor, Rudolf II, a position he held from 1604 to 1612.[93] It was in those years that the emperor worked to bring the Florentine tradition of hard-stone carving to his court in Prague. The results were displayed in the imperial Kunstkammer at Prague Castle, which included landscapes made of inlaid stones and vessels carved in jasper and chalcedony.[94] By this time, the observations of Spanish, Portuguese, and Dutch travelers were readily available to scholars who could read Latin and had access to well-stocked libraries. Boodt was familiar with Linschoten's travel narrative, probably through a Latin translation of 1599.[95] Orta's account of Asian drugs was another key source for Boodt, one that he accessed through the Latin

92. Ibid., 148–49.

93. There is no book-length biography of Boodt. Sketches of his life and work tend to focus on the *Gemmarum*: Johannes Erich Hiller, "Anselmus de Boodt als Wissenschafter und Naturphilosoph," *Archeion* 15, no. 3 (1933): 348–68; Wlodzimierz Hubicki, "Boodt, Anselmus Boetius de," in *DSB*, vol. 2, 292–93; Robert Halleux, "L'oeuvre minéralogique d'Anselme Boèce de Boodt 1550–1632," *Histoire et nature* 14 (1979): 63–78; Carlos Gysel, "A. de Boodt, lapidaire et médecine de Rodolphe II," *Vesalius* 3, no. 1 (1997): 33–42; Nicolas Zylberman, "Boece de Boodt, dernier lapidaire et premier gemmologue," *Revue de l'Association Française de Gemmologie* 177 (2011): 17–22. A few other secondary works are cited in Sinkankas, *Gemology*, vol. 1, 127–28.

94. Rudolf Distelberger, "Thoughts on Rudolfine Art in the 'Court Workshops' in Prague," in *Rudolf II and Prague: The Court and the City*, ed. Eliška Fučíková, James M. Bradburne, Beket Bukovinska, Jaroslava Hausenblasová, Lumomír Konečný, Ivan Muchka, and Michal Šroněk (London: Thames & Hudson, 1997), 188–208.

95. Boodt, *Gemmarum*, 74 (Linschoten on ballas ruby), 101 (Linschoten on emerald); idem, *Parfait joaillier*, 188 (ballas ruby), 256 (emerald). Cf. Linschoten, *Voyage to the East Indies*, vol. 2, 154 (emerald), 157 (ballas ruby).

translation compiled by Carolus Clusius, a naturalist who had worked at the court of Rudolf II's predecessor, the Holy Roman Emperor Maximilian II.[96] Finally, the emperor presided over two of the main European sites for metal mining, Saxony and Bohemia. The mining industry expanded rapidly in the sixteenth century; literature on mining and metallurgy expanded as a result. Boodt therefore had access to three types of expertise relevant to gems: Italian hard-stone carving, central European mining, and the global gem trade.

These material specialties came together in Boodt's magnum opus, the *History of Gems and Stones*. Published in Latin in 1609, this work far surpassed the lapidaries of the Middle Ages in length, breadth and accuracy. The crafts were represented by Boodt's account of the cutting, polishing, cleaving, and drilling of gems, an account that was more detailed than Cellini's and contained the first printed illustrations of the tools used in the process (fig. 1.2).[97] The global gem trade was represented by the tables of gem prices that Boodt supplied for pearl, amethyst, garnet, and diamond.[98] The new travel narratives, especially those by Orta and Linschoten, furnished Boodt with his data on the location of gem mines outside Europe. In his chapter on diamonds, for example, Boodt replaced Pliny's vague reference to "Indian" diamonds with three locations in Asia: "the province of Deccan," "the province of Bisnager," and the straits of Malacca. The first two locations correspond roughly to the sultanates of the Deccan plateau and the Vijayanagara empire.[99] Boodt would have been able to locate these places on the maps in

96. Boodt's debt to Clusius's translation is shown by a comparison of the three texts: Orta, *Colloquies on the Simples and Drugs*, 360–61 (amethyst and beryl present, cat's eye from Ceylon, ruby from Bramaa); Garcia da Orta, *Aromatum, et simplicium aliquot medicamentorum apud Indos nascentium*, trans. Carolus Clusius (Antwerp, 1567), 203 (cat's eye, not ruby, from Bramaa), 192–209 (no beryl or amethyst in chapter on gems); Boodt, *Gemmarum*, 126 (cat's eye from Bramaa), 81 (none of Orta's localities for amethyst), 107 (ditto for beryl). Boodt often cited Nicolás Monardes when he clearly meant Orta/Clusius, as noted by Adrian Toll in Boodt, *Parfait joaillier*, 148 n. b, 159 n. a; and Valentine Ball's note in Tavernier, *Travels*, vol. 2, 434. Boodt correctly cited Orta at least once: *Gemmarum*, 73.

97. Boodt, *Gemmarum*, 35–42 (gem working), 69 (diamond cutting), 96 (pearl drilling).

98. Ibid., 65–66 (diamonds), 78 (garnet), 82 (amethyst), 89–90 (pearl).

99. Diamond localities in Boodt, *Gemmarum*, 59–60. Cf. Garcia da Orta, *Colóquios dos simples e drogas he cousas medicinais da Índia* (Goa, 1563), 161. For the identification of *Malaqua* and *Bisnaguer* see Valentine Ball's notes in Tavernier, *Travels*, vol. 2, 87n1, 433, 462–64; and Arthur C. Burnell's note in Linschoten, *Voyage to the East Indies*, vol. 1, 82n3. Boodt seems not to have read Acosta: compare Boodt, *Gemmarum*, 83–92 (pearls), esp. 84–85 (localities), 100 (emeralds), with Acosta, *Natural and Moral History*, 193–95 (emeralds), 195–96 (pearls). On actual gem localities in Asia in the early modern period, see Habib, *Atlas of the Mughal Empire*, maps 15B, 16B; Sabel, "Rare Earth," 73–77, 83–92.

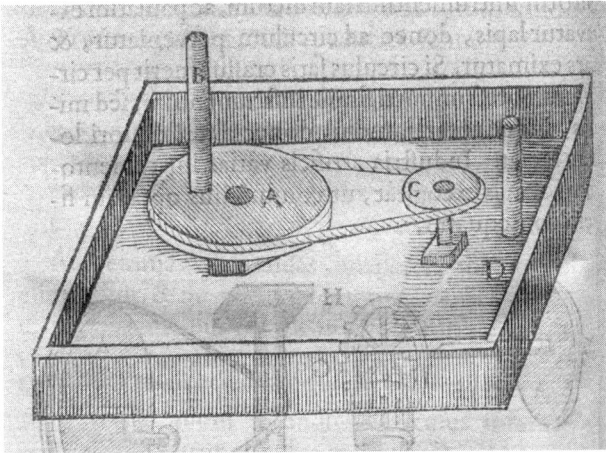

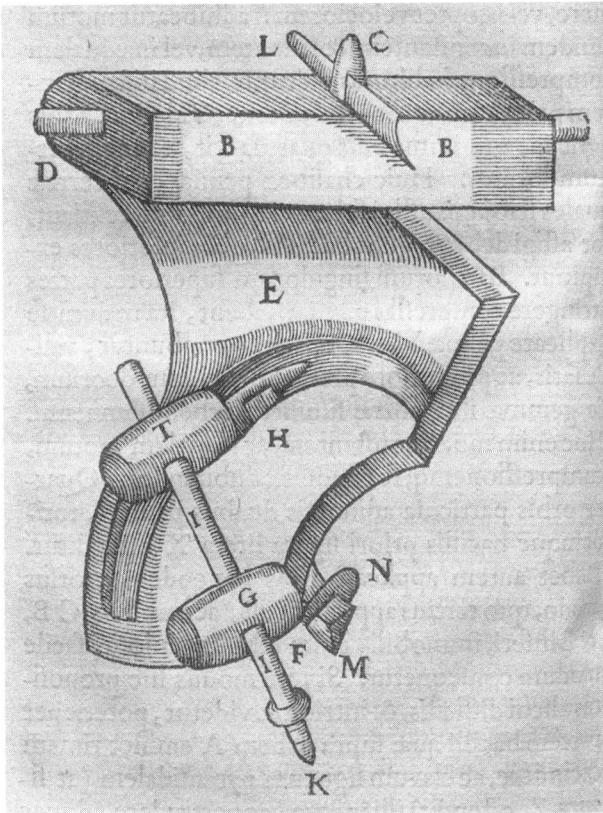

FIGURE 1.2. Gem-cutting instruments from Boodt

The top image shows a device for faceting gems. The wooden rod B is turned by hand, rotating the wooden wheel A. This drives the tin wheel C, which is covered with emery powder mixed with water. The gem is held against C with the help of the wooden rod D and the wooden "quadrant" shown in the bottom image. The quadrant holds the gem at K. The gem can be moved in three dimensions by moving the arm T through the grooved arc, and by rotating the block of wood E around the axis. From Boodt, *History of Gems and Stones*, 2nd ed. (1636 [1609]), pp. 80, 84. Courtesy of ETH-Bibliothek Zürich, Rar 1881, https://doi.org/10.3931/e-rara-15217.

Linschoten's book, along with other putative gem localities (fig. 1.3). Many of these localities were false or approximate—there were no diamond mines in the area of Linschoten's map marked "Decan," for example—but they were a better approximation than Pliny's.

These global data appeared alongside local data that were even more precise. When describing European localities, Boodt often gave the names of towns, fields, or waterways where gems were found, as in his descriptions of German agate, Silesian turquoise, and Prussian amber. Much of these data were new: Boodt mentioned opals of high quality that had recently been discovered in Hungary, as well as similar discoveries made "a few years ago" in Germany, Bohemia, and Silesia, perhaps a result of Rudolf's prospecting expeditions.[100] Boodt did some prospecting of his own, as shown by the scattered references in the *History of Gems* to stones that he found himself, such as a piece of chalcedony he spied in a field near Brussels and an Armenian stone from a mountain near Prague. He searched especially hard for garnets, measuring their sizes with the help of a device that he described and illustrated in his book.[101] The *History of Gems* was a regional mineralogy as well as a global one, owing as much to metal mining in central Europe as to gem mining in Asia and the Americas.[102]

All this contributed to Boodt's major innovation in gem classification, which was to make systematic use of the hard/soft and oriental/occidental distinctions. His description of turquoise was characteristic: "There are two kinds of turquoise," he wrote, "the oriental and the occidental." He went on to say that the oriental kind was more blue, and less green, than the occidental. This implied that the former were preferable to the latter. As he remarked a few pages later, excessively green or white turquoises were "held in contempt." Boodt's use of the term "oriental" in the rest of the treatise was uneven but extensive. Sometimes his evaluations were implicit, sometimes frank: "Among sapphires, some are oriental, the others occidental, and the latter are meaner than the oriental ones." Sometimes he substituted "European" for "occidental," though he often used the latter term for stones that he believed to occur only in Europe. Sometimes he used neither term, referring only to an "oriental" variety and leaving the reader to notice that all the other localities he mentioned were in Europe. Sometimes his use of these terms to subdivide

100. Boodt, *Gemmarum*, 14 (Hungarian opals), 125 (German agate), 134 (Silesian turquoise), 162–63 (Prussian amber).

101. Ibid., 77–78 (Bohemian garnets), 121 (Brussels chalcedony), 239 (Armenian stone).

102. On regional mineralogies in the Holy Roman Empire, see Cooper, *Inventing the Indigenous*, chap. 3.

FIGURE 1.3. Gem localities on Linschoten's map

A map from Linschoten's *Itinerario* (1596) showing part of the Indian subcontinent. The arrows indicate some of the locations named in Boodt's *History of Gems and Stones* (1609). Boodt located diamonds in "Decan" and "Bisnagar"; rubies in "Bisnagar" and "Calicut"; and sapphires in "Bisnagar," "Calicut," and "Cananor." Boodt's putative localities were somewhat fanciful, but they were more detailed than Pliny's had been. Courtesy of the David Rumsey Map Collection, David Rumsey Map Center, Stanford Libraries.

species was implicit rather than explicit, as when he referred simply to "oriental garnet" or to amethysts found in Arabia, Ethiopia, Cyprus, "and other
oriental locations." In one way or another, Boodt used the term "oriental" to
designate seventeen of the forty-one species of stone that he recognized as
"precious" in the *History of Gems*.[103] In addition, nearly all of these seventeen
species had varieties that Boodt identified as German, Selesian, or Bohemian.
Boodt was much more likely to call a variety "oriental" if he knew a nearby
variety of the same species.[104] Once again, Boodt's global geography of gems
worked in tandem with his regional geography.

Both worked in tandem with hardness. In seven cases—amethyst, emerald, topaz, asteria, sardony, chalcedony, and lapis lazuli—Boodt implied that
the oriental variety of the stone was considerably harder than its occidental
counterpart.[105] In one of these cases (amethyst) hardness was the *main* difference between the varieties. On several occasions he used hardness to separate species that were otherwise hard to distinguish, such as jasper and agate,
girasol and opal, and sardony and chalcedony.[106] Like Cellini before him, he
understood hardness in terms of the tools used by gem cutters. Hardness was
important, he explained, because hard gems can be polished by artisans, and
because this polish gives them their brilliance or glitter.[107] Elsewhere he distinguished three degrees (*gradus*) of gem hardness on the basis of the materials that scratch gems.[108] There are gems that can be scratched by steel, those
that can only be scratched by emery, and those that can only be scratched by
diamond. The steel that Boodt had in mind was not just any steel but that of
files (*lima*), probably the ones that jewelers used to distinguish real stones
from counterfeits, a procedure that he explained elsewhere in his book.[109]
The emery and diamond he had in mind was the powdered emery and pow-

103. These seventeen instances are in Boodt, *Gemmarum*, 59 (diamond), 75–76 (garnet), 80
(hyacinth), 81 (amethyst), 84 (pearls), 92 (sapphire), 99 (emerald), 104–5 (topaz), 117 (girasol),
118 (sardony), 120 (sardonyx), 121 (chalcedony), 125 (agate), 129 (jasper), 130 (heliotrope), 134
and 137 (turquoise), and 239 (lapis lazuli). Use of "European" for garnet, sapphire and topaz.
"Oriental" only for diamond and hyacinth. Implicit division for garnet, amethyst, prase, jasper,
and heliotrope.

104. The only exceptions to this rule were opal and beryl.

105. Boodt, *Gemmarum*, 81 (amethyst), 99 (emerald), 105 (topaz), 112 (girasol), 118 (sardony),
121 (chalcedony), 139 (lapis lazuli).

106. Ibid., 124–25 (agate/jasper), 112 (opal/girasol), 118 (sardony/chalcedony). Note also the
ranking of gems by hardness on 27–28: diamond, topaz, sapphire, garnet and hyacinth, in that
order, with opal the softest stone of all.

107. Ibid., 27–28.

108. Ibid., 2–3.

109. Ibid., 30.

dered diamond that cutters sprinkled on their wheels, a process he also explained elsewhere.[110] His choice of examples of the three degrees of hardness
is significant, since two of these (jasper and turquoise) were routinely used
by hard-stone carvers and the third (diamond) called for special polishing
tools that he illustrated in his book. His choice of verbs for the three degrees
is also telling. Turquoise was scratched (*radere*) by steel files, a word that suggests tests of authenticity. Jasper and diamond were worn down (*terere*) by
emery and diamond, a term that suggests the gradual abrasion achieved by
cutters and polishers. His emphasis on hardness derived from the tools of
artisans, just as his emphasis on the oriental/occidental distinction derived
from far-flung merchants and from miners closer to home. His innovation in
gem classification was to join these three traditions together and incorporate
them into the learned tradition of gem writing that stretched back to Pliny
and Theophrastus.

Classification Without Systematics

Boodt thought he had done more than this. He thought he had given one
of the first real classification of gems, maintaining that naturalists had done
no more than give alphabetical lists of gems before the sixteenth century.[111]
Historians have tended to endorse this judgment. Certainly, it is hard not to
be impressed by Boodt's learning—the lengthy account of different characters
of stones, the careful distinction between essential and accidental characters,
the clear statement of his scheme in his preface, the sheer length of the treatise, and the branching diagram of gem characters at the start of the book
(fig. 1.4).[112]

  On closer inspection, however, Boodt was working within Pliny's framework. Boodt's scheme, like Pliny's, was based on preciousness and color. He
began the treatise with the rarest and most expensive stones, on the grounds
that this arrangement was "most suited to the dignity of precious stones."
He then proceeded "by degrees to the meanest ones."[113] Like Pliny, he re-
fined this scheme so that stones of the same color were grouped together.
After diamond came the red stones, from the illustrious oriental ruby to the
humdrum amethyst—and so on for blue, green and yellow stones, in that

110. Ibid., 35.
111. Boodt, *Gemmarum*, "Ad lectorum."
112. Characters (*differentiae*) in chaps. 2 to 7, and chaps. 12 to 18. Outline, and description of
table, in "Ad lectorum." Boodt describes the table in "Ad lectorum."
113. Ibid., "Ad lectorum."

DIVISIO LAPIDUM ET
GEMMARVM.

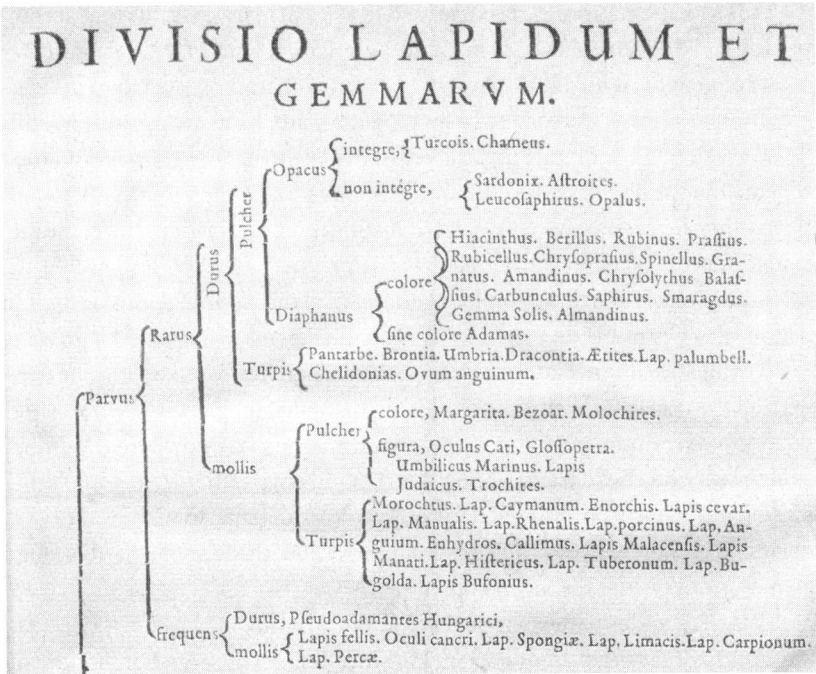

```
                            ┌ integre,┐ Turcois. Chameus.
                   ┌ Opacus ┤         ┘
                   │        └ non intégre,  ┌ Sardonix. Aftroites.
             ┌─────┤                        └ Leucofaphirus. Opalus.
             │     │
             │     │                    ┌ Hiacinthus. Berillus. Rubinus. Praffius.
       ┌─────┤     │                    │ Rubicellus.Chryfoprafius.Spinellus. Gra-
       │     │     │            ┌ colore ┤ natus. Amandinus. Chryfolythus Balaf-
       │     │     │            │        ) fius. Carbunculus. Saphirus. Smaragdus.
       │     │     └ Diaphanus  ┤        └ Gemma Solis. Almandinus.
  ┌ Rarus ┤                     └ fine colore Adamas.
  │    │          ┌ Pantarbe. Brontia. Umbria.Dracontia.Ætites.Lap. palumbell.
  │    │   Turpis ┤ Chelidonias. Ovum anguinum.
  │    │
┌ Parvus ┤
│    │                              ┌ colore, Margarita. Bezoar. Molochites.
│    │                     ┌ Pulcher┤ figura, Oculus Cati, Gloffopetra.
│    │                     │        │ Umbilicus Marinus. Lapis
│    └ mollis ┤            │        └ Judaicus. Trochites.
│             │            ┌ Morochtus. Lap. Caymanum. Enorchis. Lapis cevar.
│             │            │ Lap. Manualis. Lap.Rhenalis.Lap.porcinus. Lap. An-
│             └ Turpis     ┤ guium. Enhydros. Callimus. Lapis Malacenfis. Lapis
│                          │ Manati.Lap. Hiftericus. Lap. Tuberonum. Lap.Bu-
│                          └ golda. Lapis Bufonius.
│
│            ┌ Durus, Pfeudoadamantes Hungarici,
└ frequens ┤        ┌ Lapis fellis. Oculi cancri. Lap. Spongiæ. Lap. Limacis.Lap. Carpionum.
             └ mollis └ Lap. Perœ.
```

FIGURE 1.4. Boodt's branching diagram of gem characters
The diagram was meant to classify gems with a succession of binary distinctions. In the detail shown, the distinctions are (from left to right) small/large, rare/common, hard/soft, beautiful/mean, opaque/transparent, and whole / not whole. For example, turquoise (*turcois*) is small, rare, hard, beautiful, opaque, and whole. Though striking, this diagram was only loosely related to the classification of gems in the main text of Boodt's book. Diagram from Boodt, *History of Gems and Stones*, 2nd ed. (1636), 14. Courtesy of ETH-Bibliothek Zürich, Rar 1881, https://doi.org/10.3931/e-rara-15217.

order. The chapters on thirty-eight precious stones (*gemmae*) were followed by chapters on common stones (*lapides*), just as Pliny had followed his chapters on twenty-six principal gemstones (*principales gemmae*) with chapters on lesser stones. Moreover, Boodt's precious stones corresponded closely to Pliny's principal gemstones, insofar as the species in those categories can be identified today. The scope of the two categories may be indicated by noting that diamond, ruby, amethyst, jasper and lapis lazuli were in both, and that agate was in neither. Boodt outlined this scheme in the preface to the book; he followed it quite closely in his description of individual stones.[114]

114. Boodt outlines his scheme in "Ad lectorum." The main text is divided, more or less explicitly, as follows: chaps. 1–79 (transparent precious stones), 80–113 (semitransparent precious stones), 114–51 (opaque precious stones). There are thirty-eight species in these chapters;

This scheme was only loosely related to the learned apparatus that accompanied it. Boodt may have given a cutting-edge account of the various characters of gems, but he did not say how his list of characters related to his classification scheme.[115] As for the branching diagram, he came up with it while he was seeing the book through the press, too late to change the very different scheme that he had already used for the main text.[116] When Boodt did depart from Pliny, he rarely gave a principled justification for the departure. Indeed, sometimes his principles went against his practice, as when he rejected the oriental/occidental distinction in his theoretical discussion before using it to define varieties in his descriptions of individual gems.[117] Nor did he give any real justification for his new emphasis on transparency. Unlike Pliny, he dealt with all the transparent stones before moving to the opaque ones, with a short section on semitransparent stones between the two. Yet nowhere did Boodt explain why he chose transparency for this purpose rather than some other property. Boodt certainly helped to codify Pliny's scheme, to make it more explicit and comprehensive. But his main innovation, the idea of a hard oriental gem, was not part of his formal scheme.

This fits a wider pattern in the early modern study of gems. Agricola's *On the Nature of Fossils* is a good example, since it is often seen as a pioneering work in the classification of minerals.[118] The work was indeed new when it was first published in 1546. Agricola fitted gems into a scheme that included all minerals, not just stones and metals. Unlike Boodt, he thought carefully about how such a scheme might be crafted out of a miscellaneous list of

---

to these we can add coral, amber, and agate, stones that Boodt did not consider precious (p. 153) but which many other early modern authors did consider so. Common stones are covered in chaps. 152–282. Note also chaps. 283–98 (rocks), 303 (ancient stones unknown today), 304 (stones on the border between precious and common). Boodt is a little vague on which stones are gemmae and which merely lapides, notably in the discussion on p. 153. In particular, it's not clear from that discussion whether chapters 114–51 cover gemmae or lapides. But since he refers to many of those stones as gemmae in individual chapters, it is safe to say that the discussion of mere lapides begins in chapter 152. Boodt's French translator rendered "gemmae" as *pierres précieuses* and "lapides" as *pierres communes*: Boodt, *Parfait joaillier*, "Avertissement," 389.

115. Boodt's remarks on this (*Gemmarum*, chap. 7) are short and vague.

116. As explained in ibid., "Ad lectorum."

117. Rejection in ibid., 5, 13–14.

118. For example, in Mark Chance Bandy and Jean A. Bandy, "Foreword," in Georg Agricola, *De natura fossilium (Textbook of Mineralogy)*, trans. Mark Chance Bandy and Jean A. Bandy (Boulder, CO: Geological Society of America, 1955 [1546]), v–x, on vi. Agricola is the earliest author covered in Charles Spencer St. Clair, "The Classification of Minerals: Some Representative Mineral Schemes from Agricola to Werner" (PhD diss., University of Oklahoma, 1966). Cf. Adams, *Birth and Development of the Geological Sciences*, 170, 175, 191–95.

characters.[119] He gave tests for distinguishing between species that are otherwise easily confused. After describing the colorless gems, for example, Agricola explained that rock crystal is always angular whereas diamond is only sometimes angular, that opal changes color when inclined in light whereas *asterios* reflects a round inner light, and so on.[120] As far as formal classification was concerned, Agricola's book was more sophisticated than Boodt's. In most other respects, however, Agricola's classification of gems differed little from Pliny's. "Since gems may be classified chiefly by color," he began, "I shall speak first of the colorless ones." He then dealt with green, blue, purple, and red gems, in that order, ending with gems that have a fiery red glow, an echo of Pliny's carbunculi.[121] His debt to Pliny is especially clear in his division of species into varieties, which relied heavily on the localities that the Roman author had listed. In his description of adamas, for example, Agricola listed all and only the localities that Pliny had listed 1,500 years earlier; a close study of his description of smaragdus shows the same pattern.[122] There is little sign in Agricola's book of the merging of hardness and locality that would become so important for Boodt sixty years later.[123] Agricola's descriptions of individual gems contained few of the firsthand observations that he made on other minerals.[124] Apparently, metals were easier to study than gems in the mining towns of Bohemia and Saxony where Agricola lived while he compiled his mineralogical works.[125]

New ways of grouping gems were just as likely to appear in unsystematic works as in systematic ones. A sixteenth-century example is Jerome Cardan's

119. Agricola, *Textbook of Mineralogy*, 15–20.

120. Ibid., 124 (colorless gems), 129–30 (green gems).

121. Ibid., 118 ("chiefly by color"; colorless), 124 (green), 130 (blue; purple), 132 (red), 135 (fiery red).

122. Ibid., 121–22 (adamas), 124–26 (smaragdus); Pliny, *Natural History*, bk. 37, 207–9 (adamas), 213–25 (smaragdus).

123. Hardness is absent in Agricola, *Textbook of Mineralogy*, 5, though this is qualified on p. 11. For Agricola, hardness distinguishes gems from other classes of mineral (18, 108, 112) but rarely distinguishes one gem from another. Absence of "oriental" and "occidental" based on a search for these and cognate terms ("Indian," "Asian," "Eastern," "Western") in Bandy's English translation. Following Pliny, Agricola referred to "Indian" gems on 113, 114, 131, 147. The only hint of novelty is his reference to "European" varieties of quartz and jasper on 114, 118–21. The Latin term was "europa": Georg Agricola, *De natura fossilium*, 2nd ed. (Basel, 1558 [1546]), 273.

124. Metallic ores emphasized in Adams, *Birth and Development of the Geological Sciences*, 195.

125. Agricola's whereabouts in Pamela Long, "The Openness of Knowledge: An Ideal and Its Context in 16th-Century Writings on Mining and Metallurgy," *Technology and Culture* 32, no. 2 (1991): 335–36.

On Subtlety, first published in 1560, a few years after Agricola's book. The Italian scholar showed little interest in the formal aspects of classification, writing in a conversational style and making no effort to group gems into families. Yet he was an important precursor to Boodt in his emphasis on hardness and the Orient. He had much to say about the relative hardness of different species of gem, a property he associated with beauty, brilliance, and price. And he linked hardness to geography, giving several explanations for the high quality of "oriental" stones, a term he used in the names of four species.[126] These new approaches to gems were consolidated by writers who are not usually associated with the science of taxonomy. Renaissance naturalists such as Andrea Cesalpino and Ulisse Aldrovandi, with their prolix accounts of the mythology, etymology, and medical virtues of natural bodies, have long been contrasted with the systematizers of the eighteenth century, who had no time for what they called "useless erudition."[127] Yet erudition was compatible with classification. Aldrovandi gave as much space to the varieties (*genera*) of rubies as he did to the medical uses of rubies and to the history of the name of the species. His account of these varieties reflected the new data available from cutters and traders that had been compiled by Boodt. Data of this kind was already in Cesalpino's natural history of minerals, published a decade before Boodt's.[128] Looking ahead, the author who did the most to extend the new classification of gems in the seventeenth century was much less systematic than Boodt. Robert de Berquen, a Parisian goldsmith, had no time for learned disquisitions on *genera* and *differentiae*. There was no diagram of gem species in his *Marvels of the Oriental and Occidental Indies*, published in 1661. There was virtually no discussion of the principles of classification in the introduction to the book. There was, however, a persistent tendency to

126. Jerome Cardan, *The De Subtilitate of Girolamo Cardano*, trans. John Forrester (Tempe, AZ: Arizona Center for Medieval and Renaissance Studies, 2013), 361 (oriental topaz), 382 (oriental jasper), 386 (oriental onyx), 395 (oriental amethyst), 362, 363, 384, and 403 (explanations). Note also Cardano's knowledge of Peruvian emeralds (372) and rubies in Pegu (377). He divided gems by transparency (362) but only in passing. Hardness used to differentiate gems on 361, 363, 370, 377, 379, 380, 385, 388–89, 399. Hardness linked to other gem qualities on 363, 373, 374, 379, 380, 384, 393, 399, 422–23.

127. "Useless erudition" in Georges-Louis Leclerc, Comte de Buffon, "Premier discours," in *Histoire naturelle, générale et particulière*, vol. 1 (Paris, 1749), 27. Note, however, that Buffon appreciated Aldrovandi's "divisions" and "distributions" (26). On the distinction between Renaissance and Enlightenment naturalists, see the works cited in note 9.

128. Ulisse Aldrovandi, *Musaeum metallicum* (Bologna, 1648), 958–59 (ruby varieties); Andrea Cesalpino, *De metallicis libri tres* (Rome, 1596), 113–14. Aldrovandi died in 1605, four years before Boodt's book was published. The references to Boodt in *Musaeum metallicum* must have been added by the compilers of this posthumous work.

organize the description of gems around the idea that oriental gems are hard and occidental ones soft.[129]

It is tempting to understand this phenomenon in terms of the relationship between university-trained scholars and shop-trained practitioners. Boodt's *History of Gems* may be seen as an attempt to mediate between these two groups. Perhaps the mismatch between the formal and informal scheme in the book shows the difficulty of the task. There is certainly some truth in this view, but I want to emphasize the broader point that classification should not be equated with the formal apparatus of classification. The act of placing gems into groups was only loosely related to formal devices such as tables of species, explicit naming conventions, or theoretical discussions of the meaning of "species." This was true of learned authors such as Pliny the Elder, Albert the Great, and Boodt, as well as of practitioners such as Barbosa and Cellini. The eighteenth century may have led to new kinds of classification, or even a sharper focus on classification, but it did not invent classification tout court. Once we see this, we can ask how gem classification changed over the long term, from the first century AD to the start of the nineteenth century. To summarize the story so far: Renaissance naturalists revived and codified a scheme inherited from Pliny, while expanding that scheme to include the hard/soft distinction and the oriental/occidental distinction.

The story so far illustrates the themes and subplots mentioned in the introduction to this book. This is a transmaterialist story, in two respects. Different material substances, such as plants and gems, had different histories. And different material practices, such as trading gems and cutting them, worked together to produce new ways of grouping gems. Material evaluation is also part of the story. The new groupings went hand in hand with new forms of judgment, from Pliny's luxuries to Albert's virtues to Barbosa's prices to Cellini's tests of authenticity. The subplots have also made an appearance. The relationship between different practitioners—between traders, carvers, and miners, for example—was just as important as the relationship between practitioners and scholars (hand-hand coordination). This in turn draws attention to the geographical distribution of material expertise, such as the concentration of corundum knowledge in Ceylon and of metallic ores in central Europe (material specialties). The natural history of gems was mixed up with things now associated with experimental physics, such as magnetism, the amber effect, measurement, and powerful instruments (physics as stamp collecting). The distinction between oriental and occidental stones was a variant of the ancient practice of defining stones by their place of origin

129. See chap. 2, this volume.

(spatial species). Renaissance gem classification had a place for color, transparency, hardness, and locality, but density and double refraction would not become part of these schemes until the end of the eighteenth century (two-stage classification).

The second, eighteenth-century stage was especially pronounced in France, as we shall see in later chapters. This French bias was partly due to an accurate translation of Boodt's book published in Lyon in 1644. The translation was soon followed by many new texts on gems that allow us to look more closely at the techniques of early modern gem appraisal. How were good gems distinguished from bad in the high-stakes world of seventeenth-century gem trading? And what does this tell us about the codification of the crafts, another new science in early modern Europe?

2

Gems and Technical Writing in the Age of Louis XIV

"Sapphires, topazes, and oriental rubies are cut and formed on a copper wheel that is sprinkled with diamond powder and olive oil." The sentence is from the famous French encyclopedia edited by Denis Diderot and Jean le Rond d'Alembert and published in the middle decades of the eighteenth century. The article on the arts of the gem cutter, from which the sentence is taken, appeared in 1765.[1] Where did this precise account of the cutter's practice come from? The article is anonymous and there is no citation for this sentence, so the historian must hunt through known sources of the encyclopedia for an answer. Sure enough, an English version of the sentence appeared in the 1728 edition of the *Cyclopaedia* of Ephraim Chambers.[2] Chambers had almost certainly translated the sentence from a French commercial dictionary that was published two years earlier.[3] The author of this dictionary cited his source: André Félibien, *On the Principles of Architecture, Sculpture, Painting, and Related Arts*, an encyclopedia of the arts first published in Paris in 1676.[4] The

1. "Les rubis orientaux, les saphirs & les topases se taillent & se forment sur un rouet de cuivre qu'on arrose avec de la poudre de diamant & de l'huile d'olive." "Lapidaire," in Diderot and d'Alembert, *Encyclopédie*, vol. 9 (1765), 282.

2. "Oriental Rubies, Saphires, and Topazes, are cut and form'd on a Copper Wheel, with Oil of Olives, and Diamond Dust." Ephraim Chambers, *Cyclopaedia, or An Universal Dictionary of Arts and Sciences* (London, 1728), 430.

3. "Les rubis, saphirs & topases d'Orient se taillent & se forment sur une roüe de cuivre, avec l'huile d'olive & la poudre de diamant." Jacques Savary des Brûlons, *Dictionnaire universel de commerce* (Paris, 1726), vol. 2, 485. Savary cites Félibien in vol. 1, 1686.

4. "Quant aux Rubis, Saphirs, & Topases d'Orient, on les taille, & on les forme sur une Roüe de cuivre qu'on arrose de poudre de Diamant, avec de l'huile d'Olive." André Félibien, *Des principes de l'architecture, de la sculpture, de la peinture, et des autres arts qui en dépendent* (Paris, 1676), 361. Félibien's discussion of gem cutting and engraving is on pp. 358–81. Note the placement

paper trail ends there. Félibien, the royal historiographer in the age of Louis XIV, appears to be the ultimate source of this datum about the lapidary arts. The sentence from the *Encyclopédie* was based on a book published nearly a century earlier, at the court of the Sun King.

Félibien's book was typical of a regime that combined a fondness for gems with a commitment to documenting the techniques of merchants and artisans. Louis XIV was famously partial to gems. He made illustrious additions to the crown jewels, notably a large, blue diamond now known as the "Hope Diamond."[5] Gems were part of his public image, whether among his own courtiers, the French public, or ambassadors from gem-rich regions such as Safavid Persia and the Siamese Kingdom of Ayutthaya.[6] The king tried to secure the supply of gemstones to France, cultivating relationships with globe-trotting jewelers and setting up a hard-stone workshop in Paris with the help skilled cutters recruited from Florence.[7] In addition, Paris was home to a guild of gem cutters (*lapidaries*) and to many fine displays of gems in palaces, mansions, curiosity cabinets, shops, and workshops.[8] All this was

---

of the word "oriental" in this quote and the quotes in the three previous notes. The changing placement of the word strongly suggests that Chambers copied his text from Savary des Brûlons or Félibien, and that the author of the *Encyclopédie* article copied his text from Chambers. On Félibien, see Stefan Germer, *Art-pouvoir-discours: La carrière intellectuelle d'André Félibien dans la France de Louis XIV*, trans. Aude Virey-Wallon (Paris: Éditions FMSH-Fondation Maison des sciences de l'homme, 2016).

5. Morel, *Joyaux de la couronne*, chap. 4; François Farges, "Les grands diamants de la couronne de François I à Louis XVI," *Versalia* 16 (2014): 55–78.

6. Courtiers in Ina Baghdiantz-McCabe, *Orientalism in Early Modern France: Eurasian Trade, Exoticism and the Ancien Regime* (Oxford: Berg, 2008), 231–43. Ambassadors in ibid., 257–59; and Giorgio Riello, "'With Great Pomp and Magnificence': Royal Gifts and the Embassies Between Siam and France in the Late Seventeenth Century," in *Global Gifts: The Material Culture of Diplomacy in Early Modern Eurasia*, ed. Zoltán Biedermann, Anne Gerritsen, and Giorgio Riello (Cambridge, UK: Cambridge University Press, 2018), 235–65. French public in Charles le Maire, *Paris ancien et nouveau* (Paris, 1685), 194, 213, 214, 217; Germain Brice, *Description nouvelle de la ville de Paris* (Paris, 1698), 53, 73–76, 206.

7. Stéphane Castelluccio, *Les meubles de pierres dures de Louis XIV et l'atelier des Gobelins* (Paris: Faton, 2007); Michèle Bimbenet-Privat, *Les orfèvres et l'orfèvrerie de Paris au XVIIe siècle* (Paris: Commission des travaux historiques de la ville de Paris, 2002), vol. 1, 106–7, and vol. 2, 396 (jewelers working for Cardinal Mazarin), vol. 2, 414 (jewelers working for Louis XIV); McCabe, *Orientalism in Early Modern France*, 187, 255, 257–59.

8. See the pages from le Maire and Brice cited in note 6 as well as the following: Martin Lister, *A Journey to Paris in the Year 1698* (London, 1699), 8–9, 77, 142–44; Nicolas de Blégny, *Les adresses de la ville de Paris, avec le Tresor des almanachs* (Paris, 1692), 68–70, 93–94, 100, 139; Savary des Brûlons, *Dictionnaire universel de commerce*, especially the articles "Diamant" (vol. 1, 1684–1690), "Graveur sur pierres précieuses" (vol. 2, 273–74), "Lapidaire" (vol. 2, 486–88), and

documented in a new generation of maps, letters, inventories, printed treatises, commercial dictionaries, and guides to the city of Paris. Some of these
texts, such as merchant correspondence, were a straightforward response to
the rise of the gem trade. Others were a reflection of wider changes in French
culture and politics, such as the beginning of orientalist scholarship and the
creation of a state intelligence system under the king's most effective minister,
Jean-Baptiste Colbert.[9] Whatever the causes, the result was a flurry of primary sources that tell us a lot about the craft and commerce of gems.

These sources are especially eloquent about gem appraisal. The great question for merchants, travelers, goldsmiths, and gem cutters was how to distinguish good gems from bad. When Félibien wrote that "sapphires, topazes,
and oriental rubies are cut and formed on a copper wheel that is sprinkled
with diamond powder and olive oil," he was not only describing how to make
gems but also how to judge them. After all, cutters determined the hardness
of gems by the materials used to work them, and the hardness of gems was
one determinant of their value. As Félibien explained, gems could be cut with
steel, copper, iron, or wooden wheels, depending on their hardness and hence
their value. The point generalizes. The wealth of primary sources about gems
in seventeenth-century France allow us to look more closely at the significance of gem appraisal beyond natural history and natural philosophy.

Of course, value is never far away in the secondary literature of gems,
which are studied precisely because of their high value. Economic historians
have studied the role of trust in establishing agreement about the value of
gems in cross-cultural trade. Historians of empire and geography have noted
the association between particular gems and particular places. Historians of
jewelry have explored the symbolic value of gemstones at the royal court.
Meanwhile, the history of the goldsmiths' guild of Paris, the arbiter of the
value of gold and silver in the capital, has been extensively documented.[10] But
value is not the same thing as evaluation. There was a technical dimension

---

articles on particular species of gem. Gem cutters in Lespinasse, *Métiers et corporations*, vol. 2
(Paris, 1892), 81–95.

9. On these topics, see Nicholas Dew, *Orientalism in Louis XIV's France* (Oxford: Oxford
University Press, 2009); and Jacob Soll, *The Information Master: Jean-Baptiste Colbert's Secret
State Intelligence System* (Ann Arbor, MI: University of Michigan Press, 2011). The classic study
of information gathering in Louis XIV's reign is James E. King, *Science and Rationalism in the
Government of Louis XIV, 1661–1683* (New York: Octagon Books, 1972).

10. Economic historians: Yogev, *Diamonds and Coral*; Trivellato, *Familiarity of Strangers*;
Vanneste, *Global Trade and Commercial Networks*. Empire and geography: Warsh, *American Baroque*; Schmidt, *Inventing Exoticism*. Goldsmiths' guild: Bimbenet-Privat, *Orfèvres*. On jewelry,
see note 7 for the French case.

to the value of gems that is overlooked in most existing studies. The value of gems owed as much to the techniques applied to them—to specialized terms, instruments, and procedures—as it did to social or political forces. When these techniques have been studied, it is usually in the absence of the relevant social or political forces.[11] The challenge is to pay attention to the technical side of value without losing sight of the wider context. The study of technical writing is one way to rise to this challenge. I deal in turn with craft manuals, inventories, travel narratives, maps, and merchant correspondence.

Manuals

Robert de Berquen's *Marvels of the Oriental and Occidental Indies*, published in 1661 and again in 1669, is a good place to start.[12] There is no doubting the technical importance of this treatise. It was a self-conscious contribution to the natural history of gemstones, with the same two-part organization as Boodt's book and with copious references to the canonical authors in the genre, from Pliny to Boodt. This appears to be the first printed natural history of gems written by a practicing gem merchant or gem cutter.[13] The author proposed a new classification of gems, one centered on the hardness of stones. The basic framework was still Pliny's, with most gems grouped by color and locality; but hardness was even more important for Berquen than it had been for Boodt. Berquen repeatedly stated that hardness was the basic organizing principle of the work, and he practiced what he preached. For example, he placed topaz and sapphire in the same chapter on the grounds that these stones "do not differ in nature or hardness, but only in color."[14] The same principle lay behind his subdivision of hyacinth, one that he explicitly contrasted to Boodt's subdivision of the same stone. The principle meant that Berquen sharpened the distinction between oriental and occidental stones, since he defined the former as the hardest stones. These innovations were underappreciated by later savants—one dismissed Berquen as a naive jeweler who was "concerned only with the rarity and expense of fine stones, and did

11. For example, Ogden, *Diamonds*, chap. 11. My account builds on Ogden's chapter and on the more contextual accounts in Sabel, "Rare Earth," 187–93, 197–98, 216–21; and Warsh, *American Baroque*, esp. 227–30, 97–101.

12. This section summarizes Michael Bycroft, "Regulation and Intellectual Change at the Paris Goldsmiths' Guild, 1660–1740," *Journal of Early Modern History* 22, no. 6 (2018): 500–527.

13. Based on a study of Sinkankas, *Gemology*; and Schuh, *Biobibliography*.

14. Robert de Berquen, *Merveilles des Indes Orientales et Occidentales* (Paris, 1661), 18.

not probe their nature."[15] Berquen was certainly interested in the qualities and prices of fine stones, but he saw no inconsistency between this and the study of their natures. Indeed, he believed he had special insight into the nature of stones precisely because he was so sensitive to their price and quality. As he put it, he aimed to "order precious stones according to their degree of perfection, and principally to that of their hardness, from which derives all their luster and beauty."[16] This knowledge of gems was unique to goldsmiths such as himself, he argued, because such people "handle nothing else all their lives" (fig. 2.1).[17]

The link between natures and qualities was equally strong in a second lapidary written by a Parisian goldsmith and published in the 1660s. Pierre de Rosnel's *Indian Mercury* is notable for its division of gem species into varieties. Naturalists had been dividing gems into varieties since at least Pliny, of course, but Rosnel emphasized the practice in a way earlier writers, including Boodt and Berquen, had not. He mentioned the varieties of each species in his chapter headings, as in "On the Oriental, Carthaginian, and common amethyst." He divided his chapters into sections, each dealing with a different variety, sometimes using white space or capital letters to emphasize the passage from one section to the next. And he did very little else in his descriptions. Only once did he mention the medical or moral virtues of a precious stone (sapphire) and only rarely did he mention famous specimens of a stone (topaz, emerald, chrysolite). What mattered most was the division of each gem into *sortes* or *espèces*. Interestingly, he used these terms interchangeably with *qualités*, in phrases such as: "There is still another *qualité* of amethyst, that is to say, the Carthaginian."[18] Like Berquen, Rosnel maintained that goldsmiths were in a privileged position to discern the qualities and hence the natures of gems.[19] He even appended a guide to the appraisal (*estimation*) of gemstones to his book.[20] Here we learn, for example, that oriental amethyst is worth ten times as much per carat as Carthaginian amethyst, and that the latter stones

15. Antoine-Joseph Dezallier d'Argenville, *L'histoire naturelle éclaircie dans une de ses parties principales, l'oryctologie* (Paris, 1755), 34, cf. 14.

16. Berquen, *Merveilles des Indes*, 29, cf. 19, 36.

17. Ibid., "A Mademoiselle."

18. Rosnel, *Mercure Indien*, vol. 2, 22, cf. 13, 54, 31–32.

19. Ibid., vol. 2, "Au lecteur."

20. Berquen, "Advis aux apprentis orfèvres," in *Merveilles des Indes*, 109–12; Rosnel, "De l'estimation des pierres précieuses et des perles, ensemble des autres pierres moins précieuses," in *Mercure Indien*, vol. 2.

LES MERVEILLES
DES INDES ORIENTALES
ET OCCIDENTALES,

O V

Nouueau Traitté des Pierres precieuſes & Perles, contenant
leur vraye nature, dureté, couleurs & vertus: Chacune
placée ſelon ſon ordre & degré, ſuiuant la cognoiſſance
des Marchands Orpheures. Auquel eſt adjouſté vne pe-
tite Table fort exacte, pour connoiſtre en vn inſtant à
quel tiltre les Marchands Orpheures de Paris, & les autres dans toutes les principalles Villes preſque de toute
l'Europe, trauaillent l'Or & l'Argent.

DEDIE' A MADEMOISELLE.

Par Robert de Berqven Marchand Orpheure à Paris.

※※※※※
※※※
※

J. Chas ※ se bras.

A PARIS.

DE L'Imprimerie de C. Lambin ruë vieille Draperie,
proche le Palais, à l'Image Sainct Martin.

LES Exemplaires ſe debitent chez l'Auteur, en la ruë des
Lauandieres en la Maiſon des Marchands Orpheures.

M. DC. LXI.
AVEC PRIVILEGE DV ROY.

FIGURE 2.1. Title page of Berquen's *Marvels of the Oriental and Occidental Indies*
Berquen's 1661 book was perhaps the first printed natural history of gems written by a practicing gem merchant or gem cutter. Berquen displayed his craft prominently on the title page, calling himself a merchant
goldsmith (*marchand orphevre*) and stating that he had ranked gems according to the knowledge (*connoissance*) of goldsmiths. Copies of the book were on sale at the guildhall of the Paris goldsmiths (*maison
des marchands orphevres*). Courtesy of gallica.bnf.fr / Bibliothèque nationale de France.

are worthless if they weigh four carats or less.[21] As this examples suggest, the price of a stone depended on its variety, not just its species. Accordingly, Rosnel's appendix was organized primarily by variety. The appendix to the lapidary was continuous with the lapidary itself, with both dividing species into varieties on the basis of price and nature.

The wider context for these two manuals was the guild system, with its elaborate procedures of quality control and its competition between guilds in the same town. Berquen and Rosnel were fully integrated into this system as leading members of the Paris guild of goldsmiths.[22] In 1661, as an assistant to the warden, Rosnel carried out regular inspections of workshops in Paris where gold, silver, and gems were used. As a warden in 1662 and 1671, and senior warden in 1672, he was responsible for the assaying of gold- and silverware produced by guild members. This was a massive task, since every piece needed to be assayed under the eyes of the wardens at the guild hall, this being done twice a week throughout the year. This was also a semipublic task, mentioned in a commercial dictionary alongside the locations of goldsmiths' workshops.[23] Berquen's contribution, as the guild's clerk between 1651 and 1673, was to sign the guild's legal documents and dispose of gold- and silverware confiscated by the wardens. The regulatory duties of Berquen and Rosnel were reflected in their published writings. *Marvels of the Indies* and *Indian Mercury* were both two-part works, with one part on gold and silver and the other on gems. Both authors gave lists of the purity of gold and silver in various European polities in the hope of demonstrating that Parisian metals were the purest of all. Rosnel went into particular detail about the assaying procedures (*essais*) that enabled a Parisian goldsmith to know (*connoistre*) the purity of a piece of gold or silver.[24] Both men returned to the topic in later works, Rosnel in a summary of the guild's regulations and Berquen in a *Book of Gold and Silver Alloys*. The latter was a sort of study guide on the arithmetic of alloys, a work designed for journeymen preparing for the oral examination that preceded the submission of their masterpiece. Candidates were judged as much on the purity of their materials as the quality of their work, as we know from other sources.[25] *Marvels of the Indies* and *Indian Mercury* may be seen as

21. Rosnel, "De l'estimation," 9.

22. The system is summarized in Leroy, *Statuts et privilèges*, esp. titles 6, 7, 10, 11 and 12. For biographical details on Berquen and Rosnel, see Bycroft, "Regulation and Intellectual Change," 517.

23. Blégny, *Adresses de la ville de Paris*, 69–70.

24. Berquen, *Merveilles des Indes*, 101–6; Rosnel, *Mercure Indien*, vol. 1, bk. 2, esp. 36–37.

25. Pierre de Rosnel, *Traité sommaire de l'institution du corps et communauté des marchands orfèvres* (Paris, 1672), "Reception," 4–6; Robert de Berquen, *Le livre d'allois en or et en argent, ou brève instruction pour répondre par devant mes seigneurs de la Cour des monnoyes en*

an extension to precious stones of the goldsmith's traditional expertise in the evaluation of precious metals.

It is significant, then, that this expertise was under threat in the middle of the seventeenth century. The goldsmiths (*orfèvres*) of Paris were locked in a long-running dispute with the gem cutters (*lapidaries*) of the same city.[26] The dispute stretched back to 1584, when the cutters' guild was created in the face of fierce opposition from the long-established goldsmiths' guild. Each guild claimed the right to mount, cut, and trade gems; each tried to deprive the latter of the same rights. A compromise had been reached in 1631, with both guilds allowed to trade gems, and with the cutters having the sole right to cut them and the goldsmiths the sole right to mount them in gold and silver. But the dispute flared up again in 1665 and 1666, when two cutters were tried for selling gold and silver objects and for mounting gems in them.[27] After a similar case in 1670, the goldsmiths campaigned for a new edict to reinforce the one from 1631, something they achieved in 1673.[28]

Material evaluation was at the heart of this dispute. The cutters' main argument for stripping the goldsmiths of their right to cut gems was that the cutting and mounting of gems should be practiced by different artisans in order to deter counterfeiters.[29] The goldsmiths later made the opposite argument: frauds were so convincing that only an artisan trained in both cutting and mounting could detect them.[30] The goldsmiths had been taking action against counterfeit gems since at least the thirteenth century.[31] There were many such gems in Paris in the latter part of the seventeenth century, as commercial dictionaries and probate inventories show.[32] Goldsmiths were also concerned about the quality of gold and silver worked by cutters, a recurring

---

l'interrogatoire qui sera faitte et sur les alloiemans qui seront donné aux prétendans maistres en l'art d'orfèvrerie de Paris (Paris, 1671). Cf. Lespinasse, *Métiers et corporations*, 7n4. On quality in the exam, see Bimbenet-Privat, *Orfèvres*, vol. 1, 21.

26. For details, see Bycroft, "Regulation and Intellectual Change," 513–14.

27. Pierre Leroy, Inventaire général des archives de la Maison commune du Corps des Marchands Orfèvres Joyailliers de la Ville de Paris, documents GGG.42 to GGG.45. This handwritten inventory can be consulted in AN T/1490/11 (vol. 1, dated 1736) and AN T/1490/12 (vol. 2, dated 1741).

28. Leroy, Inventaire, documents GGG.47 to GGG.52, and HHH.53.

29. Leroy, *Statuts et privilèges*, 218.

30. Ibid., 218; "Précis sur le reglement concernant l'orfévrerie," Archives Nationales, T/1490/2. Internal evidence suggests that this undated document was written in the middle decades of the eighteenth century.

31. Leroy, *Statuts et privilèges*, 132–35.

32. For example: Blégny, *Adresses de la ville de Paris*, 70, 100; and the probate inventory of the jeweler Henri Legrand, AN MC/I/205, Jan. 23, 1697.

theme of the disputes between 1665 to 1673. Rosnel was closely involved in these disputes, as the warden of the guild in 1670 and 1671. He gave a detailed account of the *affaire des lapidaries*, as it was known, in a summary of the guild's regulations he compiled in 1672.[33] Probably he had the cutters in mind when he defined goldsmiths as "those who were the first to mount precious stones in gold and silver," and when he described the aim of the treatise as "showing the importance that gold and silver, and especially precious stones, do not fall into the wrong hands."[34] It is not surprising that Rosnel appended the summary of the regulations to the second edition of the *Indian Mercury*.[35] Both texts implied that goldsmiths had special authority in the evaluation of precious materials, whether stones or metals.

These materials had much in common in the eyes of the evaluator. The notion of connoissance, the distinction between *fin* and *faux*, and the system of guild inspections were common to both. So were precise balances—precise enough to measure quantities of considerably less than a grain, a unit roughly equivalent to the weight of a grain of barley, or about 0.05 metric grams.[36] Balances used by jewelers and assayers in early modern France were accurate to at least an eighth of a grain, about the weight of two sesame seeds; some assayer's balances from the period were much more accurate than this.[37] Portable balances were also used in the seventeenth century, for both gold and gems, with a similar design for each material (fig. 2.2). But there was an important difference between the two materials, namely that the quality of metals could be quantified but the quality of gems could not. The quality of the gold in a piece of jewelry was measured by its purity—that is, by the amount of gold that remained in the object when all impurities were separated from it. There was no such procedure for diamonds, rubies, or any other gemstones. They

33. Rosnel, *Traité sommaire*, "Recueil," 39–41.

34. Ibid., "Institution des Orfèvres," 1 (definition); "Recueil," 64.

35. Sinkankas, *Gemology*, vol. 2, 878–79.

36. Herbert Tillander, "The Carat Weight," *Journal of Gemmology* 17, no. 8 (1981): 619–23. On seventeenth-century French units, see Jean Boizard, *Traité des monoyes, de leurs circonstances & dépendances* (Paris, 1692), vol. 1, 250, 255; Jacques Savary des Brûlons, *Le parfait négociant, ou instruction générale pour ce qui regarde le commerce* (Paris, 1675), 70–71; and "Grain," in Diderot and d'Alembert, *Encyclopédie*, vol. 7 (1757), 832–33.

37. This statement is based on the numbers in jeweler's account books, such as Jean-Baptiste Caumon, "marchand joyaillier," 1748–1751, in Archives de la Ville de Paris, D5B6, reg. 1183. See also the data on assaying at the goldsmith's guild, in Berquen, *Merveilles des Indes*, 101; Boizard, *Traité des monoyes*, vol. 1, 167–68, 174–80. Balances accurate to 1/740 of a grain were available in seventeenth-century London, according to a 1679 text: Norman Biggs, "Mathematics at the Mint: A Seventeenth-Century Saga," *British Numismatic Journal* 87 (2017): 156–57. On early gold assaying in general, see introduction, note 51.

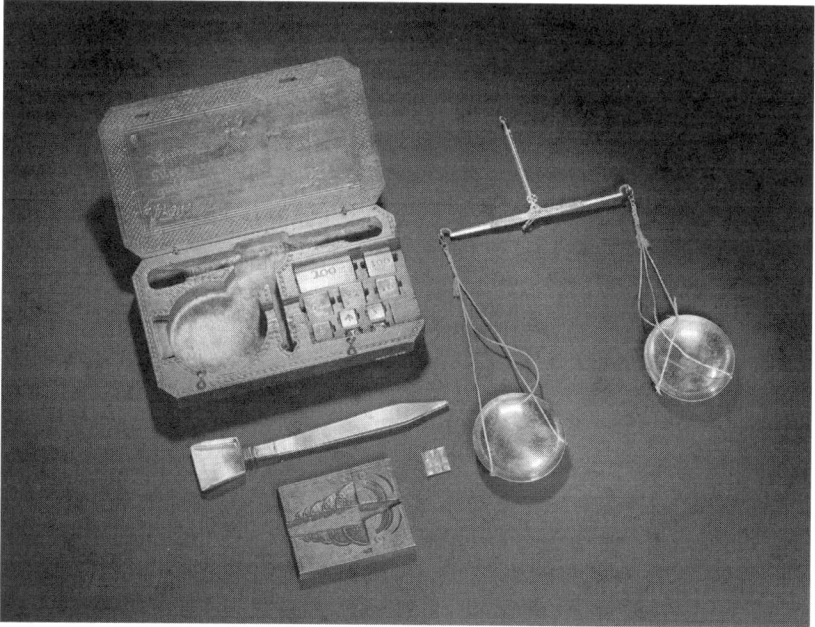

FIGURE 2.2. Portable gem balance
The balance includes a pair of scales, a set of brass weights, and a shovel for collecting small specimens. The inscription on the lid indicates that the device was made in Amsterdam in 1676. The stem of the shovel functions as a pair of tweezers. The weights range from half a grain to 200 carats, or from about 25 milligrams to about 40 grams in modern metric units. The weight shown next to the scales is marked with a maker's mark and six dots, each dot representing one grain. Jewelers' accounts show that measurements down to one-eighth of a grain (about 6 milligrams, or the weight of two sesame seeds) were routine in the eighteenth century. Courtesy of the Science Museum Group.

were too brittle to yield samples, and in any case the chemical analysis of gems lay over a century in the future. The difference between the two materials was reflected in the ambiguity of the word "carat," which referred to a unit of weight with regards to gems, and to a unit of purity with regards to metals. For example, a "22 carat" diamond was one that weighed 88 grains, whereas a "22 carat" gold ring was one that contained 88 grains of pure gold for every 96 grains of the ring.

The upshot is that gem quality was determined qualitatively, not quantitatively. Jewelers had a rich vocabulary for describing the color, shape, and transparency of gems, and other aspects of their visual appearance. They also had rules of thumb for accounting for their flaws. According to Rosnel, for example, a diamond with a few imperfections in its shape or color is a third the price of a perfect diamond of the same size and cut; if it is filled with "feathers" or other internal flaws then the price is cut in half; and so on.[38] Rosnel did his best to explain how these judgments were made, but he confessed that written instructions were no substitute for practical experience. Sometimes one simply had to "resort to the judgment of those who are versed in the knowledge (*connoissance*) of diamonds through long practice."[39] Fortunately for the historian, there are traces of this "long practice" in other sources from the period, starting with inventories.

Inventories

By an "inventory" I mean a legal document that lists a person's possessions with a view to assigning a value to them.[40] Many such documents have survived from early modern Europe. They are a rich source of information on the price of material goods, the location of these goods in private homes and other places, and the consumption habits of the owners of those goods. But they are also records of the act of evaluation itself. They tell us who did the evaluation, and they tell us something about how the evaluation was done. Like any source, they should be taken with several grains of salt—the order of the goods in an inventory does not necessarily reflect their physical location, for example. Used carefully, however, they are as good a source as any on the act of evaluation. They are particularly useful in the study of gem appraisal, for the simple reason that gems were expensive and therefore given special attention in inventories designed to evaluate a person's possessions. In seventeenth-century France, jewelers (*marchand joailliers*, members of the goldsmiths' guild specializing in gems) were often called upon to appraise the gems in larger collections. Inventories record the names of these jewelers, the prices they assigned to particular gems, and often the weight of those gems in carats or even grains. For our purposes, the interesting thing about these data is not the correlation between weight and price but the *lack* of correlation. Mismatches between weight and price show us that the quality of gems mattered as much as their weight.[41]

Consider the most famous gem inventory from seventeenth-century France, the "Inventory of the Crown Jewels" drawn up in 1666 and updated in

38. Rosnel, "De l'estimation," 2.

39. Ibid., 2–3.

40. This paragraph draws on Giorgio Riello, " 'Things Seen and Unseen': The Material Culture of Early Modern Inventories and Their Representation of Domestic Interiors," in Findlen, *Early Modern Things*, 124–50.

41. Here I build on Molly Warsh's analysis of pearl prices in seventeenth-century English inventories: *American Baroque*, 227–30.

1683 and 1691.[42] These inventories recorded the diamonds and other precious stones owned by the French crown. At the time, this was by far the largest collection of diamonds in Europe.[43] The preponderance of diamonds in the collection means that the inventories—and especially the one from 1691—give a very detailed picture of the work of the jewelers who appraised diamonds.[44] The appraisal was done by two royal jewelers, Louis Alvarez and Pierre Montarsy, in the presence of the secretary of state of the king's household (*Maison du roi*). This was an established procedure, having been followed in the inventory of 1683.[45] The 1691 inventory was a big job, involving nearly four thousand individual stones worth a total of 11.5 million livres (roughly 200 million US dollars in today's money). Most of the stones were evaluated in bulk, such as the 151 buttons for a vest, each made of five diamonds and each priced at 3,738 livres. But many stones were evaluated individually, with their price and weight recorded in the inventory along with a short description. This is true not only for the handful of world-famous diamonds at the start of the inventory—the Sancy diamond was priced at 600,000 livres, for example—but also for a large number of anonymous stones. A list of 123 diamond buttons is particularly interesting.[46] This is a large and varied set of stones that were appraised individually and at prices that were high but not spectacular, most between 5 and 50 thousand livres. This makes the inventory one of the most comprehensive datasets we have on the thought processes of working gem appraisers in early modern France.

These data show the richness of the jewelers' vocabulary for the quality of diamonds. There are numerous terms in the inventory simply to describe the color of a diamond's water (*eau*). The term referred to the color, transparency and brilliance of the internal substance of a diamond, an echo of the ancient belief that diamonds are a crystallized form of water.[47] In the inventory of the crown jewels, the term sometimes refers explicitly to quality: bad water,

42. Germain Bapst, *Histoire des joyaux de la couronne de France* (Paris, 1889), chap. 2; Morel, *Joyaux de la couronne*, chap. 4.

43. Morel, *Joyaux de la couronne*, 149.

44. The following analysis is based on the transcription of the inventory in Bapst, *Histoire des joyaux de la couronne*, 374–402, esp. 380–90.

45. The 1683 inventory is at Bibliothèque Nationale de France, Fonds Clairambault, Ms 499, p. 409. Montarsy's father had worked with Louis Alvarez and Pierre Montarsy on the 1683 inventory. On Pierre Montarsy, see Antoine Schnapper, *Curieux du Grand Siècle: Collections et collectionneurs dans la France du XVIIe siècle: Oeuvres d'art* (Paris: Flammarion, 1994), 15.

46. Here I expand upon a suggestion in Morel, *Joyaux de la couronne*, 167.

47. Argenville, *Oryctologie*, 101; "Eau, chez les joailliers," in Diderot and d'Alembert, *Encyclopédie*, vol. 5 (1755), 212.

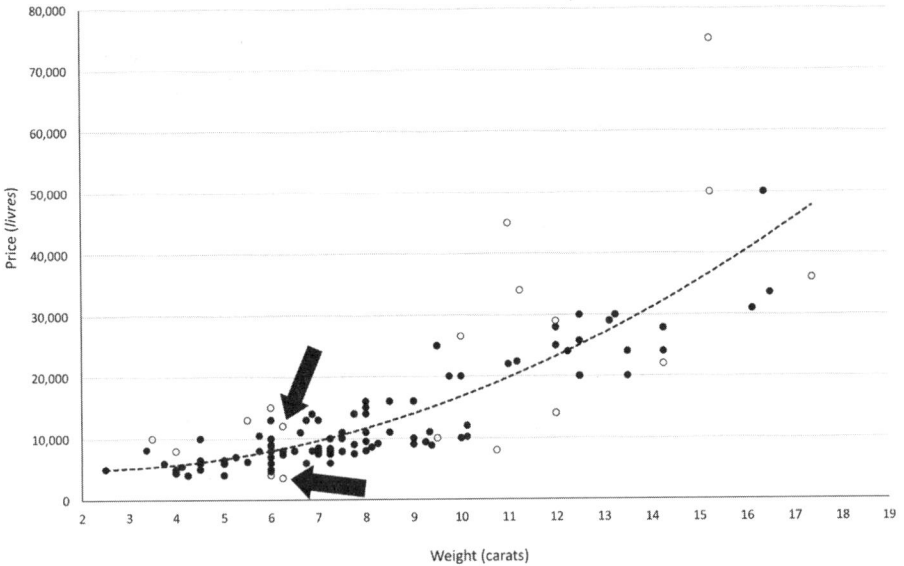

F I G U R E 2.3. Diamond quality in the French crown jewels
A chart of 121 diamonds in the crown jewels, showing that quality mattered as much as size for their price. Each dot represents a diamond. The dotted line is the parabolic line of best fit, representing the price of diamonds by weight before quality is taken into account. Note the wide variation around the parabola. The arrows point to two diamonds with the same weight but very different prices. Other such anomalies are indicated by open dots. Data from "Inventory of the Crown Jewels [1691]," AN O1/3361, transcribed in Bapst, *Crown Jewels* (1889), 380–90.

second water, good water, beautiful water, very beautiful water, crystalline water, perfect and crystalline water, crystalline and very lively water.[48] The same word could also refer indirectly to the quality of stones by noting their color: brown, blackish, straw-colored, celestial, colored like a soapy lemon.[49] There is a similar profusion of terms for different kinds of cut, degrees of clarity, and types of defect.

These judgments made a difference to the price of the diamonds, as shown by figure 2.3. The chart shows the weight and price of 121 of the royal diamonds in question, the two largest stones being omitted for the sake of clarity. The dotted line is the parabola that best fits the data points. If the size of the diamonds were the only determinant of their price, all the dots would fall on this line. There are many outliers, the most striking of which are indicated by

48. De mauvaise eau, de la seconde eau, de bonne eau, de belle eau, de très belle eau, d'eau cristalline, d'eau cristalline et parfait, d'une eau cristalline et très vive.
49. D'eau brune, d'eau noirastre, d'eau un peu foineuze, d'eau couleur de citron savonneuse, d'eau céleste.

open dots. Take the two diamonds shown by the two arrows. Both weighed 6.25 carats, but one was valued at 3,500 livres and the other at 12 thousand livres. The first of these is described as "a facetted diamond [*à facettes*], oval, well-formed, water somewhat hay-colored [*foineuze*], with a glaze [*glace*] on the crown and several small dark points [*petits points*]." The second stone is "a brown diamond, round and well-spread [*détendüe*], and very perfect as well." The difference in cut cannot explain the price difference, since there are plenty of round diamonds and diamonds *à facettes* with high prices in the inventory. Most probably it is the hay-colored water, the glaze, and the points that dragged down the price of the first diamond, and the perfect water that elevated the second. The inventory of the crown jewels was one of a kind, of course, but similar data can be extracted from the probate inventories of Parisian merchants in the same period.[50] These documents show the importance of quality in the appraisal of diamonds, and the ability of jewelers to articulate fine differences in quality.

The same documents show the significance of such judgments. For Louis XIV, the inventory of the crown jewels was part of a wider effort to decide who owned what in the royal circle. The fundamental distinction in royal collections was between the crown jewels (*joyaux de la couronne*) and the king's personal jewels (*joyaux personnels*). The crown jewels could only be sold or given away in exceptional circumstances, such as wartime; the king could dispose of his personal jewels as he wished. This distinction had existed in law since 1566, but it was strengthened in the 1660s as part of a review of all the king's precious possessions (*meubles précieux et durables*), a category that also included paintings, sculptures, tapestries, and much else. The *General Inventory of Royal Possessions* was begun around 1666 and completed in 1673, with intermittent updates thereafter.[51] As noted earlier, an inventory of the crown jewels was drawn up in 1666 and updated in 1683 and 1691. These inventories formally distinguished the king's personal possessions from crown possessions, since they only listed the latter. At least, that was the theory. In practice, Louis XIV exploited a loophole in the legal documents that codified this distinction. These documents referred only to "rare and precious" (*rare et précieux*) objects, and it was the king who decided which objects were

50. Based on a study of the probate inventories of Marguerite Le Brun, AN MC/CXV/363, Dec. 3, 1715; Charles and Claude Le Brun, AN MC/XXXIX/206, Mar. 24, 1698; and Louis-Nicolas Foucquet, AN MC/XIX/580, June 8, 1705. On the Le Brun brothers, see also Stéphane Castelluccio, *Le prince et le marchand: Le commerce de luxe chez les marchands merciers parisiens pendant le règne de Louis XIV* (Paris: SPM, 2014), 629–30, 637.

51. Stéphane Castelluccio, *Le garde-meuble de la Couronne et ses intendants du XVIe au XVIIIe siècle* (Paris: Comité des travaux historiques et scientifiques, 2004), chaps. 1 and 2.

common enough to be transferred from the crown to himself. The result was that he frequently gave away ornaments made of agate and rock crystal that were listed in the *General Inventory of Royal Possessions*. He also appropriated precious stones from the crown jewels. Indeed, this was the occasion for the inventories of 1666 and 1691. The preface to the latter inventory explained that the king wished to add some diamonds to the crown jewels "to increase the number of buttons" and to remove diamonds that "he did not find handsome enough [*assez beau*] to be included in the inventory."[52] The king himself had marked the pages of an earlier inventory to indicate the stones he wished to transfer to his personal collection. Gem appraisal served not only to decide how much a gem was worth but also to decide who owned it.

Travel Narratives

Gem appraisal was also a way of understanding the people who did the appraising. This is especially clear in French travel narratives in the middle decades of the seventeenth century. It is well known that these narratives helped to create an image of "the Orient" and "Orientals" for French readers.[53] Less well known is the role of gems—and in particular, the techniques of gem appraisal—in this early stage of European orientalism. Consider Jean Chardin's *Travels in Persia*, a classic in the genre first published in 1686. This book was awash with gems. In fact, gems were the occasion for the travels in the first place.[54] The son of a Parisian jeweler, Chardin made his first voyage to Persia in 1665–1670, learned the Persian language, worked as a gem trader for the Safavid Shah Abbas II, and returned to Europe with some 350,000 livres worth of raw diamonds. After stocking up on colored stones, he returned to Persia in 1671 with a view to selling these stones to his former patron. The

52. Bibliothèque nationale de France, Fonds Clairambault, Ms 499, p. 399; Bapst, *Histoire des joyaux de la couronne*, 369–71.

53. Recent surveys of this topic include Faith E. Beasley, *Versailles Meets the Taj Mahal: François Bernier, Marguerite de la Sablière, and Enlightening Conversations in Seventeenth-Century France* (Toronto: Toronto University Press, 2018), 3–28; and Sanjay Subrahmanyam, *Europe's India: Words, People, Empires, 1500–1800* (Cambridge, MA: Harvard University Press, 2017), 1–44. The present section builds on the account of diamond trading ca. 1700 in Diana Scarisbrick and Benjamin Zucker, *Elihu Yale: Merchant, Collector and Patron* (New York: Thames & Hudson, 2014), 151–63.

54. The standard biography of Chardin is Dirk van der Cruysse, *Chardin le Persan* (Paris: Fayard, 1998), where his movements are summarized from p. 451 onward. The identity of Chardin's parents, and the figure of 350,000 livres, are in Edgar Samuel, "Gems from the Orient: The Activities of Sir John Chardin (1643–1713) as a Diamond Importer and East India Merchant," *Proceedings of the Huguenot Society* 27, no. 3 (2000): 351–68, 352.

hitch was that Abbas II had died during his absence and been replaced by
Shah Suleiman I. Chardin's labyrinthine negotiations with the new shah are
one of the many subplots of his *Travels*.[55] For Chardin, these negotiations
were ultimately about the character of the Persian people: "everything that
happens in this country is a matter of reciprocal dissimulation."[56]

This point goes well beyond Chardin, whose *Travels* were the culmination of five decades of French traveling, travel writing, and gem trading.[57]
As early as 1631, Jean-Baptiste Tavernier, the son of a Parisian map seller, set
out to Turkey and Persia. He visited again in 1638, and for the next thirty
years he spent nearly all his time trading gems in Asia and the Middle East,
returning to France only to sell these goods or to buy European goods to
trade on his next trip.[58] Two other notable travelers of the period, Jean de
Thévenot and François Bernier, were also tied to the gem trade, though they
were not jewelers themselves. In 1666, Thévenot met Tavernier in Surat.[59] In
the same year, he spent time traveling with David Bazu, a diamond merchant
who would sell 600,000 livres worth of diamonds to Louis XIV soon afterward.[60] When Thévenot died in Persia twelve months later, he left behind
a balance for weighing diamonds (fig. 2.2) and debts to Jean Chardin and
François Bernier.[61] By this time, Bernier had spent nearly a decade as a doctor
and philosopher at the court of the Mughal emperor.[62] Bernier was in Delhi
when Tavernier arrived there in 1665 to sell European gems to Aurangzeb;
after the sale, the two spent six weeks traveling down the Ganges together.[63] A
few months later, in February 1667, Tavernier met Chardin in Bandar Abbas.[64]

55. Jean Chardin, *Voyages en Perse*, vol. 3, ed. Louis Langlès (Paris, 1811 [1711]), 37–45, 101–13,
116–18; vol. 9, 240–42, 350–58.

56. Ibid., vol. 3, 111.

57. Key dates and titles are given in Valentine Ball, "Introduction," in Tavernier, *Travels*, xi–
xxxviii; Cruysse, *Chardin le Persan*, Chronology and Bibliography; Françoise de Valence, "Introduction," in Jean Thévenot, *Les voyages aux Indes orientales*, ed. Françoise de Valence (Paris: H.
Champion, 2008), 7–23; Frédéric Tinguely, "Introduction," in François Bernier, *Un libertin dans
l'Inde moghole: Les voyages de François Bernier, 1656–1669*, ed. Frédéric Tinguely, Adrien Paschoud, and Charles-Antoine Chamay (Paris: Chandeigne, 2008), 7–34; Nicholas Dew, "Reading
Travels in the Culture of Curiosity: Thévenot's Collection of Voyages," *Journal of Early Modern
History* 10, no. 1 (2006): 39–59, 42n11, 52n46.

58. Tavernier's travels are summarized in Ball, "Introduction."

59. Ibid., xxix.

60. Valence, "Introduction," 10.

61. Christian Thévenot, *Jean de Thévenot* (Saint-Denis: Edilivre, 2012), 398, 400–401.

62. Tinguely, "Introduction," 11–17.

63. Ball, "Introduction," xxvii–xxviii, xxxiii.

64. Ibid., xxix.

This completed a remarkable eighteen months in which Tavernier met each of the three other French travelers at least once. These eighteen months overlapped with an equally intense burst of travel writing, one that ran from the first volume of Melchisédech Thévenot's compilation of travel narratives, published in 1663, to the appearance of Chardin's *Travels in Persia* in 1686. Works by Bernier, Tavernier, and Jean Thévenot appeared in the intervening years.[65]

Taken together, these narratives give a rich account of gem appraisal in the context of cross-cultural travel. Let us look more closely at Chardin's attempt to sell jewels to Shah Suleiman I. According to Chardin, the first round of negotiations took place over the course of ten days in July 1673.[66] Chardin's main interlocutors were the Nazir—the director of the Shah's household— and a jeweler whose job it was to appraise all gold, silver and gems sold at the imperial court. Chardin was dismayed to find that the Nazir would pay no more than half his asking price, and that his final offer was as low as 50 thousand livres. The Nazir argued that the largest and most expensive specimens were not "made in the style of the country." He supplemented the advice of the court jeweler with that of twenty other jewelers active in Isfahan, including Muslims, Armenians and Indians. Chardin had an assistant of his own, an Armenian merchant who was based in Isfahan whom he describes as being a "very skilled connoisseur" (*fort habile connoisseur*). The negotiations ended in stalemate, but Chardin was back two years later with a quarter of the original stones.[67] Once again the Nazir called on local jewelers, and once again they cut Chardin's price in half. When Chardin protested, the Nazir invoked the skill of these jewelers and of Persian jewelers in general. Chardin replied that his stones were so rare and beautiful that "connoisseurs from this country are unable to tell their true price." Privately, he believed the jewelers were beholden to the Nazir, and that the latter had fixed the price in advance. The Nazir eventually bought the stones for a low price. "There is no piece of trickery that he did not use to get the better of me," Chardin concluded. Tavernier told similar tales about the Mughal court, where stones brought by foreign merchants were appraised by a team of Persian and South Asian jewelers. "These Banians," he wrote, using a generic European term for South

65. For details, see note 57.

66. Chardin, *Voyages en Perse*, vol. 3, 101–3, 116–18. The nazir's role is summarized in Ronald W. Ferrier, *A Journey to Persia: Jean Chardin's Portrait of a Seventeenth-Century Empire* (London: I. B. Tauris, 1996), 82.

67. This second round of negotiations is in Chardin, *Voyages en Perse*, vol. 9, 240–42, and 350–58.

Asian merchants, "are in business a thousand times worse than the Jews."[68] Statements such as these should obviously be taken with several grains of salt. They say as much about Tavernier and Chardin as they do about the people they purport to describe. Nevertheless, they do show that gem appraisal was a basis for ethnography, a way of understanding unfamiliar people.

Gem appraisal was also involved in encounters outside the court. Tavernier's account of the Indian diamond mines showed the sorts of resource that made routine transactions possible. One resource was provided by the sultan of Golconda: a diamond weigher whose impartiality was guaranteed by the fact that he was paid by the king rather than merchants.[69] Another resource was time. Foreign merchants at the mines were given "seven or eight days or more" to examine a parcel of diamonds before they decided whether to buy.[70] "Do not trouble yourself now," a Banyan seller told Tavernier, "you will see it [i.e., a diamond] tomorrow morning at your leisure when you are alone."[71] According to Tavernier, he had a morning to examine his stones because this corresponded to one-quarter of the typical Banyan day, at the end of which (around 9 a.m.) the merchant would return to his town to bathe and pray. Tavernier was also sensitive to specific ways of handling and viewing gems. European jewelers tended to examine the water of diamonds in broad daylight, so the Frenchman was surprised to find Indian merchants studying their stones in the depths of the night, using the light of a lamp.[72] Another technique was designed to weed out diamonds with the undesirable bluish color that Tavernier called "celestial." He noted that the term itself came from Indian merchants, who detected the hue by observing a diamond in the green shadow of a leafy tree. Tavernier had seen this being done by groups of children between the ages of ten and fifteen.[73] They had a well-honed routine for evaluating the diamonds they bought and sold: weighing them with weights they carried on their belts, passing them from one member of the band to the other, and in the evening sorting the day's purchases by water, weight, and clarity. Once again, Tavernier's narrative should be treated with caution; he was a selective and slanted observer. But his account does show his sensitivity to the question of how gems were judged. It also suggests some of the ways

68. Tavernier, *Travels*, vol. 1, 134–39. Cf. Tavernier's note on the "circumspection and patience" of "Indians" in vol. 1, 395. On the European use of "banyan" and "banian," see Sabel, "Rare Earth," 78.

69. Tavernier, *Travels*, vol. 2, 68–69.

70. Ibid., vol. 2, 60.

71. Ibid., vol. 2, 65.

72. Ibid., vol. 2, 75.

73. Ibid., vol. 2, 61–62.

these judgments were embedded in his portrait of the social and religious world of South Asia.

The overall effect of these travel narratives was to reinforce the ancient idea of the Orient as the land of precious stones. There are a few exceptions to this rule, such as Tavernier's claim that the world's best opals are in Hungary.[74] But the dominant theme was the great abundance of precious stones in Asia, whether the diamonds of the sultanate of Golconda, the turquoise of Persia, or the rubies of Pegu.[75] Quality mattered as much as quantity in these images. A ruby in the Persian treasury was "the most beautiful and of the richest color that I have ever seen," according to Chardin.[76] Tavernier described a pearl "so transparent that you can almost see the light through it."[77] His drawings of Indian diamonds had captions referring to the clarity and whiteness of the drawn specimens.[78] Low-quality gems could be as evocative as high-quality ones, as shown by a story told by Bernier and Tavernier.[79] According to the story, a dispute arose about the identity of a stone sold to Aurangzeb by a courtier. The court jewelers appraised it as a ballas ruby worth 95 thousand rupees, whereas an "old Indian" dismissed it as an imitation worth 500 rupees. Aurangzeb called on his father, Shah Jahan, then under house arrest. The old emperor sided with the elderly Indian, and that was the end of the matter. "In the whole Empire of the Great Mogul," Tavernier explained, "there was no one more proficient in the knowledge of stones than Shah Jahan." The story may be apocryphal, but it is informative nevertheless. For French travelers and their readers, the ability to tell good stones from bad was as much a part of the emperor's mystique as the Taj Mahal.

Maps

This mystique was enriched by the global atlases that began to appear in Paris in considerable numbers in the same period. Maps showed where gems came

74. Ibid., vol. 2, 103. Note also his claim, on the same page, that Bohemian rubies are as good as those from Pegu, and his statement (vol. 2, 106) that there are no emeralds in Asia.

75. Representative passages are Thévenot, *Voyages aux Indes Orientales*, 233, 235–36; Bernier, *Voyages de François Bernier*, 54, 57–58; Tavernier, *Travels*, vol. 1, 395–97, vol. 2, 99–102; Chardin, *Voyages en Perse*, 488.

76. Chardin, *Voyages en Perse*, vol. 5, 431.

77. Tavernier, *Travels*, vol. 2, 110.

78. Ibid., vol. 2, 126, plate 3. The original images, with explanatory text in French, are in Jean-Baptiste Tavernier, *Les six voyages de Jean-Baptiste Tavernier, qu'il a fait en Turquie, en Perse et aux Indes* (Paris, 1676), vol. 2, 336.

79. Bernier, *Voyages de François Bernier*, 268; Tavernier, *Travels*, vol. 2, 127–28.

from, a crucial point in the classification and evaluation of gems in premodern Europe, as we have seen.[80] Moreover, atlases were linked to the travel narratives in which French jewelers recounted their experience of gem appraisal in Asia. It was the Tavernier family—especially Jean-Baptiste's father and brother—that introduced the Dutch art of cartography into France in the first half of the century.[81] The Taverniers were succeeded by the Sanson family as the kingdom's leading cartographers under Louis XIV. Nicolas Sanson drew the maps for the first French world atlas, the *General Maps of All Parts of the World*, first published in 1658. His son Guillaume oversaw the later editions of this work and drew most of the maps for a similar work published in 1681, the *New Atlas*. Guillaume's cousin, Pierre Duval, was another successful cartographer whose *Universal Geography* went through nine editions between 1661 and 1712. This work included a set of maps of modern voyages, each showing the route of a traveler indicated by a straight line.[82] Mapmakers sometimes cited travel narratives as sources of their data.[83] Nicolas Sanson's first map of Asia was dwarfed by fifty pages of written description.[84] Conversely, the authors of travel narratives sometimes drew maps themselves, such as the ones scattered through Melchisédech Thévenot's compilation of French voyages.[85] Jean-Baptiste Tavernier's voyages were maps before they were books—two maps of his travels were first printed in 1675, a year before the first edition of his travel narrative.[86]

80. See chap. 1, this volume.

81. Mireille Pastoureau, *Les atlas français XVIe–XVIIe siècles: Répertoire bibliographique et étude* (Paris: Bibliothèque nationale de France, 1984), 3–7 (survey of seventeenth-century French cartography), 6 (Taverniers and Sansons), 135–37 (Duval), 387–407 (Sansons), 395 (Nicolas Sanson's sources). See also Sabel, "Rare Earth," 109, on other jewelry-cartography connections in early modern Europe.

82. Pierre Duval, *Géographie universelle* (Paris, 1682) [Duval XI Fb], pp. 340–66. Text in square brackets shows the short title of the atlas, and the number of the map consulted, as provided in Pastoureau, *Atlas français*. Capital letters refer to editions, e.g., "Jaillot I C" refers to the third edition (C) of the earliest map (I) by Jaillot. All maps cited in this way were consulted at Département des Cartes et Plans, Bibliothèque nationale de France (Paris).

83. Pierre Duval, *Cartes géographiques* (Paris, 1688) [Duval II F, map 13] (citing Thévenot); Pastoureau, *Atlas français*, 195, referring to Nicolas Sanson, *Asie* (Paris, 1652) [Sanson III A] (citing Purchas and Linschoten).

84. Sanson, *Asie* [Sanson III A, map 9].

85. Christian Thévenot, *Melchisédech Thévenot: Bibliothécaire du roi, 1620–1692* (Saint-Denis: Edilivre, 2012), 48.

86. The maps in question were published in the 1713 Rouen edition of Tavernier's *Six voyages*. They do not appear in the copies of the first French edition of 1676 that I have consulted. Earlier versions of the two maps survive as single sheets at the National Library in Paris, and

It is not surprising, then, that maps tracked the growth of knowledge about the geographical origins of gems. Consider *Ancient Geography*, an atlas compiled by Pieter Bert and first published by Melchior Tavernier—Jean-Baptiste's brother—in 1628.[87] Bert's map of the Arabian Peninsula showed an island named Tylos on the south coast of the Persian Gulf, with a label indicating that pearls can be found on this island (fig. 2.4, top). But the Latin place-names were from Pliny's *Natural History*, and the map may have been copied from the Dutch cartographers Mercator and Ortelius, as an earlier atlas by Bert had been. Moreover, no gems were labeled on Bert's maps of India or Southeast Asia. Compare this to an atlas published in 1700, the *Curious Atlas* of Nicolas de Fer (fig. 2.4, bottom).[88] Here we again find pearls in the Persian Gulf, but they are now associated with an island labeled "Bahrem" that is located on the west coast of the gulf, rather than the south coast.[89] Also, the island on the map is in the same position as the island known today as Bahrain. "Raolgonde" and "Coulour" are marked as "diamond mines" on Fer's map of India, names that he probably borrowed from Tavernier's *Voyages*, or from a source derived from that work. These names were later identified as Rāmallakota and Kollūr, both on the Deccan plateau and both sites of real diamond mines.[90] Other maps in Fer's atlas show turquoise near "Nishapour" (Nishapur, in modern Iran), diamonds on the southwestern segment of the island of Borneo (now West Kalimantan), and pearl fisheries on the coast of what is now Venezuela.[91]

Between Bert and Fer, the standout map in terms of gem data was perhaps Nicholas Sanson's map of Asia—or rather, the text with which this map was published. This text described the pearls in the Persian Gulf and the Coromandel Coast, diamonds in the sultanate of Golconda, and rubies in Pegu, with precise place-names for each of these localities, such that the reader can find them on the adjacent maps.[92] Gems appeared on other French maps as

are dated 1675: Bibliothèque nationale de France, Département des Cartes et Plans, GE D-15505 and GE D-15506.

87. Pieter Bert, *Geographia vetus* (Paris, 1628) [Bertius A, map 15]. On Bert, see Pastoureau, *Atlas français*, 65.

88. Nicolas de Fer, *Atlas curieux* (Paris, 1700) [Fer I B, map 105].

89. Cf. Tavernier, *Travels*, vol. 2, 107–8 ("Bahren" pearls).

90. Ibid., vol. 2, 53n1, 172n1; Habib, *Atlas of the Mughal Empire*, maps 15B, 16B; Sabel, "Rare Earth," 74–75.

91. The relevant gems did indeed occur in these places in Fer's time: Khazeni, *Sky Blue Stone*, 12–13; Donkin, *Beyond Price*, chap. 10; Tavernier, *Travels*, vol. 2, 87n1; Sabel, "Rare Earth," 73–94.

92. References to gems are scattered through the text of Sanson, *Asie*, which is not paginated.

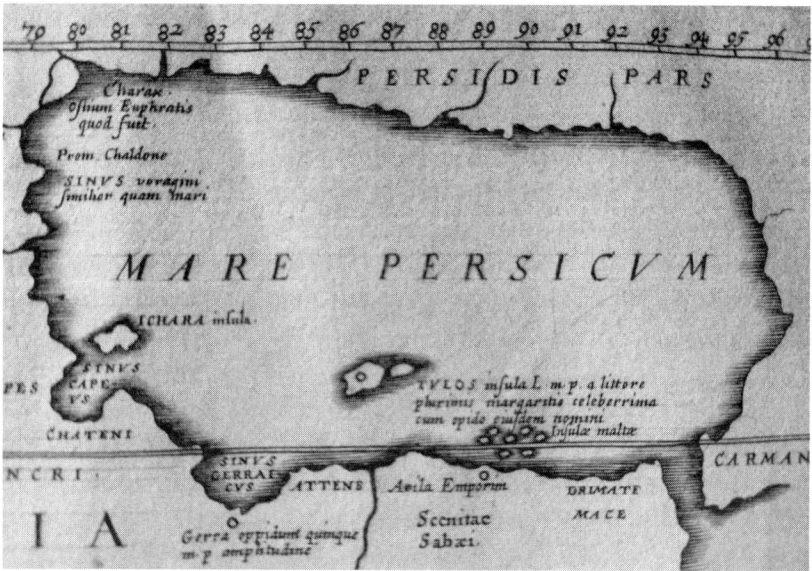

FIGURE 2.4. Pearls on Parisian maps

Two maps separated by seventy years show the growing precision of French data on gem locality. The top map is from Bert, *Ancient Geography* (1628). The Latin text attached to the island of Tylos reads, "On the shore there are many of the famous pearls as well as a town of the same name." The bottom map is from Fer, *Curious Atlas* (1700). The French text reads, "Pearls are fished in the sand around these islands." Courtesy of the Bibliothèque nationale de France; photos by the author.

well, sometimes on the marked routes of travelers (fig. 2.5).[93] The marked locations were not always accurate; on the contrary, they were sometimes seriously misleading from the present-day point of view, and even from the perspective of working jewelers at the time.[94] But the point is that they seemed increasingly accurate to European readers. They were all the more noticeable given the lack of other details on the same maps. The diamond mines of Rāmallakota and Kollūr were the only commodities marked on Fer's map of India. Indeed, they were the only things marked on the map apart from towns, kingdoms, and natural features such as rivers and mountains. Fer's map of Persia and Turkey differed only in that it labeled the sites of ancient cities such as Babylon and Persepolis. Pearls were as much a symbol of Persia as Persepolis.

The placement of gems on maps gains more significance when we see that maps were acts of classification. The distinction between Africa, America, Asia, and Europe was the basic organizing principle of the world atlas. Maps of Asia were dominated by the distinction between Turkey, Persia, India, and China, regions that formed a band across the center of these maps and were marked out with capitalized text, thick borders, and (in some editions) color-coded borders. The accompanying text reinforced these categories, whether in the form of subtitles, contents pages, or introductions.[95] Most striking are the geographical tables—structured lists of the main divisions of the earth—that helped to make Nicolas Sanson's name as a cartographer. First published on their own, they later appeared alongside Sanson's maps in the *New Atlas* (fig. 2.6).[96] Duval gave his own twist to these tables, arranging them into a grid that showed the geological features of each continent as well as their

93. The figure shows a map from Duval, *Géographie universelle* [Duval XI Fb, p. 363] with diamond mines marked on an itinerary. Sanson, *Asie* [Sanson III A, map 16], has "ubi adamantes" on a river named "Succadano" on Borneo. Jaillot, *Atlas nouveau* (Amsterdam, n.d.) [Jaillot I F, map 358] has the rivers "Lava" and "Succadano" marked as diamond localities.

94. I am grateful to Claire Sabel for pressing this point, in correspondence and at Sabel, "Rare Earth," 158–59, 232.

95. Nicolas Sanson's volumes on Europe, Asia, Africa and America were published separately in 1647, 1652, 1656, and 1657, respectively (Pastoureau, *Atlas français*, 387). Nicolas Sanson, *Cartes générales* (Paris, 1658) [Sanson V A, map 9] is divided into sections by continent; it also depicts each continent in a single map, as does Jaillot, *Atlas nouveau*. These authors referred to Africa, Asia, etc., as "parts" of the earth rather than as continents; for them, "continents" were contiguous landmasses such as the Americas or Afro-Eurasia: Duval, *Cartes géographiques*, "Introduction à la géographie." Divisions within Asia are especially clear in the map of Asia in Sanson, *Cartes générales* [Sanson V A, map 4].

96. Alexis-Hubert Jaillot's *Atlas nouveau* (Paris, 1689) [Jaillot I C] was compiled by Jaillot but included maps and tables by Sanson. On geographical tables, see Pastoureau, *Atlas français*, 37.

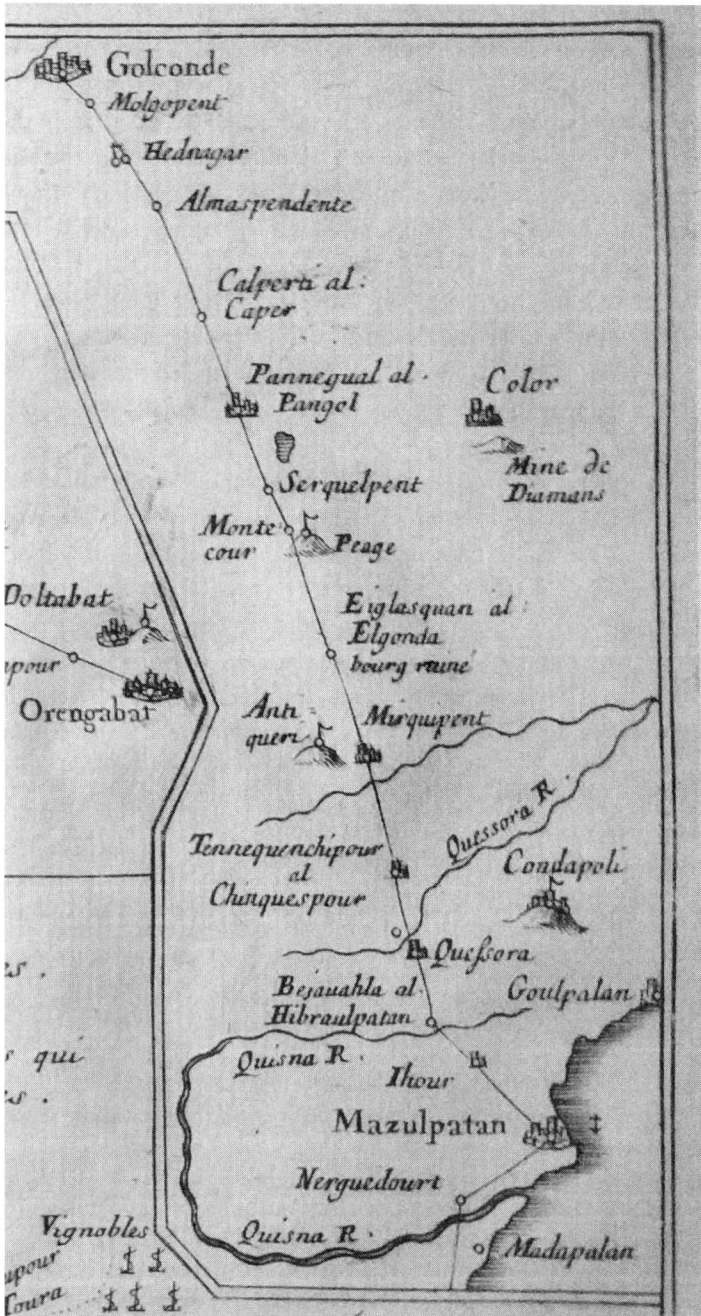

FIGURE 2.5. Diamond mine on a travel map

A diamond mine (*mine de diamans*) is marked to the right of the route from Mazulpatan to Golconda. Maps and travel narratives were overlapping genres in the age of Louis XIV, and gems were prominent in both. From Duval, *Universal Geography* (Paris, 1682). Courtesy of the Bibliothèque nationale de France; photo by the author.

TABLES GEOGRAPHIQUES,
DES DIVISIONS DE
L'ASIE,
Par le S. SANSON, *Geographe Ordinaire du Roy.*

L'ASIE, *a*

dans la TERRE FERME

{ la Turquie en Asie
la Georgie
l'Arabie
la Perse
l'Inde
la Chine
la Tartarie .

dans la MER

OCEANE { des Maldiues
de Ceylan
de la Sonde
des Molucques
des Philippines
du Iapon .

MEDITERRANÉE { de Cypre
de Rhodes
de Scio
de Metelin .

ANATOLIE { Chiutaje
Angouri
Swas
Tarabosan
Marasch
Cogni
Bursa
Marmora
Izmir
Efeso
Halicarnasse
Satalia
Tarssus
Caysairiyah
Amasia
Sinopoli
Samastro
Nicomedia .

LA TURQUIE EN ASIE, *diuise en*

TURCOMANIE { TURCOMANIE. { Erzerum
Achlat
Wan

CURDISTAN { Bitlis

DIARBECK { Diarbekir
Asanchiuf
Carsamid
Orpha
Nasibin

LA PERSE *a* { Vers la
Mer Caspie
les PROVINCES de

IRAN

LES PROVINCES DELA PERSE

Vers les
Indes, *sont*

sur les Golfes
de Balsera,
et d'Ormus,
sont

ERACK-ATZEM { Ispahan
Caswin
Soltania
Kom
Kaschan
Yest
Nehauend
Scherrimir .

ADIRBEITZA { Tauris
Ardeuil
Salmas
Merragne .
K arasbag
Eruan
Kars

SCIRVAN { Nasch Schuan .
Scamachie
Derbent .

KILAN { Resch
Kesker .

MASANDERAN { Ferabath
Sarow .

CHORASAN { Herat
Meruerud .

SABLUSTAN { Sarentz
Saruan
Bust
Memend .

SITZISTAN { Sitzistan
Chaluck
Masnih
Araba
Masurgian

FARS { Fardan .
Schiras
Firusabath
Nagira
Lar
Estaker .

CHUSISTAN { Desu
Araigan .

KIRMAN { Berdasir
Chabis
Bermasir
Bem
Tzireet

MAKRAN { Bender Kamron
Ormus .
Fihr
Kitz
Titz
Rasec .

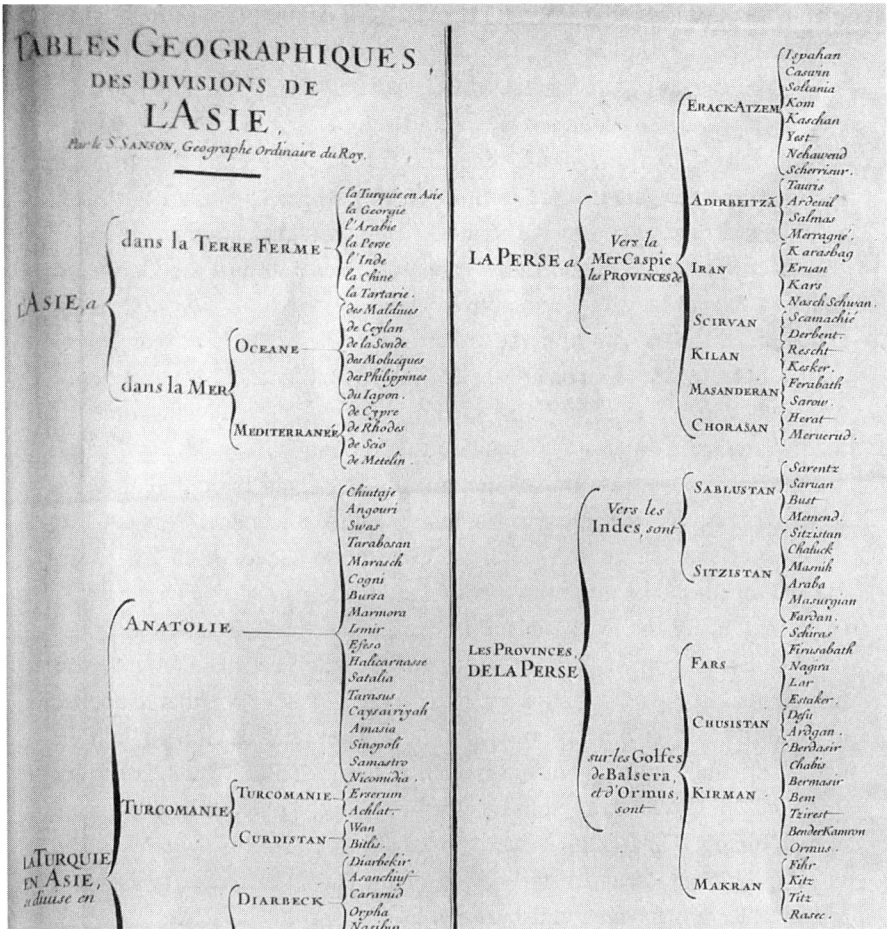

FIGURE 2.6. Geographical table of Asia
An example of geography as a classificatory project. The image is extracted from a geographical table of Asia compiled by Sanson and published in Alexis-Hubert Jaillot, *New Atlas* (1689). The table shows the broad division of Asia into earth and sea, followed by "Turkey in Asia," "Persia," and so on, each of which is divided into provinces and major cities. Courtesy of the Bibliothèque nationale de France; photo by the author.

main towns and provinces. Duval sold educational games as well as maps, and he combined the two interests in deck of cards in which each suit represented a continent. He decorated his maps with miniatures that showed a whole continent or the whole world in a single glance.[97] Readers of these atlases were conditioned to think of gems as belonging to a continent and

97. Duval, *Cartes géographiques*, map 13, "Table générale de géographie," "Jeu du monde," "Tables de géographies réduites en un jeu de cartes."

region, not just to a town or island. The pearls of Bahrain were also the pearls
of Persia and the pearls of Asia.

Finally, they were the pearls of the Orient. The terms "orientale" and "oc-
cidentale" were used widely and ambiguously on French atlases. In some con-
texts, the Orient was the part of the world that lay to the east of Europe, and
the Occident the part that lay to the west of Europe. This usage is clear in
a "universal map of commerce" published by Duval, a map that shows the
"routes toward the Oriental Indies" and the "routes toward the Occidental
Indies."[98] One of Duval's playing cards shows the "Oriental Indies," helpfully
explaining that these run from the east coast of Africa to the islands of Japan, a
region that Duval showed on a "Map of the Oriental Indies."[99] The distinction
was applied to oceans as well, the Atlantic being the "Occidental Ocean" and
the Indian Ocean and South China Sea the "Oriental Ocean."[100] But this was
not the only usage, since the distinction could be applied at different scales.
It could divide, not the whole world, but the Old World alone: "Asia takes up
the most oriental part of our continent," Sanson explained, while "Africa and
Europe together take up only the most occidental part of this continent."[101]
Other regions of the world, including Africa and the Roman empire, were
divided in the same way.[102] In sum, "oriental" could refer to a definite place,
namely Asia and part of Africa, or it could refer to any place that happened
to be to the east of some arbitrary point of reference. This chimed with the
use of "oriental" among goldsmiths such as Berquen and Rosnel. For them,
"oriental" could refer to a gem from a definite place, namely Asia, or to a gem
of high quality that could have come from anywhere. In cartography as well
as natural history, "oriental" floated free of any particular place. It was a geographical term that transcended geography.

These points about gem geography give an extra twist to the thesis, recently defended by the historian Benjamin Schmidt, that the contrast between "Europe" and "the exotic" was invented by geographers in the decades
around 1700. According to Schmidt, this is the period in which "the exotic"
acquired the connotation of charming abundance, of promiscuous variety,
whereas the term had previously referred simply to species that were not
native to Europe.[103] This new geographical aesthetic can certainly be found

98. Ibid., map 59.
99. Ibid., map 35.
100. Sanson, *Cartes générales* [Sanson V A, map 1].
101. Sanson, *Asie*, in unpaginated text.
102. Sanson, *Asie*, "Romani Imperii qua Occidens" and "Romani Imperii qua Oriens." Fer,
Atlas curieux [Fer I B, maps 46, 103].
103. Schmidt, *Inventing Exoticism*, esp. 231–57.

in maps and travel narratives I have been considering here. But these were technical texts as well as aesthetic ones. Chardin's picturesque portrayal of the court of Isfahan was woven into an informed account of cross-cultural gem appraisal. Global atlases gave increasingly precise accounts of the locations of particular gems that are in striking contrast to the "promiscuous assemblages" (Schmidt's phrase) that appeared in the cartouches and frontispieces of the same atlases. And readers did not need to wait until the 1730s to find order and method in their atlases, as Schmidt suggests—witness the geographical tables that made Sanson's fortune in the 1650s. These texts created a charming image of the world that was all the more seductive because it appeared to emerge from the expertise of goldsmiths, gem cutters, cartographers, and of observant and intrepid travelers. Technique was part and parcel of the exotic aesthetic.

Letters

Technique also helps to understand the letters of diamond merchants, a favorite source for economic historians interested in long-distance trading networks.[104] Much has been written about the role of trust in such networks.[105] But technique was as important as trust, and trust and technique were related in complicated ways. The point may be illustrated by the Chardin correspondence, a series of letters exchanged between Jean Chardin and his brother, Daniel Chardin, in the latter part of the century.[106] Jean Chardin was writing from London. He had traveled there early in the 1680s to avoid Louis XIV's growing campaign against Protestants such as himself. It was in London that Chardin published the first volume of his *Travels*, in 1686. In the same year, he and his brother went into business with two Portuguese Jewish merchants, Francis Salvador and his brother Isaac Salvador. John sent gold, coral, and emeralds to Daniel, who was based at Fort St. George, the new British trading post on the southeast coast of India. Daniel forwarded these goods to Isaac, who used them to buy diamonds in and around the Golconda sultanate.

104. Trivellato, *Familiarity of Strangers*; Vanneste, *Global Trade and Commercial Networks*; idem, *Blood, Sweat and Earth*; Yogev, *Diamonds and Coral*.

105. Trust is the main theme of Trivellato, *Familiarity of Strangers*, esp. 10–16, with technique mentioned in passing on pp. 229, 231–32, 237, 242, 246, 257–58. Trust is also the main theme of Vanneste, *Global Trade and Commercial Networks*. My account of both trust and technique builds on Sabel, "Rare Earth," 207–27.

106. The letters cited in the rest of this section are in Jean Chardin Correspondence and Documents, Beinecke Rare Book and Manuscript Library (New Haven, CT). JC = Jean Chardin, DC = Daniel Chardin.

These diamonds were then sent to John and Francis in London, who sold
them in Europe.[107]

Gem quality was a constant concern in the letters between the Chardin
brothers. Buyers insisted upon it: "diamonds of very high quality," "thin diamonds of excellent quality," "good, thin, crystalline diamonds, and not large
ones or inferior in goodness," demanded a buyer in a letter.[108] The business
partners went to considerable lengths to explain what quality meant for this
or that species. All cleaved stones are suspect with regards to water, Francis
instructed, no matter how perfect the water seems to be; a whole stone fetches
a higher price in Europe than the same stone cut in two; and so on.[109] As this
advice suggests, judgments about diamonds depended on predictions about
how they would change when cut. Chardin found this out the hard way when
he had two matching stones cut in the form of pendeloques, only to find that
one of them had acquired a brown tint in the cutting process, a defect that
halved the price of the pair.[110] To judge diamonds, one had to "reason like a
diamond cutter" (raisonner comme un diamantaire).[111] The lack of such reasoning was one source of friction between the Chardin brothers. "It is not the
done thing," John chided his brother on one occasion, "to send a parcel of
diamonds without a single word on the quality of the contents."[112]

Because diamonds were hard to judge, the Chardins sought help from
other judges. The problem was that the most skillful judges were not always
the most trustworthy ones. This was the crux of what Jean Chardin called
"the diamond affair," a problematic transaction that surfaced in his letters
early in 1697.[113] Daniel had sent a diamond to John, who showed it to Francis

---

107. The network is summarized in Samuel, "Gems from the Orient," esp. 354–55 (1686 contract), 360 (Paris buyers). There is also scattered data on the network in Cruysse, *Chardin le
Persan*, such as the discussion of the 1686 contracts on pp. 338–43. John mentions his suppliers
of coral in JC to DC, Feb. 1, 1696/7, box 1, folder 12, 7r, cf. prospective new suppliers in JC to DC,
Jan. 10, 1697/8, box 1, folder 13, 14v. John mentions an emerald supplier in JC to DC, Feb. 1, 1695/6,
box 1, folder 12, 14v.

108. Fernando Mendes to DC, Dec. 5, 1704, box 5, 77r. Fernando Mendes to DC, Dec. 28,
1707, box 5, 95r.

109. JC to DC, Feb. 1, 1696/7, 9v–10r. Cf. JC to DC, Apr. 24, 1697, 9r–v. Coral quality in JC
to DC, Feb. 1, 1696/7, box 1, folder 12, 7v; JC to DC, Jan. 2, 1708/9, box 1, folder 13, 24r. Emerald
quality in JC to DC, Feb. 1, 1695/6, box 1, folder 12, 14v; JC to DC, Nov. 15, 1700, box 1, folder 13,
44r; JC to DC, Jan. 20, 1702/3, box 1, folder 13, 60v, 61r.

110. JC to DC, Jan. 15, 1701/2, box 1, folder 10, 1r.

111. JC to DC, Jan. 2, 1700, box 1, folder 8, p. 8.

112. JC to DC, Dec. 27, 1705, box 1, folder 11, 31r.

113. JC to DC, Feb. 1, 1696/7, box 1, folder 12, 6r–v, 9r–v. A fair copy of parts of this letter,
along with more details about the affair, is in JC to DC, Apr. 24, 1697, box 1, folder 7, 6r, 8v–9r.

Salvador in the hope that the latter would buy it. John considered Francis to be a good judge of diamonds, "a more skilled connoisseur [*connoisseur*] than I." John showed the same stone to a certain Nunes, probably a jeweler or cutter, whom he also described as a "*connoisseur*" in diamonds. Both experts professed to be underwhelmed by the diamond. John suspected that Francis had exaggerated the flaws in the stone in order to drive down the price, and that Nunes was in on the game. But he could not prove this, precisely because he relied on Nunes and Francis to judge the stone. This was frustrating: "I can't stand always being under the tutelage [of Francis] for the diamonds I receive," John wrote to his brother.

One solution was to get advice from multiple independent buyers, but this raised problems of its own. John had one of his diamonds examined by at least seven pairs of eyes—two of his associates, two jewelers, Isaac Salvador, a certain Anne Maubert, and Elihu Yale, the colonial administrator who later founded a famous university.[114] The associates suggested 2,000 livres, the jewelers offered 1,500, Yale 1,400, and Maubert 1,200. Isaac Salvador offered nothing, offended by the fact that the jewelers' offer was conditional on their having the right of first refusal. Salvador's reaction illustrates one downside to the use of multiple judges, the difficulty of reconciling the demands of different judges. Another problem was the mere fact of revealing one's merchandise to a competitor. "You won't get that much if you show the stone before you sell it," a diamond cutter once said to John.[115] "[Francis] Salvador is trying to wheedle out of me the price and quality of this merchandise," John wrote to his brother in 1699, referring to a set of ten diamonds he had just received. In this case, John and Francis eventually agreed on scheme in which Francis would pay John for the privilege of seeing the stones.[116] The right to judge a stone was as much a commodity as the stone itself.

Judging diamonds was an even thornier problem in India than in London. The Chardins' solution was to hire cutters to help with the judgments, but this raised the problem of judging the cutters. The first of these assistants was Isaac Salvador, who was "reputed to be the most skilful of the two [Salvadors]

I assume that the two flawed diamonds described in a later letter relate to the same affair: JC to DC, Jan. 10, 1697/8, box 1, folder 13, 12r–v.

114. JC to DC, Jan. 20, 1702/3, box 1, folder 13, 63r–v. Maubert's first name is in Samuel, "Gems from the Orient," 64. Maubert is also mentioned in JC to DC, Jan. 20, 1702/3, box 1, folder 13, 64v. On Yale's diamond trading, see Scarisbrick and Zucker, *Elihu Yale*, esp. pt. II, chap. 1, and pp. 209–13.

115. JC to DC, Jan. 2, 1700, box 1, folder 8, p. 8.

116. Ibid., 9–10. This sale is also discussed in Samuel, "Gems from the Orient," 364.

in knowledge of [*connoissance de*] diamonds," in John's words.[117] Isaac was probably trained as a diamond cleaver. He was thoroughly immersed in Golconda society, adopting local dress, diet and language, and bearing children with Hindu women. But there was a catch: "he's a complete scoundrel," John wrote in the 1690s, around the time that Isaac began to work for Elihu Yale.[118] Isaac was soon replaced by Abraham Pluymer, a diamond cutter who joined Daniel in India in 1697 or 1698. Pluymer had worked in the Netherlands and London, had the endorsement of the jeweler's guild of London, and was reputed to be "a good *connoisseur* whose eye one can trust." He was also, in John's view, a vulgar simpleton who lacked the intelligence (*esprit*) to set himself up as a rival to the Chardins.[119] Unfortunately for the Chardins, Pluymer was not only "intelligent with diamonds" but also "sly and deceitful."[120] By 1700 he was no longer working for the Chardins. He had become very rich and was rumored to be working at the mines of Golconda. The following year he was in London, allegedly using his skills and reputation to drive down the price of Chardin's diamonds.[121]

In an ideal world, Jean Chardin would have known people who were both technically skilled and trustworthy, in both London and Fort St. George. In practice, he lived in a world where skill varied widely, where trustworthiness varied widely, and where there was no clear correlation between the two. He trusted his own brother but was continually frustrated by Daniel's inability to pick diamonds. His business partners, the Salvadors, were very good at picking diamonds but he did not trust them an inch. Chardin himself knew something about diamonds, but not enough to rely on his own judgment in difficult transactions. He looked for a second opinion—even a third, fourth, fifth, or sixth opinion—but these opinions did not necessarily agree with each other. In any case, the opinions often came from potential buyers, people who were not exactly untrustworthy but who had an interest in undervaluing Jean Chardin's wares. In all this, there was a central role for knowledge about the

117. Samuel, "Gems from the Orient," 355 and 32n. On the wider phenomenon of European merchants recruiting gem experts in the Asian trade, see Sabel, "Rare Earth," 116–29.

118. JC to DC, Apr. 5, 1693, box 1, folder 12, 12v ("complete scoundrel"). Cf. JC to DC, Feb. 1, 1695/6, box 1, folder 12, 17r ("own trap"; "license"). Yale in Samuel, "Gems from the Orient," 360.

119. JC to DC, Apr. 24, 1697, box 1, folder 7, 8v–9r. Pluymer's first name in Samuel, "Gems from the Orient," 363.

120. JC to DC, Jan. 2, 1700, box 1, folder 8, p. 13.

121. JC to DC, Nov. 15, 1700, box 1, folder 13, 43v and 44v. JC to DC, Sept. 20, 1701, box 1, folder 9, 4v. See also the critique of another cutter, a certain Glover: JC to DC, Apr. 24, 1697, box 1, folder 7, 8v–9r.

material properties of diamonds, such as their response to being cut, cleaved, or heated. Trust mattered, but technique cannot be reduced to trust.

The Mutual Influence of the Crafts

This chapter has made the case for a technical approach to value. It is a call for historians of technical writing to consider material evaluation as well as material production. It is also a call for historians interested in value to consider the technical dimension of the topic, whether in relation to guilds, trading networks, the invention of exoticism, or any other historical theme. This is one important way in which judging and knowing go together in the history of gem science.

At the same time, the sources considered here are another illustration of the importance of hand-hand coordination in craft and commerce. To begin with, there is the simple fact that different practitioners classified gems in different ways. This can be seen in table 2.1, which compares Berquen's classification with those in five other sources.[122] There are many mismatches between Berquen's classes and those of the other sources. There are even more mismatches if we consider sources that are not on the table because they are not remotely comparable to Berquen's scheme, such as Chardin's letters, with their overwhelming concern for diamonds. These differences between practitioners draws attention to the interactions between them. Interactions between goldsmiths and gem cutters were partly responsible for the natural histories of gems published by Berquen and Rosnel. Interactions between cutters and traders were at the heart of the correspondence between John and Daniel Chardin, since judgments about the value of gems depended on knowledge of markets *and* knowledge about the material properties of gems. When practitioners and scholars interacted, the effect of this interaction was often to bring together different sorts of practitioner. Félibien, the royal historiographer mentioned at the start of this chapter, could write about gem cutting because he had already written about painting, sculpting, and stone-cutting.[123]

122. The rationale for each group is as follows, reading table 2.1 from left to right and top to bottom: Félibien: cut with copper and diamond powder; cut with lead and emery; polished with wood and emery. Legrand inventory: in display case; in cabinet; and in glass-fronted cupboard, all in the probate inventory of Henri Legrand, AN MC/I/205, Jan. 23, 1697. Le Brun inventory: in section on diamants; in section on bijoux et curiosités; N/A. Crown: called "pierres de couleur" in the 1691 inventory of crown jewels; listed in section on agate vases in 1673 inventory of movable treasures; listed in section on rock crystal in the same. Tavernier: in chapter on pierres de couleur; N/A; in chapters dedicated to pearls, coral, and yellow amber.

123. This claim is based on a study of Germer, *Art-pouvoir-discours*.

TABLE 2.1. Gem taxonomies under Louis XIV

The choice of the three columns is based on Berquen's division of gems in his *Marvels of the Indies* (1661). The first row shows the species that Berquen placed in each of these groups. The remaining rows show groups of stones in other texts on gems, with these groups arranged to maximize the fit with Berquen's groups. A glance at any of the three columns shows the discrepancies between the six taxonomies. For example, over a third of the terms are underlined, meaning that they are in the wrong column with respect to Berquen's taxonomy. The three starred species are not named in Berquen's book. Diamonds are omitted for the sake of clarity.

	Greater precious stones	Lesser precious stones	Other gems
Berquen 1661	sapphire, topaz, ruby (oriental, spinel, ballas), emerald, amethyst, aquamarine, hyacinth, opal, chrysolite	iris, vermilion, garnet, carnelian, turquoise, agate, onyx, sardony, chalcedony, jasper, lapis lazuli, rock crystal	pearl, coral, amber
Félibien 1676	sapphire, topaz, oriental ruby	garnet, agate, ballas ruby, spinel, emerald, amethyst, jacinth*	turquoise, lapis lazuli, girasol, opal
Legrand 1697	sapphire, amethyst, garnet, pearl, coral	carnelian, turquoise, rock crystal, ruby, emerald	agate, counterfeit gems
Le Brun 1698	topaz, ruby, emerald, amethyst, aquamarine	agate, jasper, lapis lazuli, rock crystal	N/A
Crown 1691/1673	sapphire, topaz, ruby, emerald, pearl	agate, jasper, lapis lazuli, topaz, amethyst, jade*, prisme d'emeraude*	rock crystal
Tavernier 1676	sapphire, topaz, ruby, emerald, amethyst, hyacinth, opal, turquoise	N/A	pearl, coral, amber

Source: See note 122 in this chapter for details on the sources.

If Paris was a trading zone for gem knowledge, then trading zones are not just about the "mutual influence of artisans and the learned on each other," in Pamela Long's words.[124] They were also about the mutual influence of artisans on each other, sometimes through the mediation of the learned.

In some respects, there was surprisingly *little* mutual influence of scholars and artisans. In all the texts considered in this chapter, the most thorough discussion of the geological origins of gems was in the preface to Rosnel's *Indian Mercury*. Yet this was a distillation of Boodt's ideas that was unconnected to the descriptions of individual gems in the rest of Rosnel's book.[125]

124. Pamela Long, "Trading Zones in Early Modern Europe," *Isis* 106, no. 4 (2015): 840–47, abstract.

125. Compare Rosnel, *Mercure Indien*, vol. 2, 3–10, with Boodt, *Parfait joaillier*, bk. 1, chaps. 1–18. Rosnel described the source of his theories as "an enlightened philosopher who has studied the marvels of nature with great care" (vol. 1, p. 3).

The point is not that Rosnel was a poor natural philosopher, but that natural philosophy was not the point of his book. The converse is also true. When natural philosophers wrote about gems in this period, they did not necessarily draw on the writings of goldsmiths such as Rosnel or travelers such as Tavernier. François Bernier is a striking example, since he was a philosopher and a traveler who rarely combined these two roles his writing on gems. A reader of the *Summary of Gassendi's Philosophy*, Bernier's synopsis of the philosophical system of his master, Pierre Gassendi, would not guess from its contents that Bernier had just returned from a nine-year residence at the court of the most gem-rich kingdom on earth.[126] The strongest hint he drops is in a chapter on pearls and precious stones, where he writes that precious stones can be found "in the kingdom of Golconda."[127] But he does not mention any particular specimens he saw there, and the reference is incidental to his argument. There are scattered references to gems in the rest of the work, but these are nearly always generic: diamonds are hard, rock crystal is both dense and transparent, there is such a thing as diamond powder, and so on.[128] There is no indication in the book that Bernier spoke to jewelers or gem cutters, or that he observed them at work, even though there is evidence for these activities in Bernier's travel narratives. This is in sharp contrast to England, where Chardin made his new home in the 1680s. There, the value of gems was fast becoming a philosophical problem as well as a technical one.

126. François Bernier, *Abrégé de la philosophie de Gassendi*, ed. Sylvia Murr and Geneviève Stefani (Paris, 1992). The first edition of Bernier's work was published in 1674–1675. A second, much-revised edition appeared in 1684. The 1992 edition is based on the 1684 edition.

127. Bernier, *Abrégé de Gassendi*, vol. 5, 185.

128. Ibid., vol. 2, 93 (diamonds hard), 100 (diamond powder); vol. 3, 24 (rock crystal hard/transparent), 56 (characteristic shapes), 104 (diamonds hard); vol. 5, 72 (amber inclusions). There are also fleeting references to an experiment involving rock crystal placed in embers (vol. 3, 28), and to the shape of hyacinth (vol. 3, 29). The only discussion of a particular gem specimen in the work concerns the crystal-encrusted rock in vol. 3, 179–80.

3

Gem Collecting and Experimental Philosophy

While François Bernier was keeping natural philosophy and the arts apart in France, Robert Boyle was bringing them together in England. The result was a book that broke decisively with the medieval lapidary tradition. First published in London in 1672, Boyle's *Essay About the Origine and Virtues of Gems* (*Gems* hereafter) was unlike any other work on gems printed in Europe in the preceding two centuries.[1] The novelty lay partly in Boyle's theory that gems are formed in a liquid medium. Also new were many of Boyle's observations about gems, which included precise measurements of their specific gravity and minute studies of their crystal form. In addition, Boyle's careful comparisons between the formation of chemical substances and the formation of natural stones have impressed historians of geochemistry. Most strikingly, however, Boyle abandoned the distinction between "general" and "particular" accounts of stones that had prevailed for so long in the lapidary tradition. Gone was the partition between a general introductory discussion of stones and a detailed description of individual species of stones. After a brief open-

---

1. Robert Boyle, *An Essay About the Origine and Virtues of Gems* (1672), in Michael Hunter and Edward Davis, *The Works of Robert Boyle* (London: Pickering & Chatto, 1999), vol. 7, 3–72. In the rest of this chapter, this edition will be referred to simply as *Works*. References to Boyle's works are given as short titles only, e.g., *Gems*. These titles are based on Michael Hunter, ed., *Robert Boyle Reconsidered* (Cambridge, UK: Cambridge University Press, 1994), xiii–xvi. Works not listed there are given a short title in the same format; see bibliography for full citations and associated short titles. The main studies of *Gems* are Arthur F. Hagner's introduction to Robert Boyle, *An Essay About the Origine and Virtues of Gems* (New York: Hafner Publishing Company, 1972); and a pioneering study by Sabel, "Impact of European Trade." See also Sabel, "Rare Earth," 288–92, 293–307; Michael Bycroft, "Robert Boyle's Restless Gems," in Oosterhoff et al., *Ingenuity in the Making*, 36–49.

ing statement of his theory, Boyle embarked on a single sustained argument that combined theory and experience on every page.

How did he do this? The short answer is that he collected gems. There are nearly three hundred references to gem specimens he saw in his published and unpublished writings. These references cover most of Boyle's adult life, from his continental tour of the 1640s, to the period in the 1650s and 1660s when he made his name as an experimenter and as a founding member of the Royal Society of London, to his death in 1691.[2] The specimens Boyle saw were not a collection in the ordinary sense of the term, a set of objects that are all located in the same place at the same time. But they were certainly the product of a culture of collecting, one that included cabinets of curiosity, princely *Wunderkammer*, personal jewelry, and merchandise, as well as specialized natural history collections such as Boyle's own mineral collection.[3] Boyle observed these gems minutely. He split them open, weighed them in air and water, suspended them in a vacuum, turned them into mirrors, dunked them in water, laid in bed with them, chewed them, smelled them, and much else besides. The result was a wealth of data about gems that Boyle channeled into his philosophical theories, not only in *Gems* but also in works on color, electricity, effluvia, porosity, and more. Gems were one of the most important materials in Boyle's experimental philosophy.

There is very little discussion of these objects in the large secondary literature on Boyle. This is partly because the objects are gems, but also because they are objects. They are casualties of the tendency to think of instruments

2. Continental tour: *CPE*, 193; *Occasional Reflections*, 176–78; *Gems*, 23 and editors' note; Robert Maddison, *The Life of the Honourable Robert Boyle, F.R.S.* (London: Taylor & Francis, 1969), 41, 45; Michael Hunter, *Boyle: Between God and Science* (London: Yale University Press, 2009), 292. On the early 1650s, see "Atomical Philosophy," 234, and the observations about diamond and rock crystal in "Gems and Medicinal Stones," xiv, an early sketch of the argument in *Gems*. Key later works referring to gems were *Colours* (1664), *Electricity* (1675–1676), and *Exp. Obs. Physicae* (1691).

3. Useful overviews of early modern collecting are Paula Findlen, "Sites of Anatomy, Botany, and Natural History," in Daston and Park, *Cambridge History of Science*, vol. 3, 281–89; and Krzysztof Pomian, "Les Wunderkammer entre trésor et collection particulière," in *La licorne et le bézoard: Une histoire des cabinets de curiosités*, ed. Krzysztof Pomian (Montreuil: Gourcuff Gradenigo, 2013), 17–27. Boyle's context is covered in Arthur MacGregor, "The Cabinet of Curiosities in Seventeenth-Century Britain," in *The Origins of Museums: The Cabinet of Curiosities in Sixteenth and Seventeenth Century Europe*, ed. Oliver Impey and Arthur MacGregor (Oxford: Clarendon Press, 1985), 147–58; and Michael Hunter, "The Cabinet Institutionalised: The Royal Society's 'Repository' and Its Background," in the same volume, 159–67. The link between collections and merchandise is a theme of Smith and Findlen, *Merchants and Marvels*; Cook, *Matters of Exchange*; and Margócsy, *Commercial Visions*.

rather than collections as the key kind of hardware in the history of experimentation, especially in the history of physics. This helps to explain why Boyle's air pump has received far more scholarly attention than his mineral collection.[4] It also explains the virtual absence of natural history collecting in otherwise comprehensive studies on early modern experiment, whether in relation to Robert Boyle, the Royal Society of London, or early modern physics in general.[5] When scholars do link experiments and collections, the link is usually metaphysical or literary rather than material. It has been argued, for example, that the strange and anomalous objects in cabinets of curiosity helped to create a "pointillistic" picture of the natural world that favored experimentation.[6] Other scholars have shown that Boyle used new literary genres that originated in natural history, such as the experimental history and lists of heads and queries.[7] Yet these scholars have not argued that the objects in natural history collections were part of the routine practice of seventeenth-century experimental philosophy. On the contrary, they have sometimes denied this.[8] Separately, historians of seventeenth-century experimentation have begun to study homely instruments such as springs and kitchen utensils rather than specialized instruments such as air pumps.[9] Yet even these studies are framed in terms of instruments rather than collections. To study Boyle on gems is to study both kinds of hardware at once. And to do that is to be a transmaterialist.

4. On the air pump, see Steven Shapin and Simon Schaffer, *Leviathan and the Air Pump: Hobbes, Boyle, and the Experimental Life*, 2nd ed. (Princeton, NJ: Princeton University Press, 2011 [1985]), chaps. 2 and 6.

5. Marie Boas Hall, *Promoting Experimental Learning: Experiment and the Royal Society 1660–1727* (Cambridge, UK: Cambridge University Press, 1991); Rose-Mary Sargent, *The Diffident Naturalist: Robert Boyle and the Philosophy of Experiment* (Chicago: University of Chicago Press, 1995); Heilbron, *Electricity*; Jed Z. Buchwald and Robert Fox, *Oxford Handbook of the History of Physics* (Oxford: Oxford University Press, 2013), pt. 1, esp. chap. 7. These works cover the collection of data and of instruments, but rarely the collection of natural objects or *naturalia*.

6. Lorraine Daston, "The Factual Sensibility," *Isis* 79, no. 3 (1988): 467. Cf. Daston and Park, *Wonders and the Order of Nature*, chaps. 6 and 7.

7. Michael Hunter, "Robert Boyle and the Early Royal Society: A Reciprocal Exchange in the Making of Baconian Science," *British Journal for the History of Science* 40, no. 1 (2007): 1–23; idem, "Robert Boyle's 'Heads' and 'Inquiries,'" *Robert Boyle Occasional Papers 1* (London: University of London, 2005); Anstey and Vanzo, *Experimental Philosophy*, chap. 3.

8. Daston, "Factual Sensibility," 467; Anstey and Vanzo, *Experimental Philosophy*, chap. 3. A notable exception is Findlen, *Possessing Nature*, chap. 5, though the focus there is on anatomy and the debate about fossils, not on heat, light or electricity.

9. Domenico Bertoloni Meli, *Thinking with Objects: The Transformation of Mechanics in the Seventeenth Century* (Baltimore, MD: Johns Hopkins University Press, 2006); Werrett, *Thrifty Science*.

What about evaluation? Well, Boyle is famous not only as an experimental philosopher but also as an advocate for the utility of natural philosophy. As he put it in 1671, "the Goods of Mankind May be Much Encreased by the Naturalist's Insight into Trades."[10] This makes Boyle an important test of the thesis that useful knowledge involved evaluation as well as production. Some existing studies support this thesis, although the evidence is scattered among many different studies on different materials, as is often the case with material evaluation. The assaying of gold has received the lion's share of attention from historians, partly because it involved precise measurement and partly because Boyle used it as a metaphor for experimental inquiry.[11] But there was more to material evaluation than gold and measurement. The salubrity of air,[12] the freshness of water,[13] the richness of ores,[14] the strength of steel,[15] and the efficacy of mineral waters[16] were all part of Boyle's experimental program. In each case he took an interest in methods for "trying" or "examining" these useful properties, to use his most common terms for the practice. The terms "standard," "counterfeit," and "quality" cut across multiple materials. So did the notion that density was a mark of value and that balances were the best way to determine density. As a rule, these judgments combined natural philosophy with practical applications, or at least the promise of practical applications. Gems were no exception to this rule. Boyle was familiar with several kinds of gem appraisal, not only density measurements but also the visual acuity of merchants, the tools of cutters, and the careful observations of physicians, miners, and wearers of jewels. This expertise fed into Boyle's

10. Cited in Michael Hunter, "Boyle on the Application of Science," in *The Bloomsbury Companion to Robert Boyle*, ed. Jan-Erik Jones (London: Bloomsbury, 2021), 287.

11. Robert Boyle, "Shewing the Occasion of Making This New Essay-Instrument," *PT* 10, no. 115 (1675): 329–48; Schaffer, "Golden Means," 22–27; Sargent, *Diffident Naturalist*, 162, 171, 188; Steven Shapin, *A Social History of Truth: Civility and Science in Seventeenth-Century England* (Chicago: University of Chicago Press, 1994), 343, 347–48; Jasmin Kilburn-Toppin, "Place of Great Trust."

12. Hunter, *Boyle: Between God and Science*, 171.

13. Robert Boyle, "An Account of the Honourable Robert Boyle's Way of Examining Waters as to Freshness and Saltness," *PT* 17, no. 197 (1693): 627–41; idem, "Observations and Experiments About the Saltness of the Sea"; idem, "A Statical Hygroscope, Proposed to Be Farther Tried." The two latter texts were published in *Saltness of the Sea*.

14. *Medicina Hydrostatica*, 255–80.

15. Tawrin Baker, "Color and Contingency in Robert Boyle's Works," in *Early Modern Colour Worlds*, ed. Tawrin Baker, Sven Dupré, Sachiko Kusukawa, and Karin Leonhard (Leiden: Brill, 2016), 259, 270–71.

16. *Mineral Waters*.

natural philosophy of gems, which relied on the correct identification of gems and on a new conception of gem species.

Of course, Boyle was only one experimental philosopher in seventeenth-century England. But he was an important and influential one, widely seen at the time as an exemplary experimenter. His rich legacy of published and unpublished writings means that we can build up a detailed picture of the experimental use of gems that is hard to obtain from any surviving catalogs or inventories. Moreover, Boyle's gem collecting is a window onto wider material worlds. As Claire Sabel has shown, Boyle's work on gems was linked to the rise of London as a hub of the global gem trade.[17] London caught up with the other European centers for this trade, Lisbon and Amsterdam, in the middle decades of the seventeenth century. Much of this trade went through the English East India Company (EIC), for which Boyle served intermittently as a director from 1669.[18] The trade brought gems to London along with books about them, merchants who handled them, cutters who worked them, and travelers who had seen them drawn from the mines in Asia and the Americas. The travelers included Jean Chardin, who arrived in England for good in 1681, became a member of the Royal Society of London, and befriended Boyle and his family.[19] These events left their traces in Boyle's writings on gems. So too did the growth of metal mining in England, Wales, and Ireland. Boyle's father, the first Earl of Cork, had been a major patron of ironworks in Ireland; Boyle himself owned a lease in some English mines and was a member of the Company of Mines Royal from 1664.[20] Finally, Boyle's gem collecting was a reflection of his own movements and domestic arrangements. These included his continental tour as a young man, his spell in Oxford in the 1650s and

17. Sabel, "Impact of European Trade"; idem, "Rare Earth," 288–92, 293–307. Note also the remarks on the wider English context of Boyle's gem work, in ibid., 240–54.

18. Hunter, *Boyle: Between God and Science*, 169, 195 (EIC dates).

19. Cruysse, *Chardin le Persan*, 292–93 (Boyle's family), 275–76 and 457 (arrival in London; Royal Society of London). Chardin in *Workdiaries* 36.19, cf. 36.21, 36.78, and the editor's note on the latter entry. This workdiary is associated with the years 1685–1691. As in this note, references to Boyle's workdiaries take the form *Workdiaries* X.Y, where X is the number of the workdiary and Y the number of the "entry" cited. Where the passage cited does not have an entry number, or where a reference to its entry is ambiguous, the workdiary number is followed by a page number or folio number.

20. William Rees, *Industry Before the Industrial Revolution* (Cardiff: University of Wales Press, 1968), 25–27 and 134–37 (growth of industry), 240–47 (First Earl of Cork), 653 (Boyle's lease). Cf. Maddison, *Robert Boyle*, 112 (Company of Mines Royal), 262–63 (lease). I'm grateful to Joshua Hillman for sharing his expertise on the history of the British mining industry. For details, see "From Coallery to the Natural History of Strata: Mining and the Spatial Sciences of the Earth in Britain, 1600–1800" (PhD diss., University of Leeds, 2024), esp. chap. 2.

1660s, and his long and fruitful partnership with his sister Katherine Jones, Lady Ranelagh, with whom he lived in London from 1668 until their deaths in 1691.[21] Trade, industry, and family were all mixed up with Boyle's gem collecting. This is true whether we consider the objects themselves, their sources, their philosophical significance, or judgments of their value. I take these topics in turn.

A Virtual Collection

The obvious place to start is the mineral collection Boyle bequeathed to the Royal Society upon his death.[22] But the inventory of his collection is disappointing for anyone interested in the role of gems in his natural philosophy. There are only twenty or so gems in the document, with frustratingly short descriptions such as "Amber," "Emeralds," "Chrystalls Cornish," and so on. Only one of these entries tells us something about how Boyle acquired the specimen ("Jett taken up by Mr Longuile"). None tell us how the collection changed over time or what Boyle did with the items mentioned in it. In any case, Boyle explicitly excluded "Jewells" from the "raw and unprepared Minneralls" that he bequeathed to the Royal Society in his will. The jewels were instead to be sold off along with Boyle's lands, cattle, silver plate, and other goods.[23] The will refers to two particular jewels: "a Sardonixe Seale Ring which I usually weare on my Little Finger"; and "a Small Ring usually worne by mee on my left hand having in it Two small diamonds [with] an Emerald in the middle." This diamond ring, Boyle wrote, was "held by mee ever since my youth in great Esteeme and worne for many yeares."[24] These details tell us something about the role of jewels in Boyle's life, but nothing about their role in his philosophy.

Where historical inventories fail us, we must make our own. Boyle's gem collection can only be established by scouring his published and unpublished works for references to gems he owned, saw, or used in his experiments.

---

21. Michelle DiMeo, *Lady Ranelagh: The Incomparable Life of Robert Boyle's Sister* (Chicago: University of Chicago Press, 2021). Boyle's whereabouts in Hunter, *Boyle: Between God and Science*, 291–98.

22. Wilson, *Mineral Collecting*, 64, 163; idem, "Nehemiah Grew's Musaeum Regalis Societatis," *Mineralogical Record* 22 (1991): 337; Hunter, *Boyle: Between God and Science*, 246, summarizing Maddison, *Robert Boyle*, 201–2. The inventory is at Royal Society of London, Boyle Papers, Ms. 200.

23. "Robert Boyle's Last Will and Testament," transcribed as an appendix in Maddison, *Robert Boyle*, 261, 265.

24. Ibid., 258, 259.

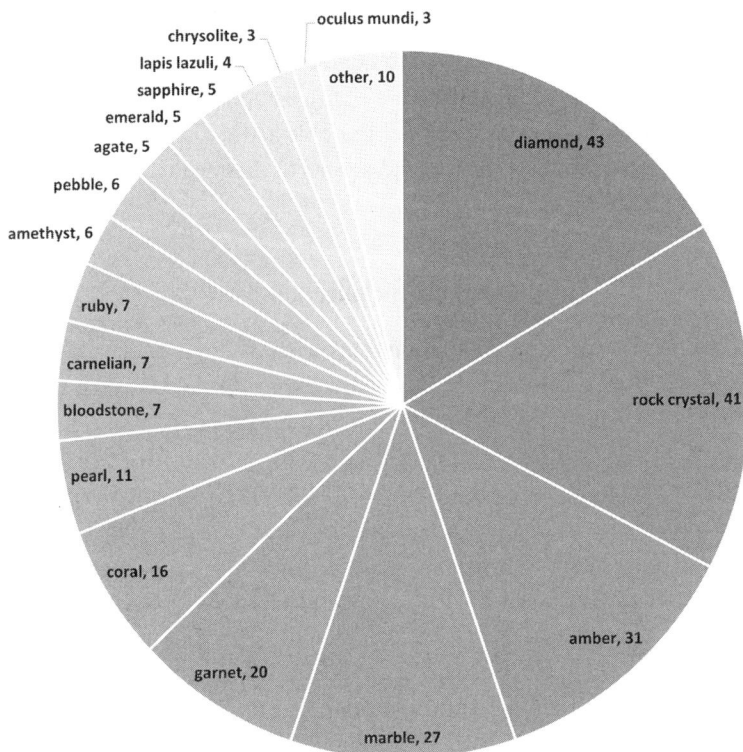

FIGURE 3.1. Gem species in Boyle's works
The chart shows the number of passages in Boyle's writings that refer to gems of different species. "Other" includes one or two references to each of the following: hyacinth, sardonyx, turquoise, cat's eye, porphyry, and topaz. The chart excludes twenty-one references to gems in which Boyle did not name the species.

These sources yield an inventory of Boyle's gems that has ten times as many entries as that of the Royal Society, with much more detail on each entry. This inventory, extracted in appendix 1, shows that Boyle referred to gems he had seen on at least 278 occasions in his writings. Many of these references were to diamond, rock crystal, amber, or marble, with a smaller number referring to ruby, emerald, garnet, agate, and a range of other species (fig. 3.1). This inventory could easily be expanded by adding references to artificial gems and to lesser species such as flints, fluorspar, and marcasites. The inventory is an underestimate in other ways as well. It excludes gems that Boyle saw but never wrote about, and it counts references to multiple stones as a single entry. There is only one turquoise specimen in the inventory, although Boyle wrote that he had "laid out for" stones of this kind.[25] The packet of "between 100 and

25. *Clayton's Diamond*, 196.

150 diamonds," seen at the headquarters of the EIC, counts as a single item; so does the "considerable number" of garnets that were "brought me out of America growing in one lump of Matter."[26] The inventory is also, in some respects, an *over*estimate of Boyle's gem collection. Boyle often referred to the same specimen in several different passages, and each of these passages is counted as a separate item—he referred to the parcel of EIC diamonds on at least five different occasions, for example (fig. 3.2).[27] The inventory also counts gems that Boyle saw but that he did not own, on the grounds that these stones were just as significant to his experimental philosophy as those he did own. In any case, the distinction between those he did own and those he did not was fuzzy, as we shall see.

The inventory not only shows the extent of Boyle's collection but also explains why it has been overlooked by historians. Simply put, it was scarcely a "collection" at all. "Virtual collection," a term that Mary Terrall has applied to some eighteenth-century collections, is a better word for the shifting and dispersed set of gems that Boyle saw in the course of his inquiries.[28] Boyle's gems have little in common apart from the bare fact that Boyle saw them and wrote about them. Certainly, they were never displayed together in the same place at the same time. It is striking that, of the 278 occasions that Boyle referred to gems he had seen, he only once stated that the gem in question ("a pretty large [diamond] that was rough") belonged to his "Collection of Minerals."[29] It is likely that there were other gems in the mineral collection and that Boyle simply did not mention their presence there. He often wrote that he had a stone "by me," or that he "keeps" a stone, without saying where he kept it.[30] However, it is equally clear that many of these specimens were not in a physical collection at the time he wrote the passages in which he referred to them. Indeed, Boyle's habits of storage and display were rather erratic. He had trouble keeping track of even his most precious specimens, such

26. "Diamonds," 387 (diamonds); *Gems*, 34 (garnets). Cf. *Electricity*, 522 (carnelians); *Gems*, 22 (rubies).

27. *Gems*, 35; "Diamonds," 387, 388; *Workdiaries* 21.631; Royal Society of London, Boyle Papers, vol. 25, p. 241, para. 2. The details of these passages strongly suggest that they all refer to the same parcel.

28. Mary Terrall, "Handling Objects in Natural History Collections," in *The Material Cultures of Enlightenment Arts and Sciences*, ed. Adriana Craciun and Simon Schaffer (London: Palgrave Macmillan, 2016), 30.

29. *Gems*, 14.

30. Boyle referred to what is almost certainly the same diamond in another place ("Diamonds," 387), without specifying that the diamond was in his collection. "By me" or "in keeping": *Gems*, 39, 47, 48, 49, 63; "Figures of Fluids," 579; *Absolute Rest*, 203; *Electricity*, 522; *Workdiaries* 32, f. 168v; *Workdiaries* 36.17.

FIGURE 3.2. Modern raw diamonds

Unpolished gem-quality diamonds, from modern mines in southern Africa. Boyle observed a parcel of such diamonds at the EIC headquarters in London, noting the semi-regularity of their shapes and colors. "Most of them [were] broken, and of very irregular Figures . . . but some few I saw that were pretty regularly figur'd." The regular ones were made up of "several Triangular surfaces that were terminated in, or compos'd, diverse solid Angles." A "very experienc'd Artificer" confirmed that regular diamonds were "six corner'd." This seems to refer to the octahedral shape of many diamonds, such as the two in the top image. Photos by John Ward, courtesy of Robert Cowley and Namakwa Diamonds Limited.

as "some number of small Diamonds" that he "lately met with among other little curiosities that lay long neglected by me."[31] There is a note of despair in his reference to a set of natural pearls that "I have somewhere yet by me" and to a geode "which I think I have not yet lost."[32]

Boyle's collection lacked coherence, then. For this very reason, however, it is a window onto the wider culture of collecting in seventeenth-century England. Static, specialized collections, such as Boyle's "Collection of Minerals," were only one part of this culture. The display of "rarities" or "curiosities," objects that excited wonder in the viewer or that confounded existing explanations, were another part of the same culture.[33] Examples from Boyle's works are an emerald "of my own . . . whose colour was so excellent, that by skilful persons twas lookd on as a rarity"; a garnet that "I kept by me for a Rarity, because of its Bigness and deep Colour"; and a pale amethyst containing brownish hairs, that "I have had my self, and shewn to others," and which was "not little wonder'd at even by curious and skilful Men" (plate 2, top).[34] As these examples suggest, Boyle sometimes showed his specimens to other people.[35] Conversely, he saw several specimens while visiting collections of rarities owned by others, including the Grand Duke of Tuscany and an unidentified "Polonian noble-man."[36] He saw at least two items that made their way into the royal collection of rarities. One was a piece of artificially colored marble that Boyle himself presented to Charles II. Another was a diamond belonging to a certain "Mr. Clayton." This was probably the lawyer and virtuoso John Clayton, knighted in 1664, whose father Sir Jasper Clayton had made a fortune as a haberdasher and had probably helped to finance the Stuart monarch's return to the throne in 1660.[37] Boyle may have had the royal collection in mind when

31. "Diamonds," 389 (small diamonds). Cf. *Gems*, 14 (rubies) and 63 (diamonds).

32. *Medicina Hydrostatica*, 227 (pearls), cf. *Gems*, 44; 55 (geode). Cf. the lost diamond in "Diamonds," 390.

33. Daston, "Factual Sensibility"; Daston and Park, *Wonders and the Order of Nature*, chap. 6; Krzysztof Pomian, *Collectors and Curiosities: Paris and Venice 1500–1800* (Cambridge, UK: Polity Press, 1990 [1987]), chaps. 2 and 3.

34. *Electricity*, 522 (emerald); *Medicina Hydrostatica*, 221 (garnet); *Gems*, 26 (amethyst). Other rarities in *Medicina Hydrostatica*, 227 (pearls); *Porosity*, 139 (marble).

35. Other examples in *CPE*, 201 (coral); *Gems*, 47 (diamond).

36. Grand Duke: *Occasional Reflections*, 176–78; *Gems*, 23 and editors' note; Maddison, *Robert Boyle*, 41. Polonian: *CPE*, 194. Cf. the "Hungarian diamond" that Boyle saw and "which the Owner would have presented me": *Absolute Rest*, 203.

37. On the fate of Clayton's diamond, see Boyle, "New Experiments Concerning the Relation Between Light and Air," *PT* 2, nos. 31–32 (1668): 581–600, 605–12, in *Works*, vol. 6, 10, 10n. Clayton is identified in Boyle, *Works*, vol. 4, 187n. On John and Jasper Clayton, see Edmund Berkeley

he referred to a stone, partly colored and partly transparent, that he saw in a ring "in a great and curious Princes Cabinet, among other rarities."[38]

The sheer variety of Boyle's specimens was also characteristic of seventeenth-century rarities. His gems ranged in size from diamonds the size of a pea to a piece of rock crystal the size of his two fists.[39] Some specimens were notable for their regularity (an "exquisitely figurd" crystal), others for their irregularity (the gravel-shaped diamonds in the EIC parcel).[40] Some came in parcels or piles, such as the diamonds and rubies and garnets mentioned earlier; others came in clusters, embedded on the internal walls of hollow rocks; others in fragments, such as a diamond cut into three pieces and a porphyry sawn in two.[41] Three of the diamonds, and two of the sapphires, had been cut or polished. Of the remaining diamonds, most were raw, "just as [they] came from the Rock."[42] Many of Boyle's specimens were notable for what they contained or for the way they were composed. They might contain insects, hairs, air, water, or wire; they might be arranged in thin parallel layers, in irregular layers, in layers made up alternately of rocks and crystals, or with one stone partially enveloped in another.[43] Boyle's collection may have been virtual, but it was part of a very real culture of collecting in seventeenth-century England.

The Jewel House

This culture can be seen, not just in the gems themselves, but also in the way Boyle acquired them. His sources were as varied as his specimens. They suggest a style of science that is more closely associated with Hugh Plat than Robert Boyle. Plat was an Elizabethan lawyer who mined the material expertise of the city of London, speaking to everyone from midwives to carpenters to apothecaries to produce his *The Jewell House of Art and Nature* (1594). For Plat, the jewel house was a metaphor for the treasures of science he found in

---

and Dorothy Smith Berkeley, *John Clayton: Pioneer of American Botany* (Chapel Hill, NC: University of North Carolina Press, 1963), 7–13. The colored marble is mentioned in *Porosity*, 139.

38. *Gems*, 23, and editors' note.

39. *Electricity*, 514 (diamond); *Gems*, 23 (two fists). Other large specimens in *CPE*, 159 (marble slabs); *Gems*, 16 (lump).

40. *CPE*, 192 (crystal); *Workdiaries* 21.631 (gravel).

41. *Gems*, 25, 62 (hollow rocks); "Diamonds," 387 (three-piece diamond); *Usefulness II*, sec. 2, 420 (porphyry).

42. *Absolute Rest*, 203.

43. *CPE*, 194; and *Gems*, 26 (fly, spider); 26 (hair); *Workdiaries* 32, f. 168v–r (water); "Figures of Fluids," 579 (air); *Gems*, 36 (wire); 23, 57 (parallel layers); 15 (alternate layers); 58 (envelope).

the capital.[44] For Boyle, jewels were real objects to be obtained in much the same way that Plat had obtained his own nuggets of knowledge—by interacting with people from many walks of life, both men and women, in the city of London and beyond.

There is a hint of such exchanges in Boyle's references to gems he had gained or lost across his lifetime. In 1669 he referred to a raw diamond "whose surface was made up of several almost triangular Planes," a specimen he had "once had"; in 1672 he referred to a "curious Agate" that "I have had"; in 1691 to a rough diamond "I had of my own."[45] Near the end of his life he wrote that he had "ben Master of several Diamonds of differing sizes, cut and uncut," implying that he was no longer master of many of these.[46] Other passages shed light on the specific forms of exchange that set these gems in motion. There were a few temporary loans, such as the diamond owned by John Clayton that Boyle subjected to numerous trials on the night of October 27, 1663.[47] On another occasion Boyle borrowed a piece of turquoise from his friend John Aubrey, observed it over several weeks, and broke it in the process.[48] A few specimens were simply bought, such as the bloodstone he purchased from a druggist. On at least one occasion, Boyle bought specimens specifically "for experiments," a phrase he applied to the "neglected" set of small diamonds mentioned earlier.[49]

More often he was "presented" with gems rather than purchasing them, though the distinction between the two was not clear-cut. A piece of topaz came from a diamond cutter, "whose Customer I had been," who had driven the stone too hard on the mill and broken it, and who "laid it aside for me, and would needs make me accept it, as a curious, though not a usefull, thing."[50] Sometimes Boyle's gems were sent to him from a distance: the set of European and American garnets that "were sent me as a Present from New England" (plate 2, bottom); another present of garnets, this time from Africa; the unspecified gems sent, "among other Rarities," by "the Governour of an American Colony"; the "flat and smooth Cornelian Stones, such as they bring

44. Harkness, *Jewel House*, chap. 6. Plat is contrasted with Boyle on 221, 251.

45. "Atmospheres," 177 (raw diamond); *Gems*, 57 (agate); "Diamonds," 387 (rough diamond); cf. "Atmospheres," 178; and *Gems*, 39.

46. "Diamonds," 389.

47. Note also the piece of amber containing a bubble that Boyle "borrow'd [from] the Burgemaster of Christian Shaven." *Workdiaries* 32, f. 169r.

48. *Absolute Rest*, 201, including editors' note. Cf. *Clayton's Diamond*, 201; *Absolute Rest*, 202.

49. *Gems*, 54 (bloodstone); "Diamonds," 389 (diamonds). Cf. "Paralipomena," 208 (bloodstone); *Gems*, 22 (rubies); *Exp. Obs. Physicae*, 425 (seed pearls).

50. *Languid Motion*, 298.

from the East-Indies, to cut Rings out of."[51] Other gems came from closer to
home: a "great number of Amethysts" that had been "taken up here in England by a Gentleman of my Acquaintance"; a rock encrusted with crystals,
found in a lake in the North of Ireland and given to Boyle by the owner of
the lake; a "diamond" found in Ireland by someone who "knew not what it
was."[52] The donors were socially varied, from an "ancient digger" who presented Boyle with soft crystals from a quarry to the "great Prince" who was
the source of the three-layered carnelian that Boyle wore in a ring.[53] A large
piece of amber, enclosing a fly the size of a grasshopper, was presented to
Boyle "by a Person no less extraordinary than it."[54]

As well as acquiring gems from travelers, Boyle encountered gems on his
own travels. Finding himself in "a very elevated piece of ground," the philosopher stumbled across a set of "exquisitely figured" crystals that he promptly
excavated.[55] Voyages allowed Boyle to identify stones that interested him and
to note the terrain in which they occurred, as when he "observed" a cluster
of rock crystals "to grow in a red Earth."[56] These trips may have been related
to his role as a member of the Company of Mines Royal: he once referred to
a set of transparent "concretions" that he came upon while "going to visit a
famous Quarry."[57] In any case, he had more than a passing familiarity with
the mining industry, as shown by his remark that clear spar is found next to
metallic veins "in most of our Western Lead-Mines," something that he knew
"partly by my own observation."[58] These mineralogical voyages probably date
from the 1660s, the decade in which Boyle became a member of the Company
and in which he compiled his unpublished writings on mineralogy and petrifaction.[59] The visit to the "elevated piece of ground" was described in a work
published in 1661; in 1670 he referred in the past tense to the time he "used to
visit mines."[60]

51. *Medicina Hydrostatica*, 227 (garnets), cf. *Gems*, 34; 35 (African garnets); 55 (unspecified);
Specific Medicines, 401 (carnelians).
52. *Gems*, 15 (amethysts); 62 (rock with crystals); 14 (Irish diamond).
53. *Gems*, 16 (digger), 23 (prince).
54. *Gems*, 26.
55. *CPE*, 192.
56. *Medicina Hydrostatica*, 272.
57. *Gems*, 16. Boyle also sometimes gave unnamed people instructions to dig out specimens
from mines: ibid., 14, 61.
58. *Gems*, 21, 40, cf. 44.
59. *Works*, vol. 7, xi–xii, vol. 13, lvii–lviii; Hillman, "Coallery to the Natural History of Strata,"
77–78.
60. *Cosmical Qualities*, 323; Hillman, "Coallery to the Natural History of Strata," chap. 2.

Gem cutters and gem merchants were another important source of Boyle's gems. His relations with these people were rich and reciprocal. He examined stones with them, bought stones from them, and received stones from them as gifts. He also observed their instruments and techniques, even introducing a new technique into a "Shop" where "I sometimes cause work to be done for me."[61] He had pieces of amber, marble and porphyry cut or ground to the correct dimensions for his experiments.[62] Boyle was especially close to a stone-cutter from Oxford who worked with him to perfect a method for staining white marble.[63] He was also close to a certain "Mr. L," the only diamond cutter he named in his published writings.[64] Mr. L was a "very skilful Cutter and Polisher of Diamonds" who had worked for twenty years in this trade in Amsterdam before moving to London.[65] This may have been the cutter "whose Customer I have been" who presented him with a cracked topaz; or perhaps the "famous and experienc'd Cutter of Diamonds" who once showed him a diamond the size of two peas; or even the "skilful Person of my acquaintance" who worked on a sapphire seal that Boyle saw.[66]

Gem merchants are equally significant and equally anonymous in Boyle's writings. The parcel of one hundred or so diamonds he saw at the headquarters of the EIC has already been mentioned. Boyle first referred to this parcel in 1672, later writing that "an Acquaintance of mine received [the parcel] lately from East India."[67] This acquaintance cannot have been Jean Chardin, who first visited England in 1680, but Chardin was almost certainly the "Gentleman eminent for his Travels into Eastern Parts, & for his skill in Jewels" who he met to discuss sapphires in the late 1680s.[68] Boyle named Chardin as his interlocutor in another passage in the same document, referring to him as "a very candid & judicious Traveller."[69] Chardin may have been the "ingenious Merchant . . . who brought many fine [diamonds] out of the East

61. *Usefulness II*, sec. 2, 478.

62. *Flame and Air*, 200; *CPE*, 152, 159; *Electricity*, 520; *Colours*, 72; *Usefulness II*, sec. 2, 420.

63. *Porosity*, 139. Cf. *Absolute Rest*, 204 ("I repair'd to an ancient Artificer"); "Diamonds," 385 (Boyle inspects the mill of a diamond cutter); "Atmospheres," 171 (firsthand observation of cutter).

64. In addition, the "Mr. R" named in an unpublished manuscript was evidently a cutter of diamonds and sapphires: "Paralipomena," 209.

65. "Mr. L" is named in "Diamonds," 386, a passage repeated in *Absolute Rest*, 204.

66. "Diamonds," 387; *Languid Motion*, 298; *Gems*, 24. Other references to diamond cutters are in "Diamonds," 388, 389; *Languid Motion*, 258, 298; *Gems*, 18, 26; *CPE*, 165; *Electricity*, 522.

67. *Workdiaries* 21.631. Cf. note 27.

68. *Workdiaries* 36.78, and editor's note. This workdiary is dated 1685–1691. The same conversation is reported in "Observations Solitary," 412.

69. *Workdiaries* 36.19.

Indies" and who showed Boyle a cracked diamond that he described in a 1685 work.[70] Boyle later referred to a "cat's eye" stone that was "taken up in the Diamond Mynes by the Person that show'd them to me," another possible reference to Chardin.[71] In addition, Boyle made tens of references to jewelers who had given him information about gems, without necessarily showing him any specimens. Some of these references may be to Chardin, such as the "Experienced Gentleman" who had read *Gems* and "had more frequented the Diamond Mynes especially those of Colconda than any I have met with."[72] Other gem merchants were certainly not Chardin, such as the "expert English jeweller" who informed him about rubies and sapphires in 1672 or earlier. There was also the "Merchant . . . whose name I know not" who was the author of a manuscript in Boyle's possession.[73] The author in question may have been Jean-Baptiste Tavernier, though there is no evidence that Tavernier and Boyle were ever in contact.[74]

In addition to rarities, gifts, voyages, merchants, and artisans, Boyle's gems came from the everyday practice of wearing jewelry. Boyle used his own jewels in his experiments, such as the "small ruby" that sometimes became electrical "when I but wore the Ring it belong'd to on my little finger."[75] Some of these jewels were designed especially to illustrate a philosophical point, such as the ring in which Boyle caused the stone to be "set rough as Nature produc'd it, because in that state the Grain is manifest to the naked Eye."[76] Boyle also cast an inquiring eye on the personal jewelry of his entourage, both men and women. He discussed a golden yellow diamond on the finger of one of his brothers; a red diamond owned by the "Dutchess of R.," probably the Duchess of Richmond, Frances Teresa; and a ruby owned by a "fair Lady" who knew Boyle well enough that she let him rub the stone in a dark room.

70. *Languid Motion*, 299.
71. "Paralipomena," 212.
72. Ibid., 212.
73. Boyle Papers 10, f. 32; *Gems*, 22. The "Merchant of Dantzik" in *CPE*, 195, is not explicitly referred to as a gem merchant.
74. Note also the information from "an expert Jeweller, that was also a Traveller" (*Gems*, 14) and a "noted Jeweller" (*Absolute Rest*, 202). Both references predate Chardin's arrival in England in 1681, but Boyle may have been in contact with Chardin (or Tavernier) by correspondence. There are other examples of conversation or correspondence with "Jewellers" (*Clayton's Diamond*, 190), turquoise sellers (*Clayton's Diamond*, 196), "the Jewellers" (*Gems*, 14), and "Jewellers" (*Gems*, 23). Some of these "jewelers" may have been gem cutters, as per Boyle's usage of the word in *Languid Motion*, 298, *Gems*, 22, and elsewhere.
75. "Atmospheres," 178 (ruby). Other rings owned by Boyle in "Atmospheres," 172; *Electricity*, 522; *Gems*, 23; *Absolute Rest*, 203; *Clayton's Diamond*, 201.
76. "Diamonds," 388.

The "observing Lady" who took a ruby ring off her finger for Boyle's inspection may have been his sister Mary Rich. A "very ingenious Lady" who told him about diamonds may have been his other sister, Katherine Jones, Lady Ranelagh.[77]

It is notoriously difficult to reconstruct Ranelagh's life, and no will or probate inventory has been found. But we can assume that she owned at least some jewels during the twenty-three years she lived with Boyle, which included the four years before the publication of *Gems* in 1672. Ranelagh moved in circles where gems were used as a form of currency, as when her father gave her 20 pounds and a diamond jewel to pay for her expenses when she left home at the age of nine.[78] It was to Ranelagh that Boyle bequeathed the diamond and emerald ring mentioned in his will, urging her to "weare it in remembrance of a Brother that truely honour'd and most dearely Lov'd her."[79] It seems likely that Boyle made observations of Ranelagh's jewels that have gone unrecorded. Women had a role, not just in showing gems to Boyle, but also making them available to be shown. There was the "very ingenious and qualify'd Lady" who traveled with her husband to a foreign monarch and brought back a piece of rock crystal containing a drop of water.[80] There was the "ancient Gentlewoman" who gave a bloodstone to a "gentleman" of Boyle's acquaintance to cure his bleeding nose.[81] Note also the piece of rock crystal, "coagulated about a kind of branching Wire," which he "observed among some Minerals left by a Gold-Smith to his Widow."[82] For all we know, women may have been owners of two other jewels Boyle saw in the possession of other people, a seal made of an oddly colored sapphire and a ring set with transparent sardonyx.[83]

All this may give the impression that Boyle had a ready supply of gems of many different kinds. Certainly, he was in an excellent position to examine these rare and expensive objects—his own wealth, the wealth and connections of his family members, his fame as an experimenter, and his proximity to cutters and jewelers in Oxford and London made sure of this. Even so, securing specimens was not an easy task. About *oculus mundi*, a variety of opal, Boyle wrote that this stone was "so rare and difficult to be got, that I had

77. "Diamonds," 388, and editors' note (brother); "Diamonds," 388, and editors' note (Duchess); *Clayton's Diamond*, 190 ("fair Lady"), probably the same ruby as on 201; *Absolute Rest*, 205, and editors' note ("observing Lady"); *Absolute Rest*, 203 ("very ingenious Lady").

78. DiMeo, *Lady Ranelagh*, 22.

79. Maddison, *Robert Boyle*, 258, and note.

80. *Gems*, 25.

81. "Atmospheres," 178. Cf. *Gems*, 70.

82. *Gems*, 36.

83. Ibid., 39, 24.

not opportunity to make upon it all the Tryals I desired."[84] And abundance
was no guarantee of success. Boyle saw many rubies, but he never saw one
glow.[85] He saw many diamonds, and he was long interested in the question
of their magnetism, but it was only in the final years of his life that he found
specimens suitable for answering this question experimentally.[86] Time was
not always on Boyle's side. John Clayton departed with his diamond a few
hours after Boyle examined it; the parcel of EIC diamonds were sold soon
after he studied them. Boyle often comes across as an anxious opportunist,
one who seized any occasion to examine an interesting specimen precisely
because those occasions were so rare.[87]

Clayton's diamond illustrates a final kind of heterogeneity in this collection, namely the extent to which Boyle experimented upon his specimens.
His operations on this diamond ran the gamut from passive observation to
intrusive experimentation. He held the stone next to his warm body, rubbed
it gently with his bare hands, placed it near a flame, looked at it under a mi-
croscope, dunked it in liquids of various kinds, and scratched it with a "white
Tile" and a "Steel Bodkin."[88] There was the same variety in Boyle's treatment
of marble. At one extreme are the marbles used in Boyle's famous experiments on air pressure, or what he called the "spring of the air." He arranged
these marbles in pairs, one on top of the other, so that the lower marble adhered to the upper; he then placed these in a glass vessel before removing the
air from the vessel with a pump. These were truly "elaborate" experiments, to
use one of Boyle's terms.[89] Boyle used at least four different pairs of marbles in
these experiments. He had them "ground very flat and polishd very carefully"
to ensure that their flat surfaces adhered to each other.[90] At the other extreme
is Boyle's observation that a piece of cold, polished marble forms drops of water on its surface in the summer months. This observation required nothing
more elaborate than placing a piece of marble on a windowsill and waiting

84. *Porosity*, 137.

85. Boyle refers to rubies he saw in *Clayton's Diamond*, 190 (failed trial at night); "Atmospheres," 178; *Absolute Rest*, 205 (stay cut short); *Gems*, 14, 22, 40; *Workdiaries* 37.93; and *Clayton's
Diamond*, 201.

86. "Diamonds," 389.

87. See esp. *Clayton's Diamond*, 188 ("I was so taken . . .").

88. Ibid., 197–203.

89. "Elaborate" in Shapin and Schaffer, *Leviathan and the Air Pump*, 38, 178.

90. *Spring, 1st Continuation*, 161; *CPE*, 158; *CPE*, 159; *Spring*, 238. Boyle mentioned pairs of
marble elsewhere, but these may have been identical to one or other of the four just cited: *Flame
and Air*, 200 ("ground very flat"); *CPE*, 152, 162; *Workdiaries* 37.52.

for droplets to form.[91] Other trials lay somewhere between the two extremes, such as the piece of black marble that Boyle ground into a mirror for the sake of experiments on color.[92] Many specimens were drastically altered by Boyle's experiments, whether by being burned, chewed, dissolved in acid, ground into powder, or colored with artificial pigments.[93] Some specimens became experimental tools, routinely used for particular purposes, such as carefully chosen pieces of amber and rock crystal that Boyle kept on hand during chemical experiments to test the strength of solutions by measuring their density.[94] Gems were fully integrated into Boyle's experimental program, whether as components of an elaborate instrument or as objects in their own right. As a piece of experimental hardware, Boyle's virtual collection of gems was as important as his air pump.

Strange Proofs

Important for what? What use could such bewildering array of objects be to an experimental philosopher? An influential answer is that seventeenth-century collections were made up of "strange facts," rare and unusual objects that could not be accommodated to any existing theories and that therefore encouraged the notion of theory-independent facts.[95] Boyle's gems were certainly rare and unusual, but they were by no means independent of his theories. For all their variety, the items in the collection did have one thing in common. They were all—or almost all—used as evidence for the corpuscular philosophy. They were used to show that many natural phenomena could be understood in terms of the sizes, shapes, and motions of small particles

---

91. *Saltness of the Sea*, 447.

92. *Colours*, 72.

93. Coral and pearl in acid: *Exp. Obs. Physicae*, 393; "Petrifaction," 391; *Producibleness*, 36; *Mechanical Origin of Qualities*, 463; *Usefulness I*, 468; "Weakened Spring," 550; *Workdiaries* 38, p. 247, p. 248; *Workdiaries* 38.86. Amber in acid: *Spring, 2nd Continuation*, 203; "Petrifaction," 396; *CPE*, 176. Amber burned: *Sceptical Chymist*, 234. Amber chewed: *Effluviums*, 280. Amber decomposed: *Sceptical Chymist*, 341; "Usefulness, Unpublished," 336; *Forms and Qualities*, 373. Various stones ground and dissolved: *Gems*, 64 (bloodstone, marble, pebbles); *Gems*, 21, 39, and *Porosity*, 138 (garnet); *Gems*, 40 (rubies); *Colours*, 69 (garnet and emerald). Colors in *Gems*, 20, 24; *Porosity*, 140. Other intrusive operations in *Languid Motion*, 258 (crystal); *Human Blood*, 49 (amber); "27th Section," 86 (rock crystal).

94. *Usefulness II*, sec. 2, 429; *Medicina Hydrostatica*, 231, 232. Note also the standard piece of electrical amber in *Exp. Obs. Physicae*, 102.

95. Daston, "Factual Sensibility"; Daston and Park, *Wonders and the Order of Nature*, chap. 6.

of matter.[96] They were part of an argument about the natural world, an argument that was persuasive precisely because the objects were so rare and heterogeneous. Gems were strange proofs, not just strange facts.

An imploding diamond illustrates the point.[97] Boyle saw the diamond in the hands of a merchant, possibly Chardin. The merchant had bought the stone at a mine in India only to find that it had developed multiple cracks a few days after the purchase. This example has all the marks of a strange fact. It originated in the East, behaved in a dramatic and unexpected way, and was described by Boyle in terms of strangeness and wonder. He referred to it as a "strange Instance" and a "Rarity"; he wrote that he "could not without wonder see so fair and hard a Stone so oddly spoil'd with Clefts." His description of the stone was very particular. He mentioned that the stone was "fair and hard," that the merchant had bought it at the mine, that it was valued at about one hundred pounds before the cracking and as worthless after the cracking, and that the clefts were so deep that the pieces could easily have been pulled apart by hand. All this may give the impression of a unique event that was hard to explain or classify.

Yet nothing could be further from the truth. The whole point of invoking the diamond, as far as Boyle was concerned, was to lend support to his thesis that natural phenomena can be understood in terms of moving matter. In his eyes, the diamond showed that there is a great deal of moving matter even in the hardest and most inert bodies. Most of the circumstantial details in the report fed into this argument. The fact that the merchant had purchased the diamond himself at the mine vouched for its authenticity, and hence its hardness. The high price of the diamond before the incident was further proof of its identity. The story of the merchant leaving the diamond to one side for "Ten Days or a Fortnight" suggests that the diamond broke of its own accord, and hence that the breakage was due to motions inside the stone rather than to an external agent. The depth of the fissures testified to the vigor of these

96. On Boyle's place in the history of the mechanical or corpuscular philosophy, see Daniel Garber, "Remarks on the Pre-History of the Mechanical Philosophy," in *The Mechanization of Natural Philosophy*, ed. Sophie Roux and Daniel Garber (Dordrecht: Springer, 2013), 5–8. Recent research has shown that Boyle's mechanism was less strict than once thought: Antonio Clericuzio, *Elements, Principles and Corpuscles: A Study of Atomism and Chemistry in the Seventeenth Century* (Dordrecht: Springer, 2000), chap. 4; Peter Anstey, *The Philosophy of Robert Boyle* (New York: Taylor & Francis, 2000), pt. II, pp. 207–8; Marina Paola Banchetti-Robino, *The Chemical Philosophy of Robert Boyle: Mechanism, Chymical Atoms, and Emergence* (Oxford: Oxford University Press, 2020). But even Clericuzio writes that Boyle "considered that explanation based on the shape, size, and motion of corpuscles were the primary, simplest and most comprehensive a naturalist could adopt" (*Elements, Principles and Corpuscles*, 107).

97. *Languid Motion*, 299.

internal movements. Even the strangeness of the event mattered for the argument, since the strangeness lay precisely in the spontaneous implosion of an apparently inanimate object. The diamond was explicable because of its strangeness, not in spite of it.

This is by no means an isolated example. The use of gems as evidence was the rule rather than exception in Boyle's writings, as a close study of the 278 references to gems in his works shows. In over two-thirds of these passages, Boyle used a gem he had seen to defend the corpuscular philosophy. The bulk of these passages appeared in *Gems*, where Boyle gave a corpuscular account of the origin and virtues of gems. On this account, gems were formed from the motions of particles of matter in a fluid medium that has since dried and hardened. Some of the original particles were derived from healing minerals, which explained the medical virtues of gems. In other works, Boyle used gems to argue for the corpuscular philosophy in several ways. Often he used the concepts of the corpuscular philosophy to explain a particular phenomenon, such as hardness, luminescence, or electricity. The air-pump experiments are in this category, since these experiments lent themselves to a corpuscular explanation even if Boyle was notoriously noncommittal about the details of this explanation. Boyle also used gems to support the central concepts of the corpuscular philosophy, such as the idea that moving matter is more widespread in nature than most people suspect. Related concepts were "effluvia," "atmospheres," and "pores": the streams of invisible matter that flow out of bodies, the halos of invisible matter that form around bodies, and the tunnels that run through bodies and give passage to fine particles of matter.

In all these passages, Boyle made it clear what he was arguing for and how his specimens fitted into that argument, either in the passage itself or in the surrounding text. There are a few other passages that contain what might be called "intertextual arguments." These are passages which do not seem to use the gem as evidence for anything, but where a comparable passage in a different text does do so. An example is "Various Observations about Diamonds," a text published in 1691 that appears to be an atheoretical list of observations, one that Boyle described as "plainly Historical." On closer inspection, most of the diamond specimens referred to in this text had already been described in *Gems*, where they fitted neatly into Boyle's argument. In addition, Boyle sometimes used gems as part of his case against traditional chemistry, as when he showed that coral can be resolved into more than five elementary substances. The remaining third of the 278 passages contained no argument for the corpuscular philosophy, intertextual or otherwise. Yet the great majority of these instances were measurements of the density of gems. They were part of Boyle's rather obsessive and unstructured efforts to measure the density of a great range of substances. If we set these measure-ments to one side, we find that Boyle used his gems as evidence for the cor-puscular philosophy in nearly 90 percent of the passages in which he referred to gems he saw.

These passages included four out of the five gems that Boyle referred to as "rarities." A carnelian, "so large and faire, that twas kept for a rarity," attracted light bodies when it was rubbed and thereby showed that electricity could be produced by mechanical means.[98] An emerald with an unusually strong color was used for the same purpose.[99] Other stones were not called rarities but had similar properties and were equally significant. The imploding diamond has already been mentioned; so has the pale amethyst that contained brownish hairs and that was "not little wonder'd at even by curious and skilful Men" (plate 2, top).[100] These examples were part of wider patterns of argument in which rare or unusual phenomena played an important role. The imploding diamond was just one example of a gem that contained unsuspected internal motions. An agate that burst of its own accord, a topaz that gaped open on the grindstone, a ruby whose electrical faculty varied from day to day, the turquoise with the migrating spots—all showed "the great effects of even lan-guid and unheeded motion," to quote the title of one of Boyle's treatises.[101] Another large class of strange proofs were gems with foreign substances vis-ibly trapped inside them. These were both a source of wonder and an argu-ment in favor of the theory of fluid origin, since it was hard to explain the presence of the foreign substances unless these gems had once been much softer than they were now.[102] Electrical and luminescent gems also combined wonder and argument. These gems were wondrous because they displayed striking phenomena (attraction, glows) with a minimum of human inter-vention. They were valuable evidence for the same reason, since they sug-gested that the striking phenomena were due to simple mechanical actions. Mr. Clayton's diamond glowed with a "very vivid Splendour" when Boyle did nothing more than press its surface with his finger; another diamond became

98. *Electricity*, 521.

99. *Electricity*, 522. The third example is the crystalline stones in *Languid Motion*, 299. The fourth is Mr. Clayton's diamond: *Works*, vol. 4, 10n. The gem called a "rarity" but not used as evidence is the garnet in *Medicina Hydrostatica*, 221.

100. *Gems*, 26.

101. *Languid Motion*, 286–87 (agate), 298 (topaz); *Absolute Rest*, 205 (ruby). See also the splintered diamond in *Gems*, 18; *Works*, vol. 10, 251–349 (unheeded motion). For details, see Bycroft, "Boyle's Restless Gems."

102. See notes 43 and 82.

electrical when he drew it from his pocket. Light and electricity seemed to be the result of mechanical processes such as friction and compression.[103]

Boyle's gems were not only rare, wondrous and striking but also varied. This variety was more grist to his philosophical mill. For example, one of his arguments for the fluid theory of gem origin was the semiregular shape of gems—the fact that some specimens had regular crystals in one part of their substance and irregular forms in another part.[104] Boyle compared these specimens to salt crystals grown in solution in the laboratory. Salt crystals had regular shapes, except in zones that had been in contact with the vessel in which they grew. The analogy suggested that the regular parts of gems were formed in a fluid medium whereas their irregular parts took on the shape of the cavities in which they grew. Boyle described multiple specimens to support these ideas. First there was a set of garnets that showed regular shapes, such as triangles and parallelograms, on their surfaces (plate 2, bottom). Boyle noted that he had taken the garnets from a lump of rock himself, a detail that mattered for his argument because it showed that the regularities were natural rather than human-made. Boyle's other examples differed from the first in telling ways. The regular shapes of diamonds and rubies showed that even the hardest stones were formed in a liquid medium. A cluster of crystals from Bristol had the same form as crystals of niter, thereby strengthening the analogy between gems and salt crystals. Another Bristol stone had crystals that were more regular and transparent at one end than at the other, suggesting to Boyle that one end had formed in contact with a stony cavity and the other in a liquid filling the cavity, just like crystals of niter in a vessel (plate 2, middle). Each of these specimens was one of a kind, but this did not harm the argument. On the contrary: it meant that they each made a one-of-a-kind contribution to the argument.

All this gives new significance to the role of cutters, merchants, family members, and other informants in Boyle's natural philosophy. Once we see that gems were theory-oriented, we can also see that Boyle's informants were theoreticians. They were not only technicians, laborers, and recorders.[105] This is reinforced by two other features of Boyle's work on gems, one material and

---

103. *Clayton's Diamond*, 201 (vivid splendor); "Atmospheres," 177 (pocket).

104. *Gems*, 14–16.

105. The following builds on earlier discussions of the role of "technicians," "interviews," and "workers" in English experimental philosophy. See Shapin, *Social History of Truth*, chap. 8; Michael Hunter, *Boyle Studies: Aspects of the Life and Thought of Robert Boyle (1627–91)* (London: Taylor & Francis, 2016), chap. 9; and Stephen Pumfrey, "Who Did the Work? Experimental Philosophers and Public Demonstrators in Augustan England," *British Journal for the History of Science* 28, no. 2 (1995): 131–56.

the other literary. The material point is that many of the phenomena that interested Boyle were too subtle, fleeting or prolonged for him to experience them directly. Rather than observing them himself, he learned about them by talking to people who had observed them as part of their daily routine. The literary point is that Boyle used words like "observation" and "circumstance" in a more theory-oriented than way we do today. In *Gems*, Boyle used both words as synonyms for "argument" and "consideration." All four words referred to general propositions that he offered in support of his hypothesis that gems are formed in a fluid medium. In this work, circumstances were not a way of separating theory from experience but a way of joining them up.[106]

Both points, material and literary, are exemplified by Boyle's passage on the "very ingenious Lady," possibly his sister Katherine Jones, who told him about the variability of diamonds:

> [She] affirm'd to me that she had divers times *observed* the like alterations in some Diamonds of hers, which sometimes would look more sparklingly than they were wont, and sometimes far more dull than ordinary. And when I objected, that possibly that dulness might be imputed to the weather, or some casual foulness of the surface of the stone, she reply'd that she had been aware of those *circumstances*, rubbing the stones clean, and otherwise taking care to secure an *Observation*, which she had made too often to have deceiv'd her self in it. (italics added)[107]

Although Boyle had made a similar observation himself, he still relied on the repeated observations of the woman in question. In addition, she had already taken her own precautions to rule out confounding factors such as dust or weather. Boyle described her work in terms of "observation" and "circumstances," words which meant much more in this case than mere data or information. The "observation" was the result of prolonged and thoughtful activity; the "circumstances" were a crucial part of the argument in favor of the observation. Moreover, the observation was direct evidence for the philosophical thesis Boyle was defending in the treatise in question, namely that absolute rest is much rarer than we imagine, and perhaps impossible. If even diamonds change over time, he reasoned, then softer bodies must be in constant motion as well. It is also worth noting that the quoted passage is a genuine dialogue, with Boyle raising an "objection" and receiving a "reply."

---

106. *Gems*, passim. This is a different view of Boyle's circumstantial details than the one in Steven Shapin, "Pump and Circumstance: Robert Boyle's Literary Technology," *Social Studies of Science* 14, no. 4 (1984): 481–520; and Shapin and Schaffer, *Leviathan and the Air Pump*, 60–65.

107. *Absolute Rest*, 203, cf. 205.

Much of the argument in *Gems* was organized in the same way, with a series of objections followed by replies, counter-replies, and so on.[108] Similar points hold for Boyle's conversations with cutters and travelers, some of whom also offered their own theories about the phenomena they described.[109] In sum, Boyle's gems were not just raw data. They were occasions for reflections about the nature of the universe, to borrow the language of Boyle's *Occasional Reflections*, a 1665 work in which he drew out the spiritual significance of everything from blossoms to insomnia. Experimental philosophy might be seen as a kind of collective meditation, with Boyle combining his own reflections on objects with those of his interlocutors.

Trials of Goodness

Evaluation lay at the heart of this meditation. The evaluation of gems was entangled with many of the people, objects, and processes we have considered so far. Evaluators included the "ingenious Lady," who tracked the dullness and sparkle of her diamonds over time. Collectors were sensitive to the problem of identifying specimens, as Boyle well knew. One of his specimens was "Rock crystall . . . taken for white Amethyst"; another was "a parcel of small and red transparent Stones, which some ghessed to be Granats; others, more probably, Rubies."[110] Boyle advised "discreet men" to seek help from "skilfull Jewelers" to decide such questions.[111] Gem appraisal raised questions of deceit and authenticity that were good topics for pious reflection. Only a skilled cutter could correctly identify a yellow diamond, as Boyle knew from his own experience; the moral he drew was that the teachings of the Church are authentic even if they appear confusing and contradictory to the ordinary mortal.[112] Gem appraisal was medical as well as moral, with Boyle giving detailed accounts of people whose ailments—a bleeding nose and palpitations of the heart—were cured by wearing bloodstone and carnelian around their necks. These accounts were couched in the language of "cases" and "tryals" that were common in medicine at the time. They even included what we would now call an experimental control, with Boyle noting that the ailments returned

---

108. *Gems*, 28–36, 38–39, 46–52, 60–63, 69.

109. Bycroft, "Boyle's Restless Gems," 42–45. Cf. *Clayton's Diamond*, 191 ("heedful Observation"); *Gems*, 25 (dialogue); "Paralipomena," 212 (confirming diamonds made of layered plates); *Saltness of the Sea*, 415 (dialogue).

110. *Workdiaries* 37.102 (white amethyst); *Gems*, 40 (garnets vs. rubies). Cf. "those pible [pebble] stones which well cut so happyly imitate Diamonds," in "Atomical Philosophy," 234.

111. Boyle Papers, vol. 38, f. 154, quoted in Shapin, *Social History of Truth*, 219.

112. *Things Above Reason*, 416. Cf. the analogy to false stones in *Occasional Reflections*, 98.

when the treatment was suspended.[113] These trials are another link to Lady Ranelagh, an avid collector, user, tester, and distributor of medical remedies, including a cure for nosebleed that she "almost certainly" shared with Boyle on an occasion when he suffered from that ailment.[114] Gem appraisal also linked Boyle to merchants and travelers, as shown by one of the items on a list of "queries about gems" that was apparently intended for people voyaging in distant countries. "By what marks & ways of trial," Boyle wanted to know, "may we best make choice, among the propos'd gems, and estimat [sic] their goodness?" As he made clear in this query, he was especially interested in qualities that were "priz'd by the Jeweller, or may be useful to the Physitian."[115]

Jewelry and medicine lay behind the hydrostatic balance, Boyle's main proposal for improving the evaluation of gems. The idea was to weigh a body in air and water and use the difference between the two weights to calculate the density of the body compared to that of water. The method was ancient. It had been used to evaluate gems by authors as varied as Abū Rayḥān al-Bīrūnī (around AD 1000) and Galileo Galilei (in 1586).[116] But it was Boyle's version of the technique that was taken up by European naturalists in the eighteenth century, as we shall see in a later chapter.[117] The technique was linked to Boyle's natural philosophy through his theory about the fluid origin of gems, which suggested that the medical virtues of gems were due to mineral impurities that entered the substance of the gem while in a fluid state. Boyle backed up this theory with measurements of the density of medical gems, arguing that these gems were denser than their nonmedical equivalents.[118] Boyle put this idea to practical use in his classic work on hydrostatics, *Medicina Hydrostatica*, which he worked on in the 1680s and published in 1690.[119] All else being equal, he reasoned, heavy gems are better for the apothecary because they contain more medically active matter. Conversely, lighter gems are better for the jeweler because they contain fewer impurities. Boyle was especially interested in using the balance to detect counterfeit gems and to distinguish between different varieties of the same gem, such as European garnets and American garnets.[120] He extended this approach to common minerals, whose

113. *Specific Medicines*, 401; "Atmospheres," 178; *Gems*, 70.

114. DiMeo, *Lady Ranelagh*, 184 ("almost certainly"), 74–83, 105, 152–58, 180–90.

115. "Queries About Gems," 30.

116. Al-Bīrūnī, *Al-Beruni's Book on Mineralogy*; Mottana, "Galileo as Gemmologist."

117. See chap. 5.

118. *Gems*, 28, 37, 38, 53, 54, 66.

119. Hunter, *Boyle: Between God and Science*, 209, 211–12; *Medicina Hydrostatica*, 209.

120. *Medicina Hydrostatica*, 219–21, 227. Cf. the table of densities in ibid., 277–80, which includes some gems.

medical virtues had been overlooked because they lacked the brilliance of rubies, rock crystal, and the like. Boyle's knowledge of rock crystal came in handy here, since he used this stone as a standard by which to judge the density of other minerals.[121] In the same work he explained how to use pieces of amber and rock crystal to measure the density of liquids—the greater the weight of these objects in a liquid, the less dense the liquid.[122] Boyle's manuscript workdiaries, now available online, are strewn with data on the density of different gems, including counterfeit gems, and on the density of liquids measured with amber or rock crystal.[123]

The hydrostatic balance is of special interest to historians of measurement, but other gem trials were equally important to Boyle's natural philosophy. Most of these trials came ready-made from jewelers and cutters. A cutter "of great Practice and Experience" told Boyle that he had often found differences of up to a carat in the weight of diamonds of the same size.[124] This, as much as the hydrostatic balance, convinced Boyle that gems contain mineral impurities that can be traced to the liquid phase of their formation. The geometrical form of gems was another quality that moved easily between commerce and philosophy. In *Gems*, Boyle used the regular shape of gems to argue for his theory about their formation. Here he cited an "expert Jeweller" who told him that "shape was a mark, by which he usually judg'd a Stone to be a right Diamond, if he had not the opportunity to examine it by the hardness."[125] Identifications such as these were a prerequisite for many of Boyle's arguments. To use a piece of rock crystal as a standard of measurement, it was first necessary to identify that specimen as true rock crystal.[126] Similarly, to show that the hardest known bodies contain unseen motions, one had to correctly identify diamonds. In such cases Boyle nearly always deferred to the judgments of cutters and jewelers. He knew, for example, that a certain green stone was a diamond rather than an emerald because "I found it among Diamonds that belong'd to Merchants too Skilful in those Gems to be impos'd

121. Ibid., 231.

122. Ibid., 231–32. Note also the use of amber to measure the strength of solutions in *Usefulness II*, sec. 2, 429.

123. Gem densities are given in abundance in *Workdiaries* 19.21, 19.89, 19.91; *Workdiaries* 29.253. Note also *Workdiaries* 37.93 (counterfeit emerald), 37.90 (counterfeit sapphire). Liquids in *Workdiaries* 37.102 (turpentine), 37.103 (spirit of salt), 37.104, 37.70, 37.88. *Workdiaries* 38.29, 38.74 (urine, brandy, claret), 38.80 (mineral water), 38 p. 246 (vinegar).

124. *Gems*, 26.

125. Ibid., 14. There is a similar claim about the regular shape of diamonds in "Diamonds," 387, this time from an "experienc'd Artificer."

126. *Gems*, 37.

upon."[127] Travelers vouched for the identity of diamonds by noting that they had personally gathered the stone from a mine.[128] Cutters tested the hardness of gems, and therefore their identity, by applying the stones to their mills.[129] The same cutters linked hardness to quality, just as Cellini and Berquen had done. One cutter told Boyle that imported diamonds had gotten "worse and worse," and were now "so soft and brittle" that he was sometimes afraid to cut or polish them lest he should "spoil" them in the process.[130] Cutters' sensitivity to hardness accounts for their acute observations about the variation in this property that Boyle relayed to his readers. Some sapphire specimens are unusually hard for that species, one cutter told him; another reported that some diamonds have zones that are harder than the rest of the stone.[131] Quality also lay behind Boyle's reports on the "unheeded motions" of gems, since these usually involved cracked, splintered or otherwise spoiled specimens.[132] All the practical marks of the goodness of gems—weight, hardness, shape, smoothness, robustness, brilliance, richness of color, presence or absence of inclusions—played a role in Boyle's philosophical arguments.[133]

These tests were all the more important given the role of the concept of a species in Boyle's philosophy of gems. Much of the argument in *Gems* hinged on this concept. Most diamonds have the same shape, Boyle reasoned. Therefore they must have formed in a liquid medium, since only in such a medium could their constituent particles congregate freely to form a regular solid.[134] The regular texture of other stones, such as New England garnets, was more

127. "Diamonds," 388.

128. *Gems*, 25; "Paralipomena," 212; *Languid Motion*, 299; "Diamonds," 389.

129. The link between the hardness and identity of gems, and the role of jewelers and cutters in judging hardness, is apparent in many passages, especially the following: *Absolute Rest*, 205 (hardness of rubies); *Electricity*, 522 (hardness of sapphire judged by jeweler); "Diamonds," 385 (hardness of diamonds demonstrated by cutter), 388 (hardness used to identify yellow diamond); *Gems*, 14 (jewelers on hardness of rubies), 21 (true stones distinguished from fluors by hardness), 22 (hardness shows rubies/sapphires same species; hardness of diamonds means they are easy to identify), 23 ("sapphirine degree of hardness"), 24 ("full hardness of a Saphire"); "Paralipomena," 209 (jeweler judges part-red stone as diamond); *Clayton's Diamond*, 201 (Clayton's diamond verified by goldsmith, based on hardness).

130. *Absolute Rest*, 204. On Cellini and Berquen, see this volume, chaps. 1 and 2, respectively.

131. "Paralipomena," 209, 218; "Diamonds," 386, 387. "Directional hardness," as it is now called, is indeed pronounced in diamonds: Read, *Gemmology*, 51–52.

132. *Gems*, 18 (splintered diamond); *Porosity*, 140 (spoiled rubies); *Languid Motion*, 298 (cracked topaz).

133. On vividness and color, as judged by jewelers, see *Clayton's Diamond*, 190; and *Languid Motion*, 298. Note also the "deep and almost dark Colour" of garnets, which suggests metallic content to Boyle in "Observations Solitary," 412.

134. *Gems*, 14.

grist to Boyle's mill (plate 2, bottom). The fact that some diamonds deviate from the norm was another argument in his favor, since this ruled out the possibility that they were generated by a "seminal principle," which would have produced a perfectly uniform species (fig. 3.2).[135] Other kinds of variation within species pointed in the same direction. Diamonds were green and yellow as well as transparent; they were more or less hard, and more or less dense; sometimes multiple colors and degrees of hardness were present in the same specimen.[136] Yet all these specimens were diamonds, as shown by their great hardness compared to other stones. All this could be accounted for, Boyle argued, by supposing that diamonds were the result of a long, contingent process during which they were imbued with other minerals. The extraneous minerals were the cause of their color and transparency, which explained why these two properties correlated so poorly with species categories such as "ruby" and "diamond." The point of these arguments was not to reject the idea of a gem species, nor to say that species were arbitrary or conventional.[137] There was such a thing as a "true diamond," a phrase that Boyle used without apology or qualification.[138] There was also a distinction between "adventitious" properties such as color and transparency, and "essential" ones such as hardness.[139] The point was rather to give a new characterization of gem species, one that is captured in one of Boyle's questions for travelers. "Whether the Gem do always, or almost always, grow pure and distinct in its kind," he wanted to know, "or do sometimes admit of mixtures and degenerations [?]."[140] He thought that gems come in kinds, but that these kinds are not always pure and distinct. This, he argued, is precisely what the corpuscular philosophy predicts. Gem appraisal was part of this argument insofar as it helped to identify the species to which a gem belonged.

This gives a new twist to the question posed at the start of the chapter: how did Boyle combine theory and experience in his study of gems? The short answer is that he collected gems. A longer answer is that he combined different kinds of theory and different kinds of experience. He combined the notion of a species with the notion of a corpuscle; the specimens of natural history with the instruments of natural philosophy; the merchandise of gem

135. Ibid., 35.

136. Ibid., 22 (variable color), 26 (variable density), 39 (transparency). Cf. notes 130 and 131.

137. Here I qualify comments about Boyle's "species skepticism" made in Shapin, *Social History of Truth*, 348; and Schaffer, "Golden Means," 22.

138. *Gems*, 22, 39.

139. *Gems*, 22 and 39 ("essential"), 37 ("adventitious"). Note also Boyle's use of the word "nature" with respect to gems, in *Things Above Reason*, 416.

140. Boyle, "Queries About Gems," 28.

traders with the tools of gem cutters. To borrow the language of economists, we might say that the "vertical integration" of theory and experience went along with the "horizontal integration" of different kinds of experience and of different kinds of theory. Experimental philosophy worked because it combined different sorts of experiment and different sorts of philosophy.

This was not a complete merger, however. Some things were left out. Nowhere in *Gems* is there a definition of "gem." Nor is there a list of known species of gems, or an attempt at a complete description of any of them. Despite Boyle's interest in the hardness of gems, he nowhere ranked all known gems in order of hardness. One needs to search long and hard in his writings for a division of gems into broad classes, and the result is scarcely worth the trouble. In an unpublished manuscript on petrifaction, in a section called "Of the Differences of Lapidescent Juices," Boyle wrote that some stones are soft and some hard, some are crystalline and others opaque, some pure and some mixed. He then brushed this division aside, "suspending the Establishment of Partitions, till further Discovery."[141] He spent much of the rest of the section giving "considerations" in favor of his theory that stones are formed from a fluid medium containing dissolved minerals.[142] Here as elsewhere, Boyle was more interested in causes than he was in classes. He invoked classes only insofar as they shed light on causes. Meanwhile, his counterparts at the Royal Academy of Sciences in Paris had begun to take a different approach. They went to great lengths to arrange materials into classes in the hope that, in the long term, causal knowledge would emerge. They made a method out of varying their matter.

141. "Petrifaction," 382.
142. Ibid., 390–98.

4

Gems and the French Origins
of Experimental Physics

There are two kinds of electricity. Two bodies with the same kind of electricity repel each other. Two bodies with different kinds attract each other. Some materials, such as glass, are easy to electrify by rubbing them. Other materials, such as metals, are hard to electrify in this way. The materials that are hard to electrify (think metals) are good at carrying electricity over long distances. The materials that are easy to electrify (think amber) are not so good at carrying electricity.

These are some of the basic rules of electrostatics, still taught to physics students today. They were clearly stated for the first time in a series of eight articles that appeared in the journal of the Parisian Royal Academy of Sciences between 1733 and 1737. These articles were written by the talented aristocrat Charles-François de Cisternay Dufay.[1] The articles are a puzzle. On the one hand, they helped to create experimental physics. They did so by establishing electricity as a coherent domain of phenomena, and by articulating general

1. The titles of these articles began with "Première mémoire sur l'électricité," "Seconde mémoire sur l'électricité," etc. I have abbreviated this to "M1," "M2," etc. The papers read to the academy in a given year were usually published in a later year. In citations, I give the year of reading, followed by the year of publication in brackets. "M1. L'histoire de l'électricité," *MAS* 1733 (1735): 23–35; "M2. Quels sont les corps qui sont susceptibles d'électricité," *MAS* 1733 (1735): 73–84; "M3. Des corps qui sont les plus vivement attirés par les matières électriques, et de ceux qui sont les plus propres à transmettre l'électricité," *MAS* 1733 (1735): 233–54; "M4. L'attraction et la répulsion des corps électriques," *MAS* 1733 (1735): 457–76; "M5. Des nouvelles découvertes sur cette matière, faites depuis peu par M. Gray," *MAS* 1734 (1736): 341–61; "M6. Quel rapport il y a entre l'électricité et la faculté de rendre de la lumière," *MAS* 1734 (1736): 503–26; "M7. Quelques additions aux mémoires précédants," *MAS* 1737 (1740): 86–100; "M8," *MAS* 1737 (1740): 307–25. An English summary of Dufay's findings appeared as "A Letter Concerning Electricity," *PT* (1734–1735): 258–66.

laws for these phenomena rather than causal explanations.[2] They exemplified a wider shift away from a deep and general natural philosophy (*physique*) and toward a flat and narrow experimental physics (*physique expérimentale*).[3] On the other hand, Dufay's articles do not fit the standard explanation of this shift.[4] French experimental physics is usually seen as an import from England and the Dutch Republic, a mixture of Boylean scrupulosity, Newtonian theory, public lectures, and elaborate instruments. Dufay did not give a course of public lectures. None of his important discoveries about electricity relied on instruments of the kind that English experimenters had made famous, such as the air pump and the electrostatic generator. His mentor was not Boyle or Newton but René-Antoine Ferchault de Réaumur, a member of the academy who is remembered mainly for his studies of the behavior of insects and the production of iron and steel. If we want to explain Dufay's laws of electricity, we need to look beyond the Dutch and English model for the emergence of experimental physics.

In particular, we need to look at the academy in the period between its founding in 1666 and Dufay's entry into the institution in 1723. We also need to look at work that seems, to the modern eye, more like chemistry, botany, or mineralogy than experimental physics. This work had various names at the time—*chimie, botanique, histoire naturelle, physique générale*—but much of it was done in the same style. Large collections were made, whether of plants, animals, mineral waters, or minerals. An operation was chosen, such

2. These are themes of past accounts of Dufay's work on electricity, including the following: I. Bernard Cohen, *Franklin and Newton: An Inquiry into Speculative Newtonian Experimental Science and Franklin's Work in Electricity as an Example Thereof* (Philadelphia: American Philosophical Society, 1956), 371–76; Duane Roller and Duane H. D. Roller, "The Development of the Concept of Electric Charge," in *Harvard Case Studies in Experimental Science*, ed. James B. Conant, vol. 2 (Cambridge, MA: Harvard University Press, 1957), 581–90; Roderick W. Home, *The Effluvial Theory of Electricity* (New York: Arno Press, 1981), xi–xiv, 47–67; Heilbron, *Electricity*, 250–60; Friedrich Steinle, "Wissen, Technik, Macht: Elektrizität Im 18. Jahrhundert," in *Macht des Wissens: Die Entstehung der modernen Wissensgesellschaft*, ed. Richard van Dülmen, Sina Rauschenbach, and Meinrad von Engelberg (Cologne: Böhlau, 2004), 515–37.

3. On the wider shift, see Pierre Brunet, *Les physiciens hollandais et la méthode expérimentale en France au XVIIIe siècle* (Paris: Albert Blanchard, 1926); Heilbron, *Electricity*, 9–19, esp. 15; Jean Torlais, "La physique expérimentale," in *Enseignement et diffusion des sciences en France au dix-huitième siècle*, ed. René Taton and Yves Laissus, 2nd ed. (Paris: Hermann, 1986 [1964]), 619–77; and Stephen Gaukroger, *The Collapse of Mechanism and the Rise of Sensibility: Science and the Shaping of Modernity, 1680–1760* (Oxford: Oxford University Press, 2011).

4. On Dufay's life and work, see Michael Bycroft, "Physics and Natural History in the Eighteenth Century: The Case of Charles Dufay" (PhD diss., University of Cambridge, 2013), and references therein.

as distillation or dissection. The same operation was performed on each item in the collection. And the outcomes of all these trials were compared among themselves. I shall call this material-driven experimentation, in contradistinction to the instrument-driven kind.[5] This is the procedure Dufay followed in the research on light and electricity that has caught the attention of historians of experimental physics. This research was done, not on a collection of plants or metals, but on a collection of gems—a large, varied, and expensive collection that Dufay experimented upon over the course of a fifteen-year career at the academy. The novelty was not the method but the extension of the method to new matter.

In other words, this chapter gives a transmaterialist account of early experimental physics. The relations between collections and instruments, between plants and minerals, and between difference classes of mineral, are the main themes. Evaluation is part of the story as well. The point of material-driven experimentation was not to entertain—many of these experiments were dull and repetitive—but to determine the value of useful materials, whether as medicines, as metallic ores, or as ingredients in the production of commodities such as lacquer and porcelain. Academicians made these judgments as representatives of the king. They were part of the machinery of an increasingly centralized French state. They codified their procedures in a new set of regulations in 1699; they led a survey of the kingdom's natural resources in the 1710s, making use of a network of provincial officials (*intendants*) established under Louis XIV; and from 1731 they were members of the Bureau du commerce, a state department responsible for commerce and industry.[6] Academicians were linked to royal power in less formal ways as well, working closely with individuals such as Philippe II, Duke of Orléans, who ruled France as the regent between the death of Louis XIV in 1715 and the majority of Louis XV in 1723. Not all academicians moved in such high circles, but Dufay and Réaumur certainly did. The point is not just that they were committed to useful

5. "Instrument-driven fact-finding" in Cohen, *How Modern Science Came into the World*, 448–62.

6. Regulations in Christiane Demeulenaere-Douyère and Eric Brian, eds., *Règlement, usages et science dans la France de l'absolutisme* (Paris: Éditions Tec & Doc, 2002). Survey in Christiane Demeulenaere-Douyère and David Sturdy, *L'Enquête du Régent, 1716–1718: Sciences, techniques, et politique dans la France pré-industrielle* (Turnhout: Brepols, 2008). Bureau du commerce in Philippe Minard, *La fortune du colbertisme: Etat et industrie dans la France des lumières* (Paris: Fayard, 1998); Harold T. Parker, "French Administrators and French Scientists During the Old Regime and the Early Years of the Revolution," in *Ideas in History: Essays Presented to Louis Gottschalk by His Former Students*, ed. Harold T. Parker and Richard Herr (Durham, NC: Duke University Press, 1965), 85–109.

knowledge, but that useful knowledge was about evaluation as much as it was about production. Gems were part of a wider pattern in which the academicians served as the king's judge of the material world. As for coffee, ginseng, Jesuit's bark, and textile dyes, so for gems: the ultimate aim was to show that French goods were at least as good as those of other kingdoms.[7]

Material-Driven Experimentation

The early academy was once thought to be a barren period for the experimental sciences, with the Parisians preferring abstract mathematics to particular experiments.[8] This view has been transformed by a generation of scholarship on anatomy, botany, and chemistry at the academy in the decades around 1700.[9] There have been various attempts to characterize this experimental program as a whole. My own view is that experiment at the early academy was often driven by materials in the form of collections.[10] The most well known example is a project on plants begun in 1667 and discontinued around 1700. This project involved a large collection of plants, most of them grown in a plot made for the purpose at the Jardin du Roi. The plants were then subjected to a standard set of chemical operations in the academy's laboratory.

---

7. Coffee in Spary, *Eating the Enlightenment*, chap. 2. Textile dyes in Agustí Nieto-Galan, "Between Craft Routines and Academic Rules: Natural Dyestuffs and the 'Art' of Dyeing in the Eighteenth Century," in Klein and Spary, *Materials and Expertise*, 321–54; and Couldardot, "Lumières au banc d'essai." Ginseng in Gianamar Giovannetti-Singh, "Galenizing the New World: Joseph-François Lafitau's 'Galenization' of Canadian Ginseng, ca. 1716–1724," *Notes and Records of the Royal Society Journal of the History of Science* 75, no. 1 (2020): 59–72. Jesuit's bark in Boumediene, *Colonisation du savoir*, bk. 2.

8. A key source of this idea was Kuhn, "Mathematical Versus Experimental Traditions," esp. 26–27.

9. Here are some notable studies: Frederic L. Holmes, *Eighteenth-Century Chemistry as an Investigative Enterprise* (Berkeley: Office for the History of Science and Technology, 1989); Alice Stroup, *A Company of Scientists: Botany, Patronage, and Community at the Seventeenth-Century Parisian Royal Academy of Sciences* (Berkeley: University of California Press, 1990); Christian Licoppe, *La formation de la pratique scientifique: Le discours de l'expérience en France et en Angleterre, 1630–1820* (Paris: La Découverte, 1996); Terrall, *Catching Nature in the Act*; Guerrini, *Courtiers' Anatomists*; Lawrence M. Principe, *The Transmutations of Chymistry: Wilhelm Homberg and the Académie Royale des Sciences* (Chicago: University of Chicago Press, 2020).

10. The rest of this paragraph summarizes Michael Bycroft, "Iatrochemistry and the Evaluation of Mineral Waters in France, 1600–1750," *Bulletin of the History of Medicine* 91, no. 2 (2017): 303–30; and idem, "Experiments on Collections at the Royal Society of London and the Paris Academy of Sciences, 1660–1740," *The Institutionalization of Science in Early Modern Europe*, ed. Mordechai Feingold and Giulia Giannini (Leiden: Brill, 2019), 236–65. Earlier characterizations of the academy's work are surveyed at the end of this chapter.

The results of these numerous trials were then compared, with a view to coming up with general rules such as the one that aromatic plants contain essential oils whereas other plants do not. The procedure also helped to discover new plant extracts and to find known extracts in new plants. All this was mixed up with the practical goal of finding new drugs—in other words, of evaluating the medical virtues of plants. Academicians studied animals and mineral waters in a similar way. A comparison between these projects and equivalent ones at the Royal Society of London, especially by Robert Boyle, shows that the academy's method was distinctive. Boyle certainly sought discoveries, generalizations, and rules, just like the academicians, but he did not use ordered collections to seek them. His collections tended to be haphazard and heterogeneous, like the gem collection discussed in the previous chapter. He systematically varied his operations, whereas the academicians systematically varied the materials to which the operations were applied. "Material-driven experimentation" therefore seems an appropriate name for this aspect of the academy's program.

The big collective projects on plants, animals, and mineral waters fizzled out in the 1690s. But material-driven experimentation did not end; it simply migrated to new materials. Mining and manufacturing were added to medicine as significant concerns for practitioners of the *sciences physiques* at the academy. Accordingly, minerals became more prominent in their research.[11] Marin de Toulon—an enigmatic figure who showed an interest in porcelain, pigments, and mineral collecting in the 1690s—was part of this shift.[12] Also important was Wilhelm Homberg, a brilliant chemist who took over the project on plants in 1691 and whose real interests lay in mining and the mineral kingdom.[13] But the crucial individual for Dufay and for material-driven experimentation was Réaumur. The latter entered the academy in 1708. His early work in natural history was on marine animals, before the study of the crafts drew his attention to the minerals used to make gold, slate and mirrors.[14] In 1710 or 1711, he took over the academy's *Description des arts et métiers*, a long-running effort to describe the arts and crafts in the kingdom. This made Réaumur an ideal candidate to lead a survey of French natural

11. Note the modest, and late, reference to mineralogy in the tables of the academy's expenses between 1666 and 1699 prepared by Alice Stroup: *Company of Scientists*, 243–62, esp. 253.

12. Principe, *Transmutations of Chymistry*, 119–20.

13. Ibid., 14 and 34–37 (mining); 97–111 (plant project); 138, 141, 144 (metals in textbook); 138 and chap. 4 (sulfur/mercury in Homberg's chemical system); 127–28 (metallic assaying).

14. Jean Torlais, "Chronologie de la vie et des oeuvres de René-Antoine Ferchault de Réaumur," *Revue d'histoire des sciences* 11, no. 1 (1958): 1–2. Cf. idem, *Un esprit encyclopédique en dehors de L'Encyclopédie: Réaumur, d'après des documents inédits* (Paris: Albert Blanchard, 1961 [1935]), 33–52.

resources carried out between 1716 and 1718. This survey has been traced to the Baconian tradition of compiling descriptions of crafts, and to the Colbertian tradition of kingdom-wide surveys (enquêtes).[15] But it was also a continuation of the academy's tradition of material-driven experimentation.

The survey was initiated by the Regent of France, Philippe II, Duke of Orléans, whose interest in mining and the chemistry of minerals had been nurtured by Homberg until the academician died in 1715.[16] One of Homberg's students, Etienne-François Geoffroy, carried forward the projects on plants and mineral waters; Geoffroy saw the survey as a continuation of the latter project.[17] The survey had the same structure as the earlier ones on plants, animals and mineral waters. A large collection of minerals was formed in Paris, drawing on the resources of the emerging state bureaucracy. The minerals were sent by the intendants, representatives of the king stationed in the French provinces.[18] They were instructed not to alter the minerals in any way, the better to compare those minerals to each other in Paris. These objects found their way into Réaumur's residences in Paris, which soon housed "samples of all the kingdom's useful earths, stones, minerals, and metals," to quote a document associated with Réaumur and written before 1722.[19] Réaumur then carried out exhaustive chemical trials of these minerals. The overall aim was to evaluate the mineral resources of France, usually with one eye on the resources of rival kingdoms. The aim was to show, for example, that the kingdom possessed "more than sixty lead mines that are as rich and abundant as those of the English."[20]

In practice, these analyses were more fragmented than the earlier ones on plants and mineral waters. They were done in different periods, on different

15. Christiane Demeulenaere-Douyère and David Sturdy, "Introduction," in their Enquête du Régent, 27–29.

16. Ibid., 11–18; Principe, Transmutations of Chymistry, chaps. 4 and 5.

17. Demeulenaere-Douyère and Sturdy, Enquête du Régent, 81. See the numerous citations under "eaux minérales (ou thermales)" in ibid., 908. On Geoffroy's role in the survey, see Demeulenaere-Douyère and Sturdy, "Introduction," 50.

18. Note the emphasis on minerals in the questionnaire sent to the intendants in 1716: Demeulenaere-Douyère and Sturdy, Enquête du Régent, 78–79.

19. René-Antoine Ferchault de Réaumur (?), "Réflexions sur l'utilité dont l'Académie des sciences pourroit être au royaume, si le royaume luy donnoit les secours dont elle a besoin," transcribed in Ernest Maindron, L'Académie des sciences: Histoire de l'Académie, fondation de l'Institut national (Paris, 1888), 108. On Réaumur's mineral cabinet, see René-Antoine Ferchault de Réaumur, "Idée générale des différentes manières de faire la porcelaine," MAS 1727 (1729): 199; idem, "Second mémoire sur la porcelaine," MAS 1729 (1731): 329; Edmé-François Gersaint, Catalogue raisonné de coquilles et autres curiosités naturelles (Paris, 1736), 31; Royal ordinance quoted in Torlais, Esprit encyclopédique, 384; Terrall, Catching Nature in the Act, 141.

20. Réaumur (?), "Réflexions sur l'utilité," 108–9.

classes of material, and with different applications in mind. The result is that the coherence and significance of the project is not easy to grasp. One historian has wondered "why so little harvest was made at the time all this information [from the survey] was gathered."[21] But much harvest *was* made, both at the time and in the decades after 1718, as a review of Réaumur's published and unpublished works in this period show.[22] At the very start of the survey, in November 1715, Réaumur read a paper on turquoise in which he thanked the regent for specimens he obtained with the help of provincial officials.[23] Once the survey was underway, he had analyses carried out on over a hundred metallic ores sent by the intendants.[24] In the next few years, up to 1722, he referred to iron ores from the survey in publications on the formation of metals, the conversion of iron into steel, the production of anchors, and the structure and classification of iron ores.[25] He was working on porcelain in the same period, using a huge number of minerals from the survey to search for French substitutes for the two main ingredients of Chinese porcelain, *kaolin* and *petuntse*.[26] As he put it, "every terrestrial material is suited to these trials."[27] A final strand of this project was a general theory of the formation

21. Roger Hahn, "Review of C. Demeulenaere-Douyère and D. J. Sturdy, *L'Enquête du Régent*," *Isis* 101, no. 2 (2010): 428.

22. The following expands on Demeulenaere-Douyère and Sturdy, "Introduction," 49–52, mainly by adding Réaumur's work on stones, earths and porcelain to the account given there.

23. Réaumur, "Sur les mines de turquoises du Royaume; sur la nature de la matière qu'on y trouve, & sur la manière dont on lui donne la couleur," *MAS* 1715 (1717): 174, 178–90. Cf. Demeulenaere-Douyère and Sturdy, "Introduction," 30–31.

24. Demeulenaere-Douyère and Sturdy, *Enquête du Régent*, 795–865; idem, "Introduction," 37–39.

25. *HAS* 1716 (1718), 76; René-Antoine Ferchault de Réaumur, "Description d'une mine de fer du pays de Foix; avec quelques réflexions sur la manière dont elle a été formée," *MAS* 1718 (1720): 139, 140; idem, *Réaumur's Memoirs on Steel and Iron*, trans. Anneliese G. Sisco, ed. Cyril S. Smith (Chicago: University of Chicago Press, 1956 [1722]), 159–61, 167–68, 171; idem, *Fabrique des ancres, lue à l'Académie en juillet 1723* (Paris, 1764). On the latter work, see Demeulenaere-Douyère and Sturdy, "Introduction," 50.

26. This work was published later: Réaumur, "Idée générale de la porcelaine"; idem, "Second mémoire sur la porcelaine"; idem, "Art de faire une nouvelle espèce de porcelaine, par les moyens extrêmement simples & faciles," *MAS* 1739 (1741): 370–88. But Réaumur was already interested in porcelain in 1717: "Idée générale de la porcelaine," 192. And he was doing trials as early as 1718: AS, Réaumur archive, dossier 55 ("Porcelain"), ff. 56–59. The manuscripts held at the Archives de la Manufacture nationale de Sèvres (Y39) are dated 1727–1729. I am grateful to Grace Chuang for providing me with electronic copies of the Sèvres manuscripts.

27. Réaumur, "Idée générale de la porcelaine," 190. Réaumur identified the survey as the basis for his search for *kaolin* in "Second mémoire sur la porcelaine," 329. It was probably also the basis for his search for *petuntse* (ibid., 330–31), since that search involved sands, earths, and

and classification of minerals. This was a substantial piece of work, yielding three articles in the academy's journal between 1721 and 1730 as well as much manuscript material.[28] Here Réaumur referred to stones, flints and crystals from various parts of France, some of which he had procured from mines and quarries that featured in his correspondence with the intendants.[29] Overall, the regent's survey was a major source of specimens for studies of minerals that Réaumur carried out over the course of more than two decades.

These projects may seem disparate, but they were joined up by versatile modes of evaluation and explanation. Fracture analysis illustrates the point. This technique involved breaking objects in two and studying the cross section of the break. Such tests were already used by artisans to determine the quality of steel bars. Réaumur himself divided iron bars into seven classes based on the appearance of their fractured surfaces. He then linked these classes to the quality of steel they produced (fig. 4.1).[30] He analyzed ceramics in the same way, with the shiny fractures of glass at one extreme and the dull, grainy fractures of earthenware at the other extreme, and with porcelain as the intermediate term.[31] He used the very same technique in a study of the formation of stones, breaking flintstones in half and studying their fractures. He interpreted these fractures as different stages in the transformation of common stones into flintstones.[32] And he interpreted the fractures of iron and porcelain in the same way. In each case, different fractures represented different stages in the progress of some substance from the perimeter of the fracture to the center. Iron became steel by absorbing "salts and sulfurs";

flints, all of which featured in the survey. See under "sable," "terre" and "caillou," in the index to Demeulenaere-Douyère and Sturdy, *Enquête du Régent*.

28. René-Antoine Ferchault de Réaumur, "Sur la nature et la formation des cailloux," *MAS* 1721 (1723): 255–76; idem, "Sur la rondeur qui semblent affecter certaines espèces de pierres, & entr'autres sur celle qu'affectent les cailloux," *MAS* 1723 (1725): 273–84; idem, "De la nature de la terre en générale, et du caractère des différentes espèces de terre," *MAS* 1730 (1732): 243–83; AS, Réaumur archive, dossier 52 (Physique: terres, sables, pierres), esp. ff. 2–8, 12–17, 31, 84–93, 103–4, 109–20. Etienne-François Geoffroy published an earlier article on the formation of stones that may have been linked to the regent's survey as well, as noted in Demeulenaere-Douyère and Sturdy, "Introduction," 50, citing Etienne-François Geoffroy, "De l'origine de la formation des pierres," AS *PV*, vol. 42 (1723), July 10, 1723, f. 213v. Cf. Fontenelle, "Sur l'origine des pierres," *HAS* 1716 (1718): 8–16.

29. Réaumur, "Nature et formation des cailloux," 265–69; idem, "Nature de la terre," 268–70.

30. Réaumur, *Memoirs on Steel and Iron*, 163–70.

31. Réaumur, "Idée générale de la porcelaine," 187; idem, "Second mémoire sur la porcelaine," 339–40; idem, "Nouvelle espèce de porcelaine," 375. Note also the gradual, outside-in movement that converted glass into glass porcelain: idem, "Nouvelle espèce de porcelaine," 387.

32. Réaumur, "Nature et formation des cailloux," 262–64.

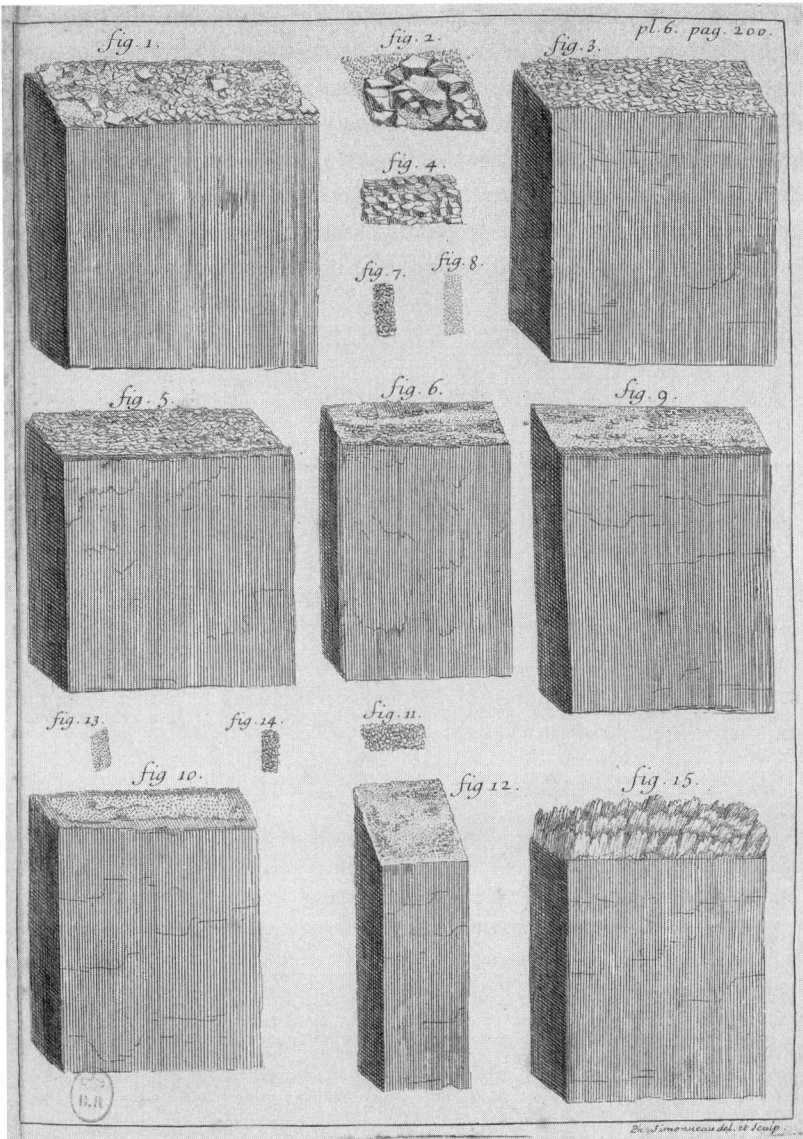

FIGURE 4.1. Réaumur's fracture analysis of iron
In a characteristic move, Réaumur divided wrought irons into seven classes according to the quality of steel they produced, as judged by the appearance of their fractured surfaces. *Figure 1* depicts the fracture of a beam of wrought iron from the first and poorest class. *Figure 2* depicts the same under a magnifying glass. Note the large, coarse, irregular platelets, which gave poor steels. Irons in the second class (*figs. 3, 4*) have smaller and more uniform platelets. Irons in the fourth class (*fig. 9*) contain fine platelets and as many grains as platelets; these irons gave the best steels. *Figures 5* and *6* show intermediate fractures. *Figures 7* and *8* relate to the specimen in *figure 6*, showing the platelets and grains, respectively. From Réaumur, *Iron and Steel* (1722), chap. 5, plate 6. Courtesy of gallica.bnf.fr / Bibliothèque nationale de France.

common pottery became porcelain by absorbing "vitrified matter"; common stones became flintstones by absorbing a "stony sugar." In Réaumur's words, these materials were "sponges" that "soak up" or "drink up" a subtle matter from their surroundings.[33] Fracture analysis was a device for classifying materials and explaining their behavior as well as for evaluating them. In sum, Réaumur was doing for minerals what earlier academicians had tried to do for plants and mineral waters—building a large collection, subjecting the collection to standard tests, and using the results to compare the behavior of different classes of material.

Assaying Gems

The mineralogical turn at the academy was also a turn toward gems. Homberg was the first academician to make a sustained study of gem-related phenomena for which substantial records survive. He came to the topic through the assaying of gold, which was sometimes done by using the putative gold to make ruby-red glass. This led Homberg to experiment with other recipes for imitating colored gems, including the engraved gems in royal collections.[34] There is some evidence for gem making at the academy prior to Homberg's arrival. There is plenty more such evidence in the work of Dufay, Réaumur, and their close collaborator Jean Hellot.[35]

33. Ibid., 258–64, esp. 256, 259; idem, "Nouvelle espèce de porcelaine," esp. 374–75, 379–83; idem, *Memoirs on Steel and Iron*, 182, 211–12, 252, 336. Réaumur's adjective was "spongieux"; his verbs were "absorber," "boire," "pénétrer," and "abbreuver."

34. Wilhelm Homberg, "Observations sur la flamme verte qui paroît lors qu'on rougit du cuivre au feu," *PV* Jan. 4, 1696, 234r–236v; idem, "Observations sur quelques effets que l'or produit seul dans le grand feu," *PV* Feb. 22, 1696, 274r–277v; idem, "Pierres factices," *PV* May 30, 1696, 74r–82v, cf. *PV* May 16, 1696, 63r; idem, "Manière de copier sur verre coloré les pierres gravées," *MAS* 1712 (1714): 187–94; Cahiers de Jean Hellot, Bibliothèque Municipale de Caen, MS 171 (hereafter "Hellot notebooks"), vol. 8, 18v–19r. On Hellot's notebooks, see Principe, *Transmutations of Chymistry*, 426–29.

35. Recipe attributed to Samuel Duclos (d. 1685) in the Hellot notebooks, vol. 3, f. 274r. René-Antoine Ferchault de Réaumur, "De l'art de faire des perles," *PV* May 23, 1711, title given in *HAS* 1711 (1713): 101; idem, "Mémoire sur la matière qui colore les perles fausses," *MAS* 1716 (1718): 229–44. For Hellot's interest in the topic, see the following passages in his notebooks: vol. 8, 118r–121v (fake pearls); vol. 1, 65v, vol. 3, 273v–274v, vol. 5, 137r, vol. 8, 260r (glass gems); vol. 1, 40v–41r, vol. 2, 92–93, vol. 5, 136r (whitening pearls); vol. 1, 81r–83r (engraved stones); vol. 3, 286v and 299r, vol. 5, 135r (dyeing gemstones); vol. 3, 119–21 (making marble); vol. 1, 122r, vol. 3, 287v–288r (foils); vol. 5, 137v (Brazilian topaz); vol. 3, 299v, vol. 5, 136v, 137r (dyeing rock crystal). For Dufay, see his "Mémoire sur la teinture et la dissolution de plusieurs espèces de pierres," *MAS* 1728 (1730): 50–67; idem, "Second mémoire sur la teinture des pierres," *MAS* 1732 (1735): 169–81;

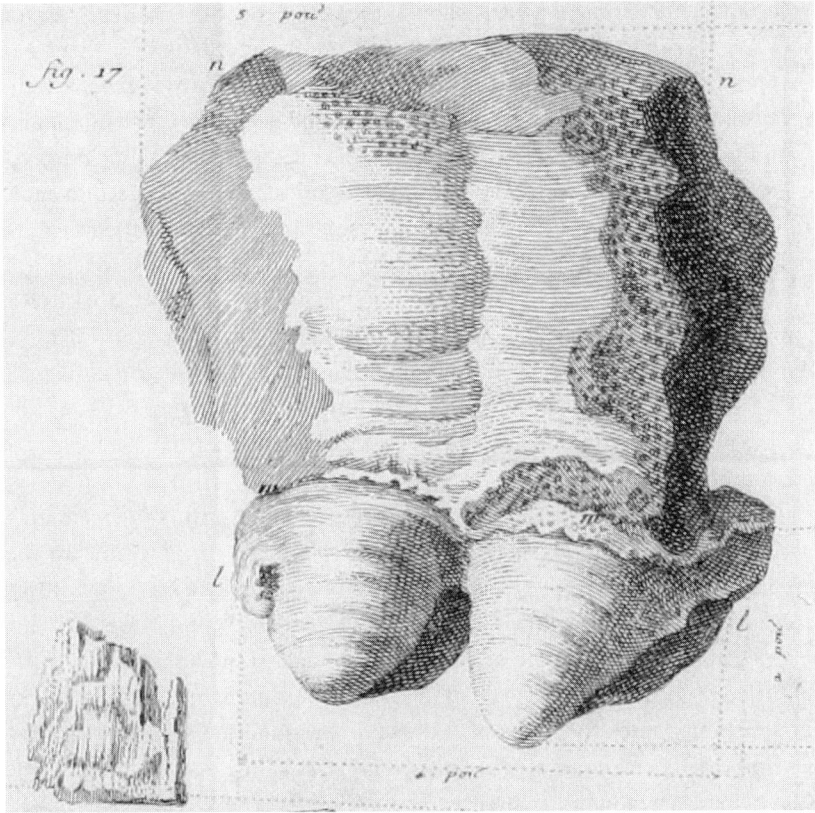

FIGURE 4.2. Réaumur's French turquoise
The main drawing shows a toothlike object found near the town of Simmore in southern France. The conical teeth are attached to a block of mineral matter that Réaumur regarded as a fossilized jawbone. The block turned blue when heated, resembling Persian turquoise. The small drawing at bottom left shows a section of this turquoise-like substance, which Réaumur believed (incorrectly) to be the same as Persian turquoise. The dimensions of the jawbone-and-tooth were 5 inches by 5.5 inches. From Réaumur, "Turquoise mines" (1715), plate 8. Courtesy of the Biodiversity Heritage Library, contributed by Natural History Museum Library, London.

But it is gem judging, not gem making, that concerns us here. The first real academic engagement with this topic appears to have been the regent's survey. Gems were part of the survey from the beginning. Already in December 1715, before the project had formally begun, Réaumur described experiments he had done on turquoise sent to him by Jean-Baptiste Imbercourt, intendant of the Montauban region (fig. 4.2 and plate 6, top). Imbercourt had

and Michael Bycroft, "Style and Substance in Rococo Science," *Journal of Interdisciplinary History* 48, no. 3 (2018): 372–76.

received a letter from Jean-Paul Bignon, the president of the academy, asking for samples of the stones and for a memorandum describing how they were extracted from a mine near the town of Simmore.[36] This was precisely the mode of inquiry that Réaumur and Bignon would generalize to all intendants with their questionnaire the following year. This questionnaire included requests for "marbles and other extraordinary stones that may be used to decorate buildings," "stones valuable either for their properties or by their color or beauty, such as jasper, crystal, precious stones of all kinds, talcs, lodestones, etc.," and precious objects found in the sea, including coral and pearls.[37] Réaumur repeated these requests in follow-up letters to the intendants.[38] He evidently saw this project as a continuation of the *Description des arts et métiers*. As he noted in the 1715 paper, his study of turquoise was part of a description of "the arts concerning gemstones [*pierreries*]" that he was in the process of compiling.[39]

This is borne out by Réaumur's correspondence with the intendants, which is replete with queries about the lapidary arts. The academician was especially interested in the jet industry in Aix-en-Province, the gem-cutting machines active around Strasbourg, and coral fishing around Marseille.[40] His queries on these topics were often specific enough to show that he had prior knowledge about them. He wanted to know, for example, whether the cutters in Strasbourg stuck their stones to the end of a small rod, "as other lapidaries do." Elsewhere, he wrote that he had seen jet being worked in Paris, and that he had "already in part written" a description of the art. The intendants responded with samples as well as answers. They sent around sixty separate parcels of gemstones to Paris over the course of the survey, many of them containing two or more samples (appendix 2). Turquoise and rock crystal were the most common species, but there were several others as well, including pearl, amber, and lapis lazuli. The transfer of these specimens from the provinces to Paris was a complicated affair that involved multiple participants at both ends. The correspondence on turquoise, for example, ran from December 1715 to at least April 1719. It involved Réaumur, the academy's

36. Demeulenaere-Douyère and Sturdy, "Introduction," 30–31, citing Réaumur, "Sur les mines de turquoises," 179–80.

37. Demeulenaere-Douyère and Sturdy, *Enquête du Régent*, 78–79.

38. Ibid., letters to the intendants of Alsace (137), Bordeaux (203), Bretagne (257), and Limoges (524), all of which ask about marble, crystal, or both.

39. Réaumur, "Sur les mines de turquoises," 179.

40. Demeulenaere-Douyère and Sturdy, *Enquête du Régent*, 108 (jet); 136 and 140 (Strasbourg); 145 ("as other lapidaries do"); 110–16 (coral). Cf. Réaumur's queries (ibid., 682, 757) about gem cutting in Rouen and Perpignan.

President Jean-Paul Bignon, three different intendants, and various informers around in Simmore—a judge, a miner, a priest, and the priest's father.[41]

The key question in these exchanges was the quality of French turquoise. Was it as good as the famous turquoise from Persia? Or was it merely a convincing imitation, like the fake pearls that Réaumur had studied a few years earlier?[42] Turquoise, like pearl, was traditionally associated with the Orient. It had acquired its European names from the (mistaken) medieval belief that the stones came from Turkey.[43] Boodt had made it clear that "occidental" turquoise was inferior to the "oriental" kind, meaning the ones from the mines near Nishapur, a town in Safavid Persia. Réaumur hoped to turn this evaluation on its head. His interest in the topic was piqued by the Persian embassy that visited Paris early in 1715. The ambassador, Mohammed Reza Beg, included 280 pieces of turquoise among the gifts he presented to Louis XIV at the Palace of Versailles in February of that year. The turquoises, like the embassy as a whole, were something of a disappointment to French observers.[44] The stones were small, whitish, and hard to polish. Réaumur thought France could do better: "It would not be difficult to send back to Persia turquoises that are more beautiful, and much larger, if we were willing to search in our mines for the stones they contain."[45]

In making this judgment, Réaumur drew on the rich seventeenth-century tradition of writing on gemstones. From Rosnel and Berquen, he learned that there was such a thing as "Languedoc" turquoise. According to Berquen, these stones were white when removed from the mine but became "Turkish blue," the same color as Persian turquoise, when they were heated. Berquen noted ironically that their only defect was their French origin, "because if they came from afar we would make a great fuss about them."[46] Tavernier's travel narrative was also useful to Réaumur. The traveler's distinction between "old rock" and "new rock" led Réaumur to believe that the ambassador's turquoises were of the latter kind, and hence that the older and more

41. Demeulenaere-Douyère and Sturdy, *Enquête du Régent*, 491, 494, 495, 497, 503, 505, 585–87, 637–53, 658–59.

42. On fake pearls, see note 35.

43. On the actual location of these stones, as reported by Persian writers, see Khazeni, *Sky Blue Stone*, chaps. 1 and 2. European traders had been locating turquoise in "Persia" since the sixteenth century: Orta, *Colloquies on the Simples and Drugs*, 358; Linschoten, *Voyage to the East Indies*, vol. 2, 141. Khazeni, *Sky Blue Stone*, chap. 4, surveys early European ideas about turquoise.

44. McCabe, *Orientalism in Early Modern France*, 254–56.

45. Réaumur, "Sur les mines de turquoises," 177.

46. Ibid., 176–76, 179. Cf. Berquen, *Merveilles des Indes*, 56; Rosnel, *Mercure Indien*, vol. 2, 26.

precious turquoise mines in Persia were exhausted.[47] Tavernier's success in selling European stones in Asia persuaded Réaumur that Languedoc turquoise could be sold in the same way.[48]

Having read up on the topic, Réaumur went on to assay French turquoise for himself. A visit to a gem cutter confirmed that the Languedoc turquoises were just as hard as the Persian ones. The same visit showed the power of the language of orientalism: The cutter called some of the specimens from Languedoc "oriental" on the basis of their hardness, and he continued to do so even after he learned their true origin from Réaumur.[49] Another apparent defect of French turquoise, besides its French origin, was the dark stripes and filaments that normally showed on its surface. Réaumur explained away these flaws with a theory about the formation of the stone. All turquoise, he maintained, is formed from the teeth and bones of dead animals; over time, the elongated "cells" of these objects are filled with the stony matter that makes up the bulk of precious turquoise. The black lines on French turquoise simply meant that their cells had not been fully filled with stony matter. The French stones were not a different species from the Persian ones, merely the same species at a different stage of its evolution.[50] The naturalist also gave an experimental confirmation of Berquen's claim that Persian turquoise turns green more easily than French turquoise. The process was too slow to test directly, so Réaumur used a chemical test to speed it up. He found that Persian turquoise turned green in distilled vinegar whereas French turquoise turned white.[51]

The ultimate test was to scale up the production of turquoise in Simmore. This was difficult work, as the increasingly pessimistic exchanges between the priest (a certain Giscaro), the priest's father, and the Parisian savants show. In an early trial, two hundred hours of prospecting yielded a mass of stones that became scaly and colorless when heated in a furnace for eight hours.[52] Clearly this would not do. Eighteen months later, the workers had found a way to produce gem-quality stones, only to learn that they were too green

47. Réaumur, "Sur les mines de turquoises," 176–77. Cf. Tavernier, *Travels*, vol. 2, 103–4.
48. Demeulenaere-Douyère and Sturdy, *Enquête du Régent*, 585–86, 647.
49. Réaumur, "Sur les mines de turquoises," 177–78.
50. Ibid., 185–86. This turns out to be false. Current thinking on French "turquoise," now known as "odontolite," is summarized in Ina Reiche, Colette Vignaud, Bernard Champagnon, Gérard Panczer, Christian Brouder, Guillaume Morin, Vicente Armando Solé, Laurent Charlet, and Michel Menu, "From Mastodon Ivory to Gemstone: The Origin of Turquoise Color in Odontolite," *American Mineralogist* 86, no. 11–12 (2001): 1519–24.
51. Réaumur, "Sur les mines de turquoises," 198–99.
52. Demeulenaere-Douyère and Sturdy, *Enquête du Régent*, 643–44.

for Parisian tastes.[53] "Help me, I beg you!" wrote the elder Giscaro in his last surviving letter, sent to Paris in March 1719.[54] The effort to replicate Persian turquoise in France had failed. But the idea of replicating oriental goods persisted. So did the wider trend of which the regent's survey was a part—the extension of material-driven experimentation to the mineral kingdom, and especially to gemstones.

Physics as Gem Collecting

These projects were taken forward by Dufay, a leading member of the academy between his entry in 1723 and his death in 1739. Like his English predecessor Robert Boyle, Dufay is usually studied as an experimental philosopher, a practitioner of la physique expérimentale. Like Boyle, however, Dufay's career shows the value of studying experimental philosophy in tandem with natural history. Dufay's links to Réaumur illustrate the point. The two men are rarely considered together, probably because Dufay is seen as a protophysicist and Réaumur as a protobiologist. In reality, their careers are hard to separate.[55] A close study of published and unpublished sources reveals the presence of Réaumur in nearly every major development in Dufay's career, from his arrival in the academy in 1723, to his appointment as a dye inspector in 1731, to his appointment as director of the Jardin du Roi the following year. The latter post was only the most visible example of Dufay's engagement with collecting, an interest that included coins and paintings, both ancient and contemporary, as well as plants, animals, and minerals. Dufay was a node in a network of collectors that included the Parisian auctioneer Edmé-François Gersaint, the writer Antoine-Joseph Dezallier d'Argenville, the British naturalist Hans Sloane, and Louis-Henri, Duke of Bourbon, a grandson of Louis XIV.[56] Like Réaumur, Dufay tried to replicate Asian goods with European materials,

53. Ibid., 658–59.

54. "Secouré moy, je vous prie": ibid., 659.

55. This paragraph summarizes Bycroft, "Physics and Natural History," chaps. 2 and 3. This in turn builds on hints in Heilbron, *Electricity*, 251, 255; Licoppe, *Formation*, 113–24; and Terrall, *Catching Nature in the Act*, 49, 52.

56. The standard accounts of early French natural history collecting are Edouard Lamy, *Les cabinets d'histoire naturelle en France au XVIIIe siècle et le Cabinet du Roi (1635–1793)* (Paris: E. Lamy, 1930); and Yves Laissus, "Les cabinets d'histoire naturelle," in Taton and Laissus, *Enseignement et diffusion*, 659–712. On Gersaint and Argenville, see Guillaume Glorieux, *A l'enseigne de Gersaint: Edme-François Gersaint, marchand d'art sur le Pont Notre-Dame, 1694–1750* (Seyssel: Éditions Champ Vallon, 2002); Anne Lafont, *1740, un abrégé du monde: Savoirs et collections autour de Dezallier d'Argenville* (Paris: Fage, 2012). The classic study of the shift from antiquities to natural productions is Pomian, *Collectors and Curiosities*, chap. 4.

translating a major work on Chinese lacquer into French.[57] Also like his mentor, Dufay studied the arts of gem making and gem cutting in France and abroad.[58] Both men had a broad conception of la physique, one that included everything from salts to salamanders. They also had a common interest in magnetism, electricity, and luminescence. In unpublished notes linked to his work on steel and porcelain, Réaumur anticipated one of Dufay's most important findings about the Bologna stone, a mineral that could be made to glow in the dark. Réaumur hinted that many bodies "have the privileges and properties that we have accorded to [the Bologna stone] and that we have considered unique to her alone." Dufay confirmed this in a 1730 article, showing that a great range of materials could be made to glow in the dark just like the Bologna stone.

Behind this result lay a wider project on gems. Gems were among the materials that Dufay tried to transform into phosphors, both in the 1730 article and in earlier piece on phosphors, where he mentioned jasper, jacinth, ruby, opal, and garnets, among other "precious stones."[59] The wider significance of gems is evident in the following passage in the 1730 paper:

> Some years ago, I formed the intention of examining, by every means I could think of, the nature of all fine stones (*pierres fines*). Calcination was one of the main tests I used. Since I aimed to omit no stones in the class (*classe*) of fine stones, I examined those that are placed in that class only because they cannot be easily placed elsewhere. Common topaz is one such stone, and as there are many kinds (*sortes*) it is important to note that the topaz in question is known only in medicine. . . . It is a very soft stone, yellow, heavy, and chalky; and when I came to describe it, it reminded me immediately of the Bologna stone, from which it differs only by its external form, which is usually a little rounded and rough, while the topaz is more often cubic, or at least almost always terminates in parallel faces. Without pausing to make a more detailed comparison, I calcined this topaz in a crucible. . . . I [then] exposed it to the

57. Michael Bycroft, "What Difference Does a Translation Make? The *Traité des vernis* (1723) in the Career of Charles Dufay," in *Translation and the Circulation of Knowledge in Early Modern Science*, ed. Sietske Fransen and Niall Hodson (Leiden: Brill, 2017), 66–90.

58. Dufay, "Teinture des pierres," 175; idem, "Second mémoire sur la teinture des pierres"; idem, "Lumière des diamants," 362–65; Hellot notebooks, vol. 3, 299v, vol. 8, 193; AN F/12/992, "Charles Bro."

59. Bernard le Bovier de Fontenelle, "Sur une pierre de Berne qui est une espèce de phosphore," *HAS* 1724 (1726): 58–71, esp. 60–61 on the Bologna stone. Dufay's paper on this topic was never published, but traces of it survive in the academy's archive: *PV* 1725, Mar. 10 (78r), June 20 (141r), Aug. 29 (223v), Sept. 5 (227r). The paper was based on a letter sent to the academy by the Swiss naturalist Louis Bourguet; an extract of this letter survives in the Hellot notebooks, vol. 8, f. 207r–v.

light of the day, then carried it into a dark room, and found it to be the equal of the best Bologna stones.[60]

The first line of this passage shows that gems were the basis for a general program of material-driven experimentation that Dufay had begun "some years ago," meaning some years before 1730. The remainder of the passage gives some idea of the nature of the program. It was chemical, involving operations such as calcination. It was also a form of natural history, involving the careful differentiation of "classes" and "kinds" mineral and the close observation of the color, texture, weight, hardness, and shape of minerals. Observations of this kind led to the discovery of a new phosphor (common topaz) that triggered a search for many other new phosphors. The new phosphors were a by-product of the project on gems, not the other way round.

The quoted passage also shows that Dufay had amassed something close to a complete collection of "fine stones" by 1730. This was no doubt the beginning of the "collection of precious stones" that Dufay bequeathed to the Jardin du Roi when he died in 1739.[61] This bequest is well known, having been mentioned by the academy's secretary, Bernard le Bovier de Fontenelle, in his *éloge* of Dufay; but details about the collection are hard to come by. The only thing approaching a catalog of Dufay's collection is a manuscript list of the pieces of rock crystal he owned when he died, a list written down by Dufay's friend and literary executor, Jean Hellot (see fig. 4.3). This list gives some idea of the extent of the collection—Hellot mentioned a dozen specimens from France, Austria, Switzerland, Italy, Germany and India—but it is evidently a partial list.[62] In the absence of catalogs, the best place to turn is the detailed record of Dufay's experimental research. The articles he published in the academy's *Mémoires*, along with unpublished notes related to these articles, are the best sources we have on the contents of his gem collection. Taken together, these sources show that he had access to most of the "fine stones" listed by Argenville in his mature treatise on mineralogy (appendix 3).

A closer look at these sources shows how Dufay wove gems into his work on light and electricity. A few years after generalizing the phenomenon of phosphorescence, Dufay carried out an equally comprehensive study of the

60. Dufay, "Mémoire sur un grand nombre de phosphores nouveaux," *MAS* 1730 (1732): 526.

61. Bernard le Bovier de Fontenelle, "Éloge de M. Dufay," *HAS* 1739 (1741): 81. See also the brief references to gems at the Jardin du Roi in Louis-Jean-Marie Daubenton, "De la connoissance des pierres précieuses," *MAS* 1750 (1754): 30; and Antoine-Joseph Dezallier d'Argenville, *L'histoire naturelle éclaircie dans deux de ses parties principales: La lithologie et la conchyliologie* (Paris, 1742), 199.

62. Hellot notebooks, vol. 8, 193v.

Aux depens de M.^r le Président Henault de l'Académie Françoise.

FIGURE 4.3. Engravings of crystals from Argenville

Figures 1 and *2* are described by Argenville as being "faceted by nature," an analogy to cut gems. *Figures 3* and *4* are "true matrices," in his words, showing the matter in which crystals form. *Figure 5* is a single crystal containing elongated objects that the author compared to straws. The collection of Argenville's contemporary, Dufay, was especially rich in crystals such as these. From Argenville, *Oryctology* (1755), plate 3. Courtesy of gallica.bnf.fr / Bibliothèque nationale de France.

luminosity of diamonds.[63] The main finding of this study, as of his work on phosphors, was that an effect once considered rare and capricious was in fact a reliable feature of many materials. The effect was the ability of diamonds to glow in the dark after being rubbed, heated, or exposed to light. The effect itself was not new—Robert Boyle had generated glows in John Clayton's diamond in many different ways. Dufay's contribution was not to invent new techniques but to apply the old techniques to many specimens. These included two white diamonds weighing nearly 20 carats each; green, blue, yellow, pink, and violent specimens; a set of four hundred yellow diamonds that Dufay "once had the occasion to have to hand"; and many others.[64] Georges-Louis Leclerc, Comte de Buffon, would later say of Dufay that "perhaps no-one else has handled as many rough and cut diamonds."[65]

Buffon also dropped a hint about the origin of these stones: the naturalist "had borrowed the diamonds of the crown, and those of our princes, for his experiments."[66] There is a glimpse of this kind of aristocratic exchange in Hellot's note on Dufay's rock crystals: "The piece of cut Rhine crystal, white and with a handsome water, belonging to the Duke of Antin and previously to Mr. the Duke [i.e., Louis-Henri, Duke of Bourbon], in which one can see a well-preserved piece of bulrush, was at Dufay's place in August 1737."[67] Dufay himself was coy about the sources of his stones in his published works. Fortunately, he left behind a set of laboratory notes in which he named several of these sources.[68] These names were a microcosm of the emerging community of Parisian mineral collectors: the Duke and Duchess of Bourbon; the academicians Jean-Paul Bignon, Claude-Joseph Geoffroy, and Henri-Louis Duhamel du Monceau; and a certain "Mr. Dénery" and "Me. Durf," two names that later appeared in Argenville's list of owners of mineral and shell collections in Paris.[69] The bulk of the stones in these notes came from the duke and

63. Dufay, "M6. Électricité et lumière," 512–17; idem, "Recherches sur la lumière des diamants et de plusieurs autres matières," MAS 1735 (1738): 347–72, esp. 353, 355–56.

64. See the passages cited in the previous note, especially Dufay, "Lumière des diamants," 355 (twenty carats), 356 ("once had the occasion").

65. Georges-Louis Leclerc, Comte de Buffon, Histoire naturelle des minéraux, vol. 4 (Paris, 1786), 267.

66. Ibid.

67. Hellot notebooks, vol. 8, 193v.

68. These notes are scattered across five sides in "Notes sur l'électricité," AS, Dufay dossier, in folder entitled "Donner à l'abbé Nollet."

69. The full list of donors named is "M. le Duc," "Me. la Duchesse," "Mr. Decleves," "Mr. Deymout," "Mr. Philippe," "Mr. Dénery," "Me. Durf," "Baron," "Bignon," "Geoffroy," and "Duhamel." Dénery and Durf are listed in Antoine-Joseph Dezallier d'Argenville, La conchyliologie, ou histoire naturelle des coquilles de mer, d'eau douce, terrestres et fossiles (Paris, 1780), 797, 808;

duchess, including a single jewel containing multiple diamonds of different colors. Dufay's trials on this jewel give a flavor of his experiments: "The fleece (*toison*) of Mr. le Duc being exposed [to the sun], the blue diamond, the large white one and the yellow sheep (*mouton*) were luminous in the dark. Some of the white ones and the small red ones were [luminous] as well, but two or three in a peculiar way, and much more than the others. The red one and three-quarters of the white ones were not [luminous]." The references to the "fleece" and "sheep" must refer to the insignia of the Order of the Golden Fleece, the prestigious order of chivalry to which the duke belonged. A wedding ring set with diamond and amethyst was borrowed from another of Dufay's acquaintances. He also used his own jewels: "my rock crystal buckles, my whole case of rings (*tout mon baguier*), and some alabaster buckles, gave nothing" when exposed to the sun and then observed in the dark. He even experimented on a stone that he called "my best diamond," exposing it to the sun then enclosing it in soft black wax to see how long its glow lasted.

As these examples indicate, there was no single location for Dufay's experimental gems. They were just as likely to come from personal collections of jewelry as from natural history cabinets. Dufay did have such a cabinet, one that he described briefly in his 1736 article on the luminosity of diamonds. He had been "working for several years to form a cabinet of fine stones, of jaspers, agates, prisms, and singular crystallization," all of which were "arrayed in drawers divided into compartments."[70] He added that he had tested the luminosity of "a great number of stones" from his cabinet. He also named those that gave the brightest glows when exposed to sunlight. These were four "singular crystallizations," or metals encrusted with crystals of different forms and colors; false emeralds, rubies, and topaz; and assorted other stones, namely rock crystal, talc, selenite, marble, lapis lazuli, and a Bologna stone. This cabinet presumably bore some resemblance to the collection of Dufay's somewhat older contemporary, John Woodward, a collection that is still extant and that contained rings, agates, and rock crystal (plate 3).[71] Dufay did not mention any diamonds in his cabinet, however, nor any emeralds, rubies or sapphires. He almost certainly owned these species, but he probably kept

---

Laissus, "Cabinets d'histoire naturelle," 685, 698. "Duhamel" was surely Dufay's collaborator Henri-Louis, Duhamel Du Monceau. "Geoffroy" probably referred to Claude-Joseph Geoffroy, and not his elder brother Etienne-François, since the latter died in 1731. I have not been able to identify "Mr. Decleves," "Mr. Deymout," or "Mr. Philippe."

70. Dufay, "Lumière des diamants," 357.

71. On Woodward on gems, see Hillman, "Invisible Labour in the Woodwardian Collection," 50–53; and James Evans, "The First Identification of Spinel," *Gemmology Bulletin* (2020): 2–7.

them elsewhere, as in the "case of rings." It is hard to know exactly where they were kept, since there are no surviving inventories of Dufay's possessions.

One inventory that does survive is that of Claude-Joseph Geoffroy, a friend of Dufay's and the owner of a considerable number of gems. An inventory of Geoffroy's possessions was drawn up on the occasion of his death in 1753. In this document, gems crop up in several places in the "cabinet of natural history," a room in Geoffroy's house dedicated to the plant, animal and mineral kingdoms.[72] This room housed a wooden cabinet made up of at least eighteen drawers, one of which contained "several precious stones," including diamond, ruby, sapphire, topaz, opal, turquoise, cat's eye, and emerald.[73] This was probably the drawer that Dufay had in mind when he wrote that "only one of M. Geoffroy's pieces of emerald was luminous," and that "all the other stones that were in the sixth drawer were not [luminous] at all."[74] In the same room, a shelf held 108 jars containing pieces of "jacinth, garnet, pebbles from Medoc, amethyst and various crystals."[75] Another cabinet in the room housed what appears to be a collection of hard stones, including agate, onyx, carnelian, rock crystal and lapis lazuli.[76] Pieces of agate, coral and jet were scattered through other shelves and drawers in the room. All this was distinct from Geoffroy's silverware (*argenterie*), which had its own section in the inventory.[77] There we find a rich collection of rings mounted in precious stones, including diamonds, sapphire, opal, hyacinth, and garnet. These were accompanied by a pair of diamond earrings worth 1,400 livres, two-thirds the annual salary of senior members of the academy at the time. Geoffroy owned many gems, but the phrase "gem collection" is misleading insofar as it suggests a single set of objects housed in the same piece of furniture. The same probably applies to Dufay's gem collection.

This culture of collecting supplied more than just raw materials for Dufay's experiments. It supplied other kinds of hardware, such as the drawers in Dufay's cabinet. These were "very useful for doing a large number of

---

72. Probate inventory of Claude-Joseph Geoffroy, AN MC/CVIII/505, June 25, 1753. This document was brought to light by David Sturdy in *Science and Social Status: The Members of the Académie des Sciences, 1666–1750* (Woodbridge: Boydell Press, 1995), 339.

73. Probate inventory of Claude-Joseph Geoffroy, item 52 under "Cabinet d'histoire naturelle."

74. Dufay, "Notes sur l'électricité."

75. Probate inventory of Claude-Joseph Geoffroy, item 54 under "Cabinet d'histoire naturelle."

76. Ibid., items 69, 70, and 72 under "Cabinet d'histoire naturelle."

77. Ibid., under "Sur l'argenterie."

experiments simultaneously, and without losing time."[78] Apparently, Dufay removed each drawer successively from the cabinet, exposed the contents of the drawer to the sun's rays, then carried the drawer into a dark room to compare the glows of the objects in the compartments. This was quicker than testing each sample separately; it was also more telling, since each sample was treated in the same way. This method could also lead to unexpected discoveries. As Dufay remarked, he only tested lapis lazuli for luminosity because it happened to be in the same drawer as the hard, transparent stones that seemed more likely candidates. Testing the whole drawer at once, he was surprised to find that lapis lazuli was as luminous as any of the others.[79] The practice of keeping multiple samples of the same species was another aid to experimentation, since it allowed Dufay to determine the effect of variables that were otherwise hard to isolate. He owned a series of rock crystals that differed only in their color, allowing him to test the effect of the color of a gem on the intensity of its glow.[80]

Finally, Dufay's connoisseurial eye meant that he was a keen observer of the visual properties of gems. His published and unpublished writings are strewn with remarks on the qualities of the gems he tested: "a very beautiful peach-blossom diamond," "a blue diamond that is fairly well-spread but full of points," "a small white diamond that has a rose cut but is rather mean-looking (*vilain*)."[81] These remarks would not have been out of place in the inventory of the crown jewels compiled under Louis XIV and discussed in chapter 2. Nor were they out of place in Dufay's experimental articles, where the shape, color, and visual subtleties of gems were all-important. He paid close attention to the color changes undergone by gems heated in a crucible.[82] He kept track of the cut, color, hardness, and geographical origin of the gems he tested for luminosity, thereby showing his knowledge of the jargon of jewelers. He distinguished oriental stones from occidental ones, high-quality lapis lazuli from the common variety, and Auvergne topaz from the oriental, Indian, and Bohemian varieties of the stone.[83] He was a skilled observer of the faint glows of diamonds and other precious stones. He maximized his chances of seeing them by closing one eye when exposing a diamond to the

78. Dufay, "Lumière des diamants," 356–57.
79. Ibid., 358–59.
80. Ibid., 367.
81. Dufay, "M6. Électricité et lumière," 516; idem, "Notes sur l'électricité."
82. Fontenelle, "Pierre de Berne," 59–60.
83. Dufay, "Lumière des diamants," 358–59, 371.

sun; he then opened that eye, and closed the other, when he observed the diamond in the dark.[84]

This was an ingenious precaution, as historians have recognized, but we should not ignore the visual expertise that Dufay acquired simply by observing many gems for the purpose of pleasure. According to Fontenelle, he had "a great appetite for purely ornamental things [*choses de pur agrément*]."[85] He may be compared to Antoine de la Roque, a prominent writer in Paris who built a dazzling collection of colored stones during Dufay's lifetime. La Roque had "every merit as a connoisseur," Gersaint wrote in his catalog of the collection, published in 1745.[86] Gersaint went on to paint a portrait of the gem connoisseur in his cabinet: "All these stones, even the smallest, have a distinctive value that is relative to their quality. He was so sensitive to their beauties that he never dared to admire them by wearing them on his finger. . . . It was enough to examine them at home, among his cases of rings (*baguiers*), where he admired the pleasant variety and harmony that he found in their rich colors . . ."[87] Of course, Dufay not only wore gems on his fingers but also rubbed them, exposed them to the sun, and roasted them in a crucible. But it is easy to imagine him admiring the gems in his baguier in much the same way as la Roque.

Dufay continued to experiment on gems until the last few months of his life, when he made a remarkable discovery concerning double refraction. As naturalists had known for several decades, crystals of Iceland spar divide rays of light in two.[88] One of the rays obeys the ordinary law of refraction; the other ray, known as the "extraordinary" ray, does not. The effect had also been observed in rock crystal, as Dufay reminded the academy in a paper he began reading in July 1738. He continued reading for another five sessions, indicating that this was a substantial piece of work.[89] Unfortunately, the text was

84. Ibid., 353–55.

85. Fontenelle, "Éloge de Dufay," 79.

86. Edmé-François Gersaint, *Catalogue raisonné des différens effets curieux & rares contenus dans le cabinet de feu M. le Chevalier de la Roque* (Paris, 1745), vii.

87. Ibid., 111.

88. On early studies of double refraction see Leó Kristjánsson, *Iceland Spar and Its Influence on the Development of Science and Technology in the Period 1780–1930* (Reykjavík: University of Iceland, 2010), chaps. 3 and 4; Jed Buchwald, "Experimental Investigations of Double Refraction from Huygens to Malus," *Archive for the History of Exact Sciences* 21, no. 4 (1980): 311–73; and Hélène Metzger, "Une théorie curieuse de la double réfraction chez Buffon," *Bulletin de la société minéralogique* 37 (1914): 162–76.

89. *PV* 1738, July 16, 161r; July 19, 162r; July 23, 163r; July 30, 167r; Aug. 9, 171v; Aug. 13, 173r. Cf. Geoffroy to Sloane, Aug. 26, 1738, HS Ms. 2055, ff. 372–73.

not published in the academy's journal; nor was it recorded in the minutes of the meetings. The only record of Dufay's findings is a few lines in Fontenelle's éloge. Yet this is enough to show that Dufay had stumbled upon something important. He had looked for double refraction in "all transparent stones." He had found it in many of these stones. And he had reduced his findings to a rule: "all right-angled transparent stones have only one refraction, and all those whose angles are not right have a double [refraction] whose size depends on the inclination of [the stone's] angles."[90]

In a sense, this result was ahead of its time. Dufay "had taken the first steps toward a correlation of crystal form and optical properties," in the words of a modern mineralogist.[91] In another sense, the result was very much a product of its time. It was the culmination of an experimental program that Dufay had been pursuing since at least 1724: "examining, by every means I could think of, the nature of all fine stones." This program used the material culture of natural history—gem collections, orderly cabinets, and fine visual judgments—to study phenomena that we now associate with experimental physics. Réaumur had adapted material-driven experimentation to gems; Dufay adapted gem-driven experiments to the study of light and electricity.

Electricity as a Science of Materials

What does this have to do with the laws of electricity summarized at the start of this chapter? One answer is that gems cropped up in key passages in Dufay's articles on electricity. They were part of his electrical research from the start, as amber had been recognized since antiquity as a body that attracts light bodies when rubbed. This property was still the defining feature of electricity in the 1730s—batteries, current electricity, and the concept of electric charge were several decades away.[92] In addition, Dufay used diamonds to investigate the link between light and electricity; rock crystal to establish the distinction between two kinds of electricity; and a ruby as a bearing in a device designed to look for connections between electricity and gravity.[93] But there is a deeper point here, one regarding the structure of the investigation rather than the use of this or that material. Dufay studied electricity in the same way that Réaumur had studied steel and porcelain. He took a large collection of

90. Fontenelle, "Éloge de Dufay," 81.
91. Adolf Pabst, "Charles-Francois du Fay, a Pioneer in Crystal Optics," *American Mineralogist* 17 (1932): 570.
92. Note Dufay's definition of electricity in "M1. L'histoire de l'électricité," 23.
93. "M6. Électricité et lumière," 512–17; "M4. Attraction et répulsion," 464–69; "M8," 321.

materials, gems among them; he did the same thing to each material; and he looked for patterns in the results. To see this, we need to take a fresh look at the articles in question.[94]

There are six points to make here. The first is that *Dufay relied on classification to show that electricity is a general property of matter.* He reached this conclusion after taking various bodies, rubbing them and observing whether they attracted pieces of brass or gold leaf.[95] It has been said that he tried everything, but that is not literally true. He was well aware that he could not test every object on the surface of the earth. Instead, he divided bodies into classes and tested a few items from each class. He knew exactly what he was doing: "I was content to try a certain number of each species [*espèce*], and we can reasonably suppose that the same goes for all the others."[96] His classes were a microcosm of his career in chemistry and natural history (table 4.1). The distinction between salts, metals, stones, and bituminous or resinous substances, was common in the natural history of minerals at the time; so was the distinction between opaque gems and transparent ones. Dufay was familiar with shellac, gum copal, Chinese varnish, and several kinds of gum and resin, as a result of his work on lacquer.[97] The woods and vitrifications are a reminder of his directorship of the Jardin du Roi and of Réaumur's efforts to make porcelain. All these bodies could be made to attract light bodies when rubbed, with the exception of metals and bodies that were too soft to rub. Dufay suspected that even these bodies would, with the right preparation, behave in the same way, and hence that electricity was "a property common to all bodies."[98] Further trials convinced him that the capacity to become electrical by communication—that is, by being placed near a body already electrified by friction—was also a general property of matter.[99] His confidence was based on the heterogeneity of the bodies he tried, which was in turn based on his ability to place them into classes. Classification was an aid to generalization.

Having made this generalization, Dufay drew a fundamental distinction between two kinds of material. This was an early version of our distinction between conductors and insulators. The point here is that *the distinction between conductors and insulators began as an act of gem classification.* Dufay

94. This section summarizes Bycroft, "Physics and Natural History," chap. 4.
95. "M2. Corps susceptibles," 74–81.
96. Ibid., 78.
97. For details, see Bycroft, "What Difference Does a Translation Make?"
98. "M2. Corps susceptibles," 74.
99. Ibid., 81–84.

TABLE 4.1. Dufay's electrics

Classes of material that Dufay heated and rubbed in the hope that they would become electric, such that they attracted light bodies such as gold leaf. They all became electric in this way except those in classes 11 and 13. Note also the gems in classes 1, 2, 3, 5, and 6. The practice of performing the same operation on a large and ordered collection of objects was widespread at the early Paris Academy.

1. *Materials already known to be electric*
amber
resins
bituminous substances
precious stones
carnelian

2. *Resinous, bituminous, and fatty materials*
amber
jet
asphalt
gum copal
shellac
pine resin
plant resin
common sulfur
white wax
Chinese varnish

3. *Transparent precious stones*
white diamond
colored diamonds
garnet
peridot
cat's eye
sapphire
ruby
topaz
amethyst
rock crystal
emerald
opal
jacinth

4. *Vitrifications*
white and transparent glass
porcelain
earthenware
varnished earth
lead glass
antimony
copper
other vitrifications

5. *Transparent stones*
Venetian talc
Muscovy talc
Berne phosphor
gypsum
transparent selenite
all transparent stones of whatever nature

6. *Opaque stones*
carnelian
agates and jaspers of all kinds
porphyry
granite
marbles of all colors and degrees of hardness
lodestone
sandstone
slate
building stones

7. *Thread-like materials*
silk
wool
thread
cotton
feathers
head hair
hairs of all animals dead or alive

8. *Miscellany, already tried by others*
paper
parchment
leather

9. *Miscellany, not already tried by others*
straw
all dried herbs
ivory
bones
antlers
fish scales
whale bone
shells of all kinds

10. *Wood*
boxwood
ebony
lignum vitae
sandalwood
oak
elm
ash
limewood
fir
wicker
cork

11. *Bodies that melt during rubbing*
aqueous gums
strong glue
fish gum

12. *Salts*
alum
rock candy

13. *Metals*

Source: Data from Dufay, "M2. Quels sont les corps qui sont susceptibles d'électricité" (1735).

stated the distinction for the first time in a study of hard stones. The passage is worth quoting in full:

> by this means [heating followed by rubbing] I electrified all the species [*es-pèces*] of agates and jaspers that I tried, as well as porphyry, granite, marbles of all colors and all degrees of hardness, the lodestone, sandstone, slate, and building stones; so that I believe it would be very hard to find a species of stone that cannot be rendered electric by this means. It is true that one can identify two classes [*classes*] into which all stones can be placed; some are electric without any further preparation than rubbing, while others, such as jaspers, opaque agates, and the hardest marbles, need to be heated in advance, sometimes very vigorously; these need to be very hot and rubbed for a long time, and [even then] the electricity they acquire is weak: It seemed to me that the hardest stones needed more heating, and became less strongly electrified, than the others; black marble, for example, is less electric than white, and white marble less so than building stones.[100]

The distinction drawn here was between bodies that can be electrified by rubbing alone, and those that cannot be electrified in this way. Dufay usually called the former bodies *électriques*, so I shall refer to them as "electrics" and to the other category as "nonelectrics." For Dufay, nonelectric bodies included those that need to be heated in advance (such as jasper) and those that cannot be electrified by rubbing even when hot (such as metals). As the quoted passage shows, Dufay introduced this as a distinction between two classes of mineral. Just as stones could be hard or soft, precious or non-precious, opaque or transparent, they were now either easy or difficult to electrify by friction. Moreover, Dufay developed this distinction by analogy to hardness, noting that soft stones were easier to electrify than hard ones. His distinction was not entirely new—a similar distinction had long been implicit in the assumption that electricity was confined to a small number of bodies. But Dufay's finding that nearly all bodies can be made electrical called for a new distinction, not between those that can and cannot be made electrical but between those that are easy to electrify and those that are not. Hence the importance of the quoted passage.

Even more important was Dufay's use of this distinction. He immediately used it to state what he called "one of the general laws of electricity."[101] This brings us to a third point: *the most important law in Dufay's electrical research has been misidentified by historians.* The law emerged from his efforts to electrify bodies by communication. He placed these bodies on little stands, noting

100. Ibid., 77.
101. Ibid., 84.

that bodies are easier to electrify when the stands are made of electrics such as glass rather than nonelectrics such as copper. This has recently been called the "rule of Dufay."[102] But it was not the "general law" that Dufay had in mind. There was nothing very surprising about the fact that electrics promoted electricity when they were used as stands—they promoted electricity almost by definition. And Dufay certainly found the "general law" surprising: "This observation is not the kind that one can foresee, and it derives from experiments alone."[103] The surprise did not lie in the material of the stands but in the materials that stood upon them. One might expect that electrics such as amber would be easier to excite by communication than nonelectrics such as copper. But Dufay found the opposite to be the case. Copper became *more* electric than amber in the presence of a rubbed glass tube, when the same stand was used for the two materials. As he later wrote, "I constantly remark'd, that such Bodies as of themselves were least Electrical, had the greatest Degree of Electricity communicated to them at the Approach of the Glass Tube."[104] In the original paper, Dufay referred to "contraction" (*contractor*) rather than "communication." I shall therefore refer to his discovery as the "contraction law." This law, he wrote, "may shed more light than any other on the nature of electricity, as we shall see from the use I shall make of it in what follows."[105]

And use it he did. Hence the fourth point: *Dufay repeatedly invoked the contraction law to explain the contrasting behavior of different materials.* He began with the materials he used as stands. Why are bodies harder to electrify by communication when they are placed on a copper chandelier than when they are placed on a block of glass?[106] Because, Dufay explained, the copper soaks up the electricity that would otherwise be available to the body on top of it.[107] In other words, because the copper "contracts" electricity and retains

102. Ibid., 82–83; Heilbron, *Electricity*, 253; Pieter Present, "Petrus van Musschenbroek (1692–1761) and the Early Leiden Jar: A Discussion of the Neglected Manuscripts," *History of Science* 60, no. 1 (2022): 106.

103. "M2. Corps susceptibles," 83.

104. Dufay, "Letter Concerning Electricity," 259. Dufay wrote the same thing, though in a somewhat garbled manner, in his 1733 paper: "M2. Corps susceptibles," 83.

105. Dufay, "M2. Corps susceptibles," 83.

106. Ibid., 82.

107. This is implicit in Dufay's emphasis on the volume of supports in "M2. Corps susceptibles," 82. It is explicit in Dufay, "M8," 325, where he explained why a cork ball contracts more electricity when hung from a silk thread than when hung from a length of string. There is a similar explanation in "M4. Attraction et répulsion," 475. Dufay sometimes wrote about the "dissipation" of electricity through a room, but only when he was considering the volume rather than composition of his materials; an example is in "M3. Corps attirés," 250.

it. More explanations followed in Dufay's third paper. Why does an electrified tube attract gold leaf when a glass pane is interposed between the tube and the leaf, but not when the glass pane is replaced by a sheet of metal? Again, because the metal soaks up electricity whereas the glass does not.[108] The sensitivity of metals to the glass tube could be explained in the same way—they acquired more electricity from the tube and therefore were more strongly attracted to it.[109] "We see here that the same laws apply everywhere," Dufay concluded, "and the bodies that are electric on their own [*électriques par eux-mêmes*] are those that are least inclined to block, retain, or absorb the flow of electricity [*les écoulements électriques*]."[110] Similar reasoning accounted for the ease with which metals drew light from excited bodies and the ease with which they delivered sparks to nearby bodies.[111] The contraction law lay behind six of the sixteen principles of electricity that Dufay summarized at the end of his sixth article on electricity.[112] He reminded readers of the law in the very last paragraph he published on electricity, where he summed up his experiments on cork balls suspended by silk lines. Silk, "being a material more apt than string to become electric on its own, is for this reason less disposed to receive [*recevoir*] electricity from the ball, as I have demonstrated by a large number of experiments."[113]

What about the ten other principles on Dufay's list? In particular, what about the discovery of shocks, sparks, and the two electricities, the achievements that take up most space in existing summaries of Dufay's electrical research? These were genuine achievements, but they have overshadowed Dufay's own vision of a science of electricity based on the contraction law. The irony is that *Dufay's well-known discoveries were all by-products of the contraction law*. Consider the two electricities.[114] This discovery can be traced back to Dufay's experiments on colored ribbons, which had shown that nonelectrics are more strongly attracted to a glass tube than electrics. This was a fairly straightforward application of the contraction law. The nonelectric ribbons contracted more electricity from the tube than the electric ribbons and were therefore more strongly attracted to it. Dufay then wondered how the electrified tube would affect a body that it had electrified by contact rather

---

108. Dufay, "M3. Corps attirés," 236, cf. 239–42.

109. This is implicit in ibid., 236.

110. Ibid., 243.

111. "M6. Électricité et lumière," 509, 518, 520–23.

112. Ibid., 523–25, principles 3, 4, 5, 7, 8, 11.

113. "M8," 325; italics added.

114. For a similar analysis of Dufay's other discoveries, see Bycroft, "Physics and Natural History," 137–39.

than at a distance. Suppose a piece of gold leaf is dropped onto the tube—would the leaf stick to the tube or be repelled by it? Dufay found that it was repelled.[115] Next, he wondered whether other materials would behave in the same way as the glass tube.[116] He was looking for a distinction analogous to the one between electrics and nonelectrics, asking: "Do electrical bodies differ among themselves only by their various degrees of electricity?"[117] He reached for one of the ingredients for Chinese varnish, gum copal. To his surprise, he found that copal *attracted* the gold leaf that had been *repelled* by the glass tube. There seemed to be two classes of material, such that materials of one kind attract bodies that are repelled by materials of the other kind. This would later be understood as a distinction between "positive" and "negative" electricity, but it began as a distinction between two kinds of material. Hence Dufay's names for the two "species" of electricity he had identified, "vitreous" and "resinous."[118] In subsequent work, Dufay treated this distinction in the same way as the distinction between electrics and nonelectrics. He added new materials to each class, including rock crystal on the list of vitreous substances and wax and amber on the list of resinous ones.[119] And he observed these materials in new situations, looking for differences between the two classes. When he rubbed diamonds to generate light, for example, he compared the behavior of these vitreous bodies to resinous ones such as gum, wax, amber, and sulfur.[120] He was looking for "yet another difference between the two electricities."[121] He searched for differences between amber and glass in the same way that he had earlier searched for differences between amber and copper.

The upshot was a set of empirical rules concerning the contrasting behavior of different groups of material. But these rules were not merely empirical. *The contraction law was mixed up with a mental picture of electricity that differs from the modern one.* Today we think of metals as "conductors" of electricity, as bodies through which electricity flows easily. By contrast, we say that glass is an "insulator" through which electricity flows with difficulty or

---

115. "M4. Attraction et répulsion," 458.

116. The following expands on accounts of the discovery of the two electricities in Friedrich Steinle, "Exploratives experimentieren: Charles Dufay und die Entdeckung der zwei Elektrizitäten," *Physik Journal* 3, no. 6 (2004): 47–52; and John L. Heilbron, "Dufay, Charles-François de Cisternai," in *DSB*, vol. 4, 215.

117. "M4. Attraction et répulsion," 464.

118. Ibid., 467, 469; "M6. Électricité et lumière," 524.

119. "M4. Attraction et répulsion," 464–69.

120. "M6. Électricité et lumière," 504–20.

121. "M5. Nouvelles découvertes," 355.

not at all. But this is not what Dufay had in mind. For Dufay, electricity does not flow *through* a piece of metal; it flows *into* the metal and stays there. That was the crux of the contraction law, which stated that metals (and other non-electrics) contract electricity more easily than glass (and other electrics). It is significant that he used the verb "contractor" interchangeably with the verb "absorber." It is also significant that he referred to phosphors as "sponges" that soak up light.[122] True, he did not use this metaphor explicitly for electrics. But that usage was attributed to him after his death.[123] And the metaphor makes sense given the starting point of his investigation. His paradigm case of electricity was a piece of amber sitting on a stand, not a piece of wire or thread hung from a ceiling. So, it is not surprising that he saw contraction, rather than long-distance transmission, as the fundamental electrical phenomenon. This is just what one would expect from a gem collector whose mentor, Réaumur, had an abiding interest in steel, porcelain, and other "spongy" materials. More generally, Dufay's material-driven study of electricity is just what one would expect at an academy that had been building collections and experimenting on them since 1667.

The Varieties of Matter

This chapter has taken us well beyond gems, but this is precisely because gems were so thoroughly integrated into the academy's activities in the early decades of the eighteenth century. Gems are therefore a window onto the academy as a whole. The view from this window may be compared to other perspectives on academic experimentation that historians have offered recently. One view is that natural history and experimental physics went together at the academy. "Physics should not be separated from natural history," as one of Dufay's biographers, the physicist Jean Becquerel, put it two centuries later.[124] This is certainly a good lesson to draw from Dufay's career, but we need to be clear about what it means. It sometimes meant using instruments such as microscopes and thermometers to study plants, animals, and minerals, as Mary Terrall has shown.[125] But it also meant using the tools of natural history to study light and electricity, tools such as large and well-organized

122. Phosphors in "M2. Corps susceptibles," 74. Verbs in ibid., 83; "M3. Corps attirés," 243, 254.

123. Jean-Antoine Nollet, "Observations sur quelques nouveaux phénomènes d'électricité," *MAS* 1746 (1751): 12. The "able man" referred to here was almost certainly Dufay.

124. "La physique ne doit pas être séparée des sciences dites 'sciences naturelles.'" Jean Becquerel, "Chaire de physique appliquée à l'histoire naturelle," *Archives du Muséum national d'histoire naturelle* (1935): 99.

125. Terrall, *Catching Nature in the Act.*

collections.[126] Partly for this reason, the academy's investigations were not loose or unstructured or miscellaneous, connotations that historians often attach to the phrase "experimental history."[127] They were highly structured, precisely because they were based on collections that were organized into classes. They were not quite "exploratory experiments," in the sense that the historian and philosopher Friedrich Steinle has given that phrase.[128] They could be used to test and explain regularities and not just to discover them. Insofar as they were exploratory, they were a particular kind of exploration in which materials were the source of variation, in contrast to instrument-driven experiments such as Francis Hauksbee's on electricity and Boyle's on air.[129] Furthermore, material-driven experimentation was not a novelty in the eighteenth century. It therefore cannot be explained in terms of the ejection of wonders and marvels from respectable science around 1800.[130] The novelty in the eighteenth century was to extend this method to minerals, including gems. It was an old method applied to new materials.

Moreover, these new materials were themselves heterogeneous. Gems, steel, earths, porcelain, metallic ores and flintstones had different properties and uses and associated arts. Techniques originating with one material could migrate to other materials, as when Réaumur took fracture analysis from iron-working and applied it to porcelain and flintstones. This was not just a matter of improving the techniques of the arts, nor of communicating them to the public. It was also a matter of using those techniques as investigative tools, especially evaluative techniques such as assaying, gem appraisal, fracture analysis, and the decomposition of mineral waters.[131] This process is only partly captured by the thesis that academicians mediated between natural

126. Ibid.

127. For example, Kuhn, "Mathematical Versus Experimental Traditions," 12; and Klein and Lefèvre, *Materials in Eighteenth-Century Science*, 22–28.

128. Steinle, "Exploratives experimentieren"; idem, "Entering New Fields: Exploratory Uses of Experimentation," *Philosophy of Science* 64 (1997): S65–S74. See also Present, "Musschenbroek and the Early Leiden Jar," esp. 127–28. When exploratory experiments focused on materials, they had much in common with what I am calling material-driven experimentation. Here is an example: Pieter T. L. Beck, "Strong Foundations: Petrus van Musschenbroek's Experimental Research on the Strength of Materials," *Historical Studies in the Natural Sciences* 53, no. 2 (2023): 109–46.

129. For the contrast between Dufay and English electricians, such as Hauksbee, see Bycroft, "Physics and Natural History," 139–43.

130. Licoppe, *Formation*, 113–16; Lorraine Daston, "The Cold Light of Facts and the Facts of Cold Light: Luminescence and the Transformation of the Scientific Fact, 1600–1750," *Early Modern France* 3 (1997): 1–27; Daston and Park, *Wonders and the Order of Nature*, 352.

131. Cf. Marc J. Ratcliff, "Experimentation, Communication and Patronage: A Perspective on René-Antoine Ferchault de Réaumur (1683–1757)," *Biology of the Cell* 97, no. 4 (2012): 231–33.

philosophy and the mechanical arts.[132] And it is only partly captured by the thesis that they became true philosophers by liberating themselves from this or that art, such as medicine.[133] Rather, their philosophy consisted in mediating between different arts and their associated materials. They were transmaterialists, specialists in exploring the varieties of matter.

One thing they did not do is write natural histories of minerals along the lines of Boodt's *History of Gems* or Berquen's *Marvels of the Indies*. It is true that Réaumur sketched out a classification scheme for minerals, based on his study of flintstones, but this remained a sketch rather than a treatise. Dufay took for granted the received classification of gems in his articles on electricity. He made no effort to work out a new classification of gems based on the phenomena—electricity, phosphorescence, double refraction—he had explored so thoroughly. This effort began after Dufay's death in 1739, using two of Dufay's legacies: the collection of precious stones at the Jardin du Roi and the field increasingly known as la physique expérimentale.

My point is that "communication" of trade secrets and rigorous "experimentation" were part of the same process.

132. As argued in Charles C. Gillispie, "The Natural History of Industry," *Isis* 48, no. 4 (1957): 398–407; and Paola Bertucci, *Artisanal Enlightenment: Science and the Mechanical Arts in Old Regime France* (New Haven, CT: Yale University Press, 2017).

133. As suggested in Principe, *Transmutations of Chymistry*, 413.

5

Precision and Preciousness in
Enlightenment Mineralogy

Are emeralds precious? Not for the French naturalist Georges-Louis Leclerc, Comte de Buffon. Genuine precious stones, he wrote in 1785, are those made of a simple and homogeneous material. The measure of the simplicity of a substance, he continued, is the optical phenomenon known as double refraction. Since emeralds have two refractions, they cannot be perfectly homogeneous and therefore cannot properly be called precious stones (*pierres précieuses*).[1] This conclusion was idiosyncratic, but Buffon's approach to precious stones was typical of European naturalists in the second half of the eighteenth century. They believed that precious stones were as real as any other category of mineral, and that preciousness was as real as any other property of minerals. But they hoped to improve judgments of preciousness by making them more exact. The qualities of gems were still tied to their natures, but now both were tied to numbers. The qualification of gems was now a form of quantification.

Gems were therefore part of the "quantifying spirit," the wave of quantification that spread across the natural sciences in the latter part of the eighteenth century. Historians have shown that the quantifying spirit was normative: numbers were used to evaluate everything from air to brandy to territory, not to mention humans and human societies.[2] Gems confirm this finding. They also show that natural history and experimental physics continued to

1. Georges-Louis Leclerc, Comte de Buffon, *Histoire naturelle des minéraux*, vol. 3 (Paris, 1785), 558; vol. 4 (1786), 247, 250–51, 626.

2. Here are some representative examples: Ashworth, "British Alcohol Standards"; Schaffer, "Measuring Virtue"; Sumner, *Brewing Science*; Arne Hessenbruch, "The Spread of Precision Measurement in Scandinavia 1660–1800," in *The Sciences in the European Periphery in the Enlightenment*, ed. Kostas Gavroglu (New York: Springer, 1999), 179–224; John L. Heilbron, "The Measure of Enlightenment," in Frängsmyr et al., *Quantifying Spirit*, 207–44; Svante Lindqvist,

interact in the latter part of the century, even once la physique expérimentale had emerged as a well-defined field. Natural history not only helped to create this field, as we saw in the previous chapter; it also helped to make the field numerical. To see this, we need to bring together two Enlightenment worlds that historians usually study separately. On the one hand, there is the world of precise instruments and numerical tables, with its links to artisans and public spectacle. On the other hand, there is the world of specimens and classification schemes, with its links to trade, empire, and connoisseurship in the fine arts.[3] Both worlds were involved in the quantification of gem quality. Concepts such as "nuance," "structure," "variety," and "correlation" helped to fuse the two worlds, because they applied equally to classes and to numbers.

The end result was a reclassification of gemstones. The old properties of color, locality, and transparency gave way to new criteria for placing gems into groups. Density, refraction, and crystal form were prominent among the new criteria, with electricity and magnetism playing supporting roles. These properties were not new in any absolute sense. They had been studied for centuries, if not millennia, including by naturalists interested in gems. They had been studied by experimenters like Boyle and Dufay in the decades around 1700, as we saw in the last two chapters. The novelty was to move these properties from the margins of gem classification to the center. The most visible outcome was the creation of a science of crystallography in France in the last quarter of the century. But this science did not come out of nowhere. At least in France, it was the culmination of several decades of gem quantification, starting with the experiments on double refraction done by Dufay in the 1730s and noted in the previous chapter. The key site was the Jardin du Roi, with its large gem collection and its proximity to the experimenters at the academy; private collectors and enterprising cutters also played a role. All this was bound up with the invention of mineralogy, a science that is usually traced to Swedish botany and German mining but which owed just as much to French jewelry.[4]

"Labs in the Woods: The Quantification of Technology During the Late Enlightenment," in Frängsmyr et al., *Quantifying Spirit*, 291–314.

3. These two worlds are surveyed in Curry et al., *Worlds of Natural History*, pt. 2; Buchwald and Fox, *History of Physics*, pt. 2. On connoisseurship and natural history, see Bycroft and Wragge-Morley, "Science and Connoisseurship."

4. Swedish botany in Burke, *Origins of the Science of Crystals*, 57–59; Laudan, *Mineralogy to Geology*, 71–78. German (and Swedish) mining in Theodore Porter, "The Promotion of Mining and the Advancement of Science: The Chemical Revolution of Mineralogy," *Annals of Science* 38, no. 5 (1981): 543–70; Fors, *Limits of Matter*, chap. 5; Oldroyd, *Sciences of the Earth*, passim. The medical side of Enlightenment mineralogy is covered in Matthew Eddy, *The Language of Mineralogy: John Walker, Chemistry and the Edinburgh Medical School, 1750–1800* (London: Routledge,

From Lapidaries to Mineralogies

It may seem odd to say that mineralogy was a new science in the eighteenth century. Minerals had been studied since ancient times, in Europe and elsewhere; they had attracted the attention of Parisian savants from around 1700, as we saw in the previous chapter. But there is a real sense in which mineralogy did not exist before about 1750. Only in 1785 did the Paris Academy of Sciences created a class of researchers dedicated to "minéralogie," as opposed to "botanique" or "histoire naturelle."[5] To continue with the French case, the first major work with the word "minéralogie" in the title appeared in Paris in 1762.[6] Earlier works had classified minerals, but they had not made classification their main priority. Even *The Perfect Jeweler*, the French translation of Boodt's *History of Gems and Stones*, was as much about the uses, virtues, and origins of gems as about their division into classes. In addition, earlier French works had tended to deal with particular regions of the mineral kingdom, such as stones or metals, rather than covering the mineral kingdom as a whole.[7] The 1762 work, by contrast, was dedicated to arranging the entire mineral kingdom into classes—or rather, into classes, orders, genera, subdivisions, species, and varieties, to use the author's technical terms. The scheme was not only described in the text but also summarized in ten "synoptic tables," each of which showed the members of an entire class on a single page. The classification scheme was what the author called the "systematic part" (*partie systematique*) of the book.[8] All the rest, including the practical uses of minerals, recent discoveries concerning them, and theories

2016). Other works do consider the French examples but without much attention to gems: Spencer St. Clair, "Classification of Minerals"; Arthur Birembaut, "L'enseignement de la minéralogie et des techniques minières," in Taton and Laissus, *Enseignement et diffusion*, 365–418. My own approach is closest to that in Jonathan Simon, "Mineralogy and Mineral Collections in 18th-Century France," *Endeavour* 26, no. 4 (2002): 132–36; idem, "Taste, Order and Aesthetics," 97–112; idem, "The Values of the Mineral Kingdom and the French Republic," in *Ordering the World in the Eighteenth Century*, ed. Donald Diana and Frank O'Gorman (Basingstoke: Palgrave Macmillan, 2006), 163–89.

5. Roger Hahn, *The Anatomy of a Scientific Institution: The Paris Academy of Sciences, 1666–1803* (Berkeley: University of California Press, 1971), 99.

6. Jacques-Christophe Valmont de Bomare, *Minéralogie, ou Nouvelle exposition du règne minéral* (Paris, 1762). Earlier works with the word in the title, listed on the online catalog of the Bibliothèque nationale de France, are translations of German and English works: Jean Gotschalk Wallerius, *Minéralogie, ou Description générale des substances du règne minéral*, trans. Paul-Henri Thiry, Baron d'Holbach (Paris, 1753 [1747]); Diederich Wessel Linden, *Lettres sur la minéralogie et la métallurgie pratiques* (Paris, 1752).

7. Earlier French treatises that deal with minerals, without being mineralogical in the sense I use here, are listed in Argenville, *Oryctologie*, 1–36.

8. Bomare, *Minéralogie*, vol. 1, vi–vii.

about them, was confined to footnotes. This approach to minerals was widely disseminated in France by Jacques-Christophe Valmont de Bomare, the author of the work in question. As well as writing books on minerals, Bomare gave an annual series of public lectures on the topic that was based on his collection in Paris, and that ran from 1756 to 1788.[9]

Where did Bomare's style of mineralogy come from? Chemistry, botany and Sweden are certainly part of the answer—Bomare's book was derived from a work by Johan Gotschalk Wallerius, a professor of chemistry at the University of Uppsala and a colleague of the great botanist Carl Linnaeus.[10] But equally important was the long tradition of gem science, both in France and elsewhere. Bomare recognized gems as a genuine category, calling it "precious stones and crystals." He also gave a disproportionate amount of attention to this category, giving thirty-five pages to "precious stones and crystals" and another eighteen pages to "agates." Taken together, this is more than he spent on most of the ten classes of minerals he considered in the work, the only exceptions being salts (sixty-five pages), semi-metals (seventy-two) and metals (152). The class of stones, in which most of the gemstones appeared, received more pages than another other class. Even within that class, "precious stones and crystals" received more attention than any other genus. The only genus that came close was the flintstones (*cailloux*), and that only because the chapter on flintstones contained many pages on agates.[11] Bomare's *Mineralogy* was essentially a book about metals and stones, with the stones dominated by gemstones.

9. Jacques-Christophe Valmont de Bomare, *Catalogue du cabinet d'histoire naturelle de M. Bomare de Valmont comprenant les minéraux, végetaux, animaux, & quelques productions, tant de la nature que de l'art* (Paris, 1758); John G. Burke, "Valmont de Bomare, Jacques-Christophe," in *DSB*, vol. 13, 565–66. Bomare's was the earliest French work listed in the surveys in Torbern Bergman, *Manuel du minéralogiste, ou Sciagraphie du règne minéral distribué d'après l'analyse chimique*, ed. and trans. Jean-André Mongez (Paris, 1784 [1782]), xiii–lxxix (survey), esp. xxix (Bomare); and Jean-Baptiste Louis Romé de l'Isle, *Des caractères extérieurs des minéraux* (Paris, 1784), 57–82, esp. 66.

10. St. Clair, "Classification of Minerals," 87n148; Bomare, *Minéralogie*, vol. 1, xiii. Wallerius's work was *Mineralogia eller Mineralriket* (Stockholm, 1747), translated into French as Wallerius, *Minéralogie, ou Description générale des substances du règne minéral*, by Baron d'Holbach. On the linked careers of Wallerius and Linnaeus, compare Sten Lindroth, "Linnaeus, Carl," in *DSB*, vol. 8, 375–76; and Uno Boklund, "Wallerius, John Gottschalk," in *DSB*, vol. 14, 144.

11. Page counts from Bomare, *Minéralogie*, vol. 1, 109–287. The only major difference in the second edition of the work is that gems took up eighty-six pages rather sixty, consistent with the general expansion of the work. The category of precious stones was still intact in Jacques-Christophe Valmont de Bomare, *Dictionnaire raisonné universel d'histoire naturelle*, 2nd ed. (Paris, 1775 [1764]), vol. 6, 684–87.

Moreover, Bomare and his French contemporaries continued to classify gems in much the same way as before. This can be seen by a three-way comparison between Bomare's *Mineralogy*, a similar work by Antoine-Joseph Dezallier d'Argenville from 1755, and Rosnel's *Indian Mercury* from the previous century (appendix 4). On the face of it, Rosnel's book was very different from those by Argenville and Bomare. The goldsmith had focused on gems at the expense of other minerals. And he eschewed taxonomic devices such as synoptic tables and multilayered taxonomies. These differences in form are misleading, however. They conceal a great deal of continuity in content. All three authors drew a basic distinction between two kinds of gem. Rosnel's distinction between "true precious stones" and "agates" resembled Argenville's between "transparent crystalline stones" and "semitransparent crystalline stones." Both resembled Bomare's distinction between "precious stones and crystals" and "agates, or semitransparent pebbles." These categories contained a core set of species that persisted across the three authors. Stones such as diamond, ruby, and garnet populated the first category; stones such as agate, chalcedony, and carnelian populated the second. There were certainly differences between the authors, especially when we consider categories further down the table in appendix 4. Marble, amber, and pearl are interesting cases. Long treated as gemstones, these species were classified by Bomare as a calcareous stone, a flammable substance, and a "fossil foreign to the earth," respectively.[12] Overall, though, there was remarkably little change between Rosnel and Bomare, given the century-long gap between them and the formal differences between their books.

In fact, in some respects the emergence of mineralogy helped to strengthen, rather than weaken, the traditional classification of gemstones. The distinction between hard oriental gems and soft occidental ones is a case in point. This distinction had emerged in the sixteenth century and been consolidated in the seventeenth, as we saw in earlier chapters.[13] It became even more pronounced in the eighteenth century, as a comparison between Rosnel and Bomare shows. Rosnel had mentioned "oriental topaz," "Indian topaz," and "German topaz"; Bomare took this scheme a step further by enumerating just two varieties, "oriental topaz" and "occidental topaz." Whereas Rosnel mentioned two kinds of garnet, oriental and occidental, before going on to describe three other kinds of the same stone, Bomare's systematic approach forced him to clear up the ambiguity, and he did so in favor of the oriental/occidental distinction. Of the eleven precious stones in Bomare's book, all but three were

12. Bomare, *Minéralogie*, vol. 1, 153–62 (marble); vol. 2, 268–72 (amber), 322–23 (pearl).
13. On the earlier history of the distinction, see chaps. 1 and 2, this volume.

divided into oriental and occidental varieties. The distinction circulated widely in France: in dictionaries, encyclopedias, academic articles, manuals for gem connoisseurs, tables of mineral densities, and even in the form of colored illustrations (plate 4).[14] These sources constantly associated the word "oriental" with hardness and "occidental" with softness, just as Boodt, Berquen, and Rosnel had done in the seventeenth century. Eighteenth-century authors were aware that this association was misleading—they were aware, for example, that there were soft stones in Ceylon and Pegu, and hard stones in Brazil and Peru.[15] But they did not abandon the notion of an "oriental" stone. They simply insisted that they were using the term as a proxy for the hardness and quality of gems and not as a literal guide to their geographical origin.[16] In doing so, they were following the language of the cutters and jewelers on whom they relied for data on gem hardness. As Argenville and Bomare explained, "oriental" was the jewelers' term for a hard stone, "occidental" their term for a soft stone.[17]

Some of these ideas were copied from earlier texts, or from recent texts published outside France. Seventeenth-century lapidaries had a long afterlife in the eighteenth century—Berquen and Rosnel were still being cited with approval as late as 1801, in no less a work than René-Just Haüy's *Treatise on Mineralogy*.[18] But Enlightenment naturalists also drew their ideas on hardness from firsthand interaction with gems and with the cutters and merchants who handled them. As we have seen, academicians such as Dufay and Réaumur combined gem collecting with the study of gem cutting and gem making.[19] Argenville's mineralogical treatises emerged from the same culture. Indeed, Argenville came to minerals by way of gemstones, publishing a *Lithology* in 1742 that resembled the *Oryctology* he would publish the following decade. The earlier work focused on stones rather than minerals in general;

14. Dufay, "Lumière des diamants," 358; Daubenton, "Connoissance des pierres précieuses," 35–36; Argenville, *Lithologie et conchyliologie*, 53; Argenville, *Oryctologie*, 180; Brisson, *Pesanteur spécifique*, vi–vii, xvi–xviii; Bomare, *Dictionnaire raisonné*, vol. 6, 685; Diderot and d'Alembert, *Encyclopédie*, vol. 12 (1765), 593–95, on 594, and vol. 11 (1765), 644; Louis Dutens, *Des pierres précieuses et des pierres fines*, 2nd ed. (Florence, 1783 [1776]), 19–20.

15. The examples are from Dutens, *Pierres précieuses*, 19–20. Cf. Sinkankas, *Gemology*, vol. 1, 291–92, on 291.

16. For example, after calling the distinction misleading, Dutens then used it to divide most precious stones in his book into varieties. See his chapters on ruby, sapphire, topaz, amethyst, aquamarine, chrysolite, garnet, hyacinth, agate, and sardonyx.

17. Argenville, *Oryctologie*, 180; Bomare, *Dictionnaire raisonné*, vol. 6, 685. Cf. Dutens, *Pierres précieuses*, 19–20.

18. Haüy, *Traité de minéralogie*, vol. 2, 489n1, 490n3.

19. See chap. 4, this volume.

it focused especially on precious stones, which Argenville described before the other stones, at greater length, and with capitalized headings.[20] In addition, the *Oryctology* came with a detailed account of the lapidary arts. This contained much that was absent in other published sources of the time, suggesting that Argenville had observed cutters and jewelers at work.[21]

These links between naturalists, collectors, and practitioners persisted into the latter part of the century. Jewelers themselves owned notable collections of gems, as shown by the lists of leading collections that Argenville published in 1767 and 1780.[22] The catalogs of such collections sometimes amounted to miniature treatises on mineralogy, with lengthy descriptions of the species to which each item belonged—the catalog for Antoine de la Roque, drawn up by Edmé-François Gersaint in 1749, is a case in point.[23] In the last quarter of the century, the collection of colored gems owned by the tax farmer Alexandre Estienne d'Augny was a key source for a natural history of gems written by Louis Dutens.[24] D'Augny was later described as a distinguished connoisseur (*illustre amateur*) who was sought out by gem cutters for his fine specimens and his advice on how to cut them.[25] This description comes from a work by Antoine Caire-Morant, a cutter from Briançon who worked closely with collectors, connoisseurs, and academicians in Paris in the last third of the century.[26]

20. Argenville, *Lithologie et conchyliologie*, 43–75.

21. Argenville, *Oryctologie*, 172–82. Note also the glossary (95–112), where nearly a quarter of the terms (thirty-one out of 304) are related to the lapidary arts. Compare this account with Boodt, *Parfait joaillier*, 90–103, 173–75; Berquen, *Merveilles des Indes*, 12–15; Rosnel, *Mercure Indien*, vol. 2, 12; and Félibien, *Principes de l'architecture*, 358–81, 459–70. Note, for instance, Argenville's description of the scoring of diamond (177–78), and terms such as "labora" and "grasse," neither of which appear in Félibien's glossary. Nor do they appear in the *Encyclopédie* of Diderot and d'Alembert, as shown by a word search on https://enccre.academie-sciences.fr/encyclopedie/. Cf. Argenville, *Lithologie et conchyliologie*, 42–43, 52–54.

22. The jeweler Pierre-André Jacquemin had a collection in which "the suite of precious stones is very considerable": Antoine-Joseph Dezallier d'Argenville, *Conchyliologie nouvelle et portative* (Paris, 1767), 316. Cf. Wilson, *Mineral Collecting*, 177; Laissus, "Cabinets d'histoire naturelle," 688. The jeweler Fagnier had "a considerable number of fine and precious stones" in a collection noted in Argenville, *Conchyliologie* (1780), vol. 1, 265. Cf. Laissus, "Cabinets d'histoire naturelle," 686. Another jeweler, named Fouché, had some agates and jaspers but apparently no precious stones: Argenville, *Conchyliologie* (1780), 686. Cf. Laissus, "Cabinets d'histoire naturelle," 686.

23. Gersaint, *Catalogue raisonné du cabinet de la Roque*, 112–35, 139–40, 142–43. Cf. Dutens, *Pierres précieuses*, 6–7.

24. Dutens, *Pierres précieuses*, 3–4, 60, 68, 113–14, 117, 119, 128.

25. This is in the anonymous preface to Antoine Caire-Morant, *La science des pierres précieuses, appliquées aux arts* (Paris, 1826), iv.

26. On Caire-Morant, see chap. 7, this volume.

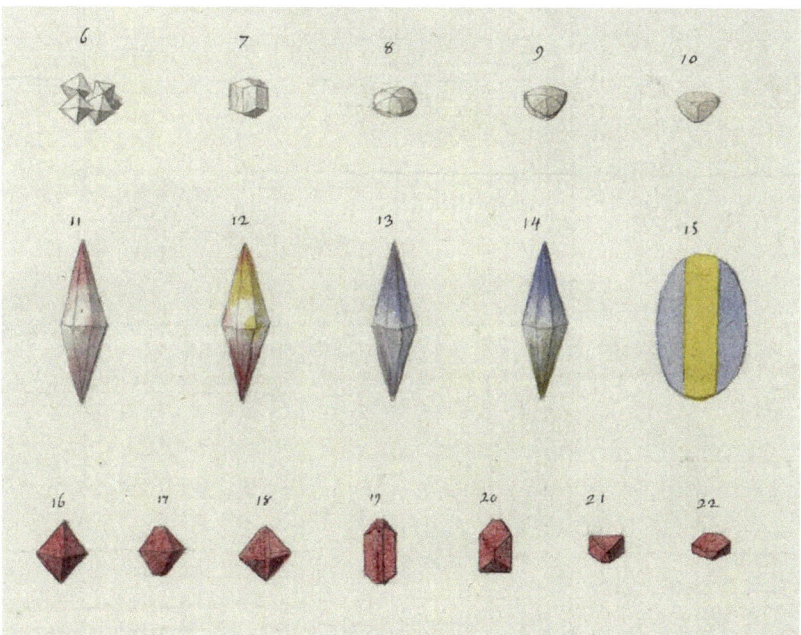

PLATE 1. The term *saphirus* was applied only to blue stones in medieval Europe. Europeans were later exposed to stones of other colors that resembled saphirus and to cultures that treated all these stones as one species, known today as corundum. *Top*: A modern collection of raw corundum specimens. The specimens have different colors, but all come from the same place, the Ratnapura region in Sri Lanka. *Bottom*: Diagrams from Swebach Desfontaines, *Mineralogy* (1790), vol. 3, plate 99. Figures 11 through 14 show uncut bicolored sapphires. Figure 15 depicts a cut bicolored sapphire that was part of the French crown jewels from at least 1774. Figures 6 through 10 show crystal forms of diamond, figures 16 through 22 crystal forms of spinel. The artist drew on the work of Romé de l'Isle. *Top*: © Tony Gill. *Bottom*: Courtesy of the Library and Archives, Natural History Museum, London.

PLATE 2. Modern specimens similar to minerals described by Boyle. *Top*: Transparent, polished amethyst with included goethite needles. Boyle owned a "pale amethyst" that contained "hairs of a brownish colour." *Middle*: Rock crystal from Bristol. Boyle was struck by the regularity of "Bristol diamonds" (a six-sided prism terminated by a six-sided pyramid) and by their irregularity (opposite sides of different lengths, some terminations ending in a line rather than a point, etc.). *Bottom*: Garnet crystals in a sericite-schist matrix. Boyle took garnet crystals "out of their Wombs" himself, to show that their regular shapes were natural rather than artificial. Objects from the collections of the Natural History Museum, London, BM.58341, BM.48054, BM.1911,34. Photos by the author.

PLATE 3. *Top*: A drawer from Woodward's collection showing specimens of rock crystal. *Bottom*: A detail of another drawer from Woodward's collection, showing two rings set with gems among pieces of agate. The drawers and specimens date from the early eighteenth century and therefore give some idea of the appearance of Dufay's mineral collection. Photos © 2026 Sedgwick Museum of Earth Sciences, University of Cambridge. Reproduced with permission.

PLATE 4. Colored images of gem specimens, based on Romé de l'Isle's crystallography. The title signals the taxonomic importance of *pierres précieuses* as late as 1789. Some panels show the same stone in multiple colors; others show a single characteristic color. Note also the grouping of stones into "Oriental" and "Peruvian" species, each with a characteristic geometry. From Swebach Desfontaines, *Natural History* (1789), plate 3. Courtesy of the Library and Archives, Natural History Museum, London.

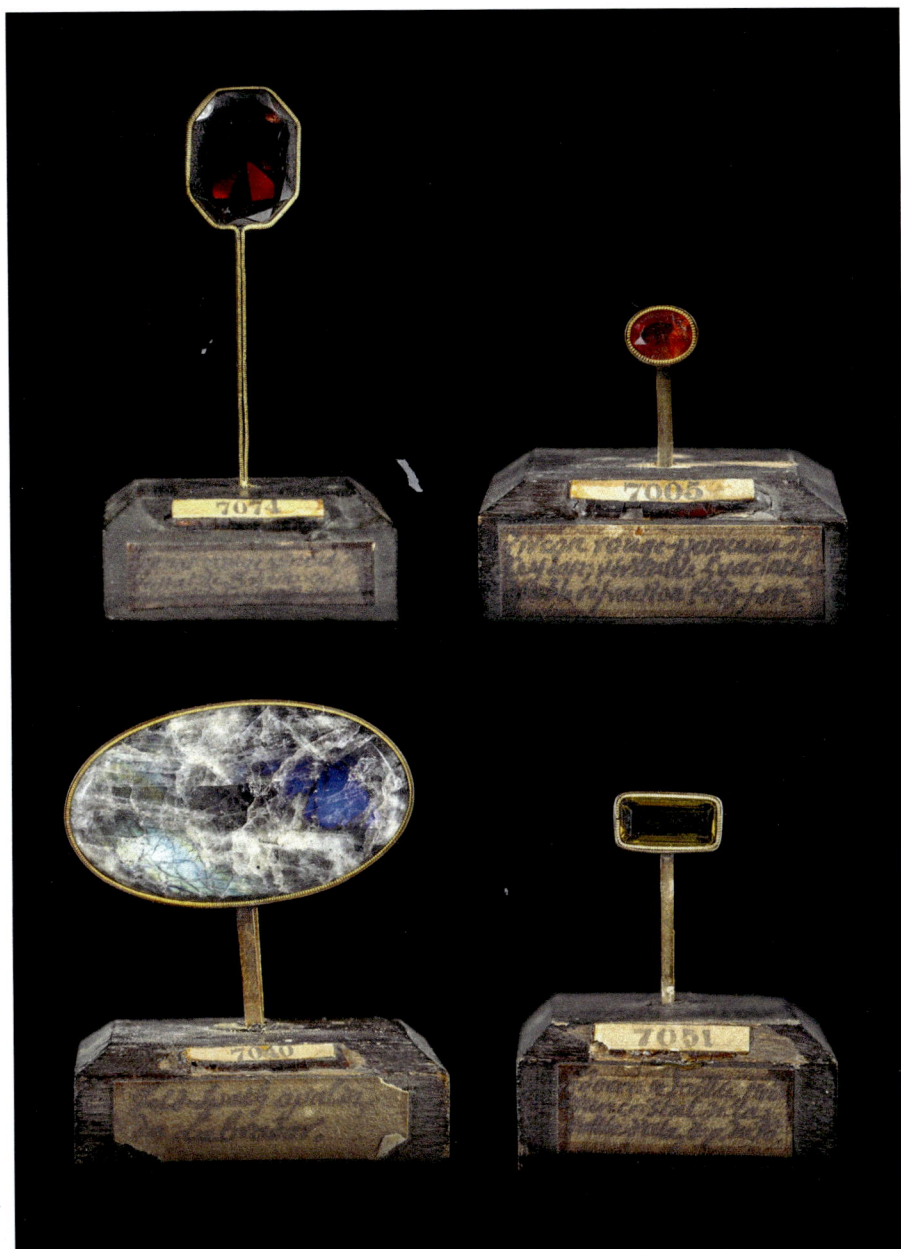

PLATE 5. Items from Haüy's collection of cut gems. *Top left*: A garnet given to Haüy by a certain Achard, probably the jeweler who advised Haüy for his *Treatise on Precious Stones* (1817). *Top right*: Red zircon, or Ceylonese hyacinth, a stone with a color similar to garnet but very different chemistry and crystallography. This specimen was described by Haüy as having "very strong double refraction." *Bottom right*: Greenish-yellow zircon, the same species as the previous stone but of a different color. *Bottom left*: Haüy used Newtonian optics to explain the iridescence of opal and similar stones, such as the opaline feldspar shown here. Photo by François Farges, © Muséum national d'histoire naturelle, Paris.

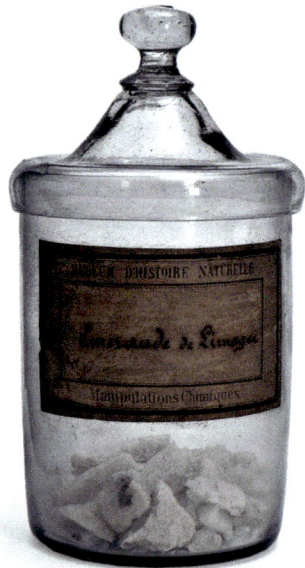

PLATE 6. Enlightenment savants sought domestic alternatives to foreign gems. In France, Réaumur argued that stones from the town of Simmore are superior to Persian turquoise. The material was in fact fossilized mastodon ivory that turns blue when heated (*top*). Soon afterward, German chemists argued that "Saxon topaz" had the same qualities as its Asian counterpart. Saxon topaz from the period was found in the Schneckenstein site, as was the specimen shown here (*bottom left*). Later, Vauquelin found beryllia in stones from Limoges, France, which suggested that these stones were a variety of emerald. The flask (*bottom right*) shows residue from Vauquelin's analysis. *Top*: Photo by François Farges, © Muséum national d'histoire naturelle, Paris. *Bottom left*: © Jeff Skovil, courtesy of the Mineralogical Collection of the TU Bergakademie Freiberg. *Bottom right*: © Muséum national d'histoire naturelle, Paris, rights reserved.

PLATE 7. Varieties of beryl from Haüy's research collection. Haüy argued on chemical and crystallo-graphic grounds that these stones all belonged to the same species, despite their different forms, colors, and names, which included "emerald" and "aquamarine." Haüy called the inclusive species "*émeraude*," but it later became known as "beryl." From left to right: hexahedral prism, partly honey-colored and partly pale yellow; cylindroidal form, greenish-blue; peridodecahedral form, green; hexahedral prism, yellowish-green. The geometrical terms are Haüy's. The hexahedral prism was the "primitive form" of emerald in Haüy's scheme. Photo by François Farges, © Muséum national d'histoire naturelle, Paris.

PLATE 8. Chart from Mawe's *Treatise on Diamonds and Precious Stones* (1813). Mawe insisted that hue is a reliable guide to gem identity at a time when many mineralogists said otherwise. There are red diamonds and red rubies, Mawe argued, but the redness of diamond is different from that of ruby. The chart was meant to illustrate this point using a range of gem species. Courtesy of Bodleian Libraries, Oxford.

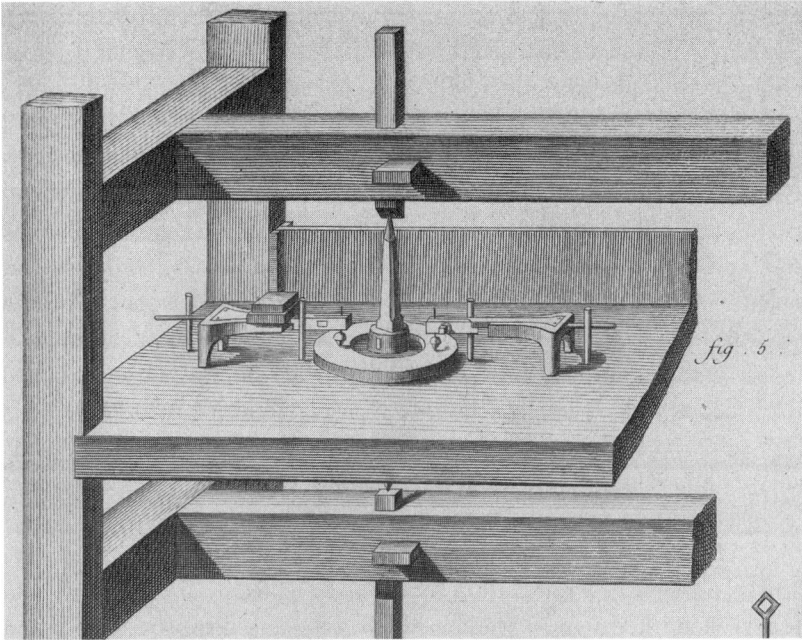

FIGURE 5.1. Lapidary's wheel with weight
The lead wheel (center) is shown with two diamonds that are each held by a rod clasped by a pair of pincers. The pincers on the left are charged with lead weights. The weights speed up the cutting process by pressing the diamond against the wheel. The gem cutter Caire-Morant used an arrangement like this to measure the hardness of gems based on the size of the weights required to cut them. From Diderot and d'Alembert, *Encyclopédie* [plates], vol. 3 (1763), "Diamantaire," plate 1 bis. Image © Bibliothèque Mazarine, 2° 3442-24.

A characteristic product of these interactions was Caire-Morant's effort to quantify the hardness of gems. The project drew on the ideas of the mineralogist Jean-Etienne Guettard, the diamond cutters' practice of pressing their gems against the grinding wheel, and the physicist's fondness for precise balances and tables of weights (fig. 5.1). The project was cut short by the French Revolution, but it was part of a wider pattern in which the old practice of ranking gems by hardness led to the hardness scales of modern mineralogy.[27] This is yet another example of the persistence of gems as a basis for the study of minerals. It also hints at how the worlds of natural history and experimental physics could combine to measure the value of gems. There was a similar convergence around color, double refraction, and density, and ultimately around

27. Caire-Morant, *Science des pierres précieuses*, 24. For details, see Michael Bycroft, "The Hand of the Connoisseur: Gems and Hardness in Enlightenment Mineralogy," *History of Science* 60, no. 4 (2022): 517–21.

crystal form. I take these topics in turn, keeping one eye on the conceptual terms—from "nuance" to "correlation"—that made the convergence possible.

Color and Nuance

Color was the first gem property to be quantified at the Jardin du Roi. This is not surprising, given the ancient tradition of classifying gems by color and the more recent tradition, dating from Isaac Newton's *Opticks* in 1704, of quantifying color. The key work was done by Louis-Jean-Marie Daubenton while he was the caretaker (*garde et démonstrateur*) of the natural history collection at the Jardin.[28] In this post, which he held from 1745, Daubenton had ready access to the gem collection that Dufay had bequeathed to the king. "For several years," Daubenton wrote in 1750, "I have had the habit of seeing, at every hour of the day, so to speak, the prolific collection of precious stones in the Cabinet du Roi."[29]

Daubenton had a connoisseur's eye for these stones. Witness the twelve articles on gems he wrote for the encyclopedia edited by Denis Diderot and Jean le Rond d'Alembert. These articles were published between 1751 and 1755.[30] The dominant theme was color. The "hues and shades" of these stones can "vary almost indefinitely," as Daubenton wrote in the article on agate. Despite this variety, indeed because of it, color was a reliable guide to the identity of stones. Aquamarine is easily identified by its color, Daubenton argued, because it is a mixture of blue and green, whereas the comparable blue stones (sapphires) have no trace of green and the comparable green stones (emeralds) have no trace of blue. He differentiated gems not only by their hues but also by other color-related properties, such as the beauty (*beauté*) of red rubies, which sets them apart from red diamonds, and the clarity and vividness (*vif et net*) of the orange sardonyx, which distinguishes these stones from orange agates. Color could also distinguish between varieties of the same species, as between red carnelian and milky-white carnelian. The color of some

---

28. Camille Limoges, "Daubenton, Louis-Jean-Marie," in *DSB*, vol. 3, 111.

29. Daubenton, "Connoissance des pierres précieuses," 30. Dufay's bequest is noted in Fontenelle, "Éloge de Dufay," 81–82. Cf. Franck Bourdier, "Origines et transformations du cabinet du Jardin Royal des Plantes," *Revue générale des sciences pures et appliquées*, 18 (1962): 40.

30. The articles are on amethyst, chrysolite, aquamarine, diamond, emerald, beryl, carnelian, agate, chalcedony, carbuncle, coral, and amber. These are easy to find at https://enccre .academie-sciences.fr/encyclopedie/. See esp. vol. 1, 167 (agates), 357 (amethyst), 199 (aquamarine); vol. 4, 940 (diamond), 244–45 (carnelian). My account of these articles expands on Jeff Loveland, "Louis-Jean-Marie Daubenton and the *Encyclopédie*," *Studies on Voltaire and the Eighteenth Century* 12 (2003): 187.

carnelians, Daubenton noted, is midway between those of red carnelian and orange sardonyx. He compared these intermediate carnelians to the point on the solar spectrum where red shades into orange.

This reference to the solar spectrum was more than a metaphor. While Daubenton was observing the gems in the royal cabinet, he was also experimenting with light. His aim was to perfect the kind of visual judgments that he communicated so vividly in his articles in the *Encyclopédie*. The problem was not the weakness of the eye but its strength: the eye was able to make distinctions that were impossible to convey in words. The result, according to Daubenton, was a confused nomenclature that was exploited by unscrupulous jewelers, who sold cheap gems for the price of expensive ones by using the same name for both.[31] Daubenton's solution was to define gem species by locating them on the solar spectrum.[32] After all, Newton had divided the spectrum into seven principal colors. And a study of the gems in the royal collection convinced Daubenton that each of these colors corresponded to an established species of gem. Other species seemed to lie at the boundaries of these colors: ballas rubies were a mixture of red and orange, Brazilian topaz a mixture of orange and yellow, and so on. The seven solar colors corresponded to seven gem genera.

It remained to divide these genera more finely. To this end, Daubenton invented a device for turning the color of a gem into a number (fig. 5.2).[33] First, he placed the gem next to a piece of colorless rock crystal. Next, he cast part of the solar spectrum onto the crystal, using a glass prism to divide the sun's light into seven colors. He then moved the rock crystal through the spectrum until it had the same hue as the gem. The distance traveled by the rock crystal measured the hue of the gem. Though simple in principle, this arrangement was hard to implement. The first problem was to find a ray of light to cast upon the prism and thereby generate the spectrum. This could be done, Daubenton advised, by piercing two holes in the shutters of a room and closing the shutter. One then had to "mount an equilateral prism in the usual way to receive the ray of light that passes through one of the holes."[34] Evidently, Daubenton was drawing on his knowledge of the sort of optical experiments that Newton had described in his *Opticks*, in which rays of light are passed through the windows of dark rooms and refracted through

31. Daubenton, "Connoissance des pierres précieuses," 28–29.
32. Ibid., 29–31.
33. Ibid., 31–34.
34. Ibid., 31.

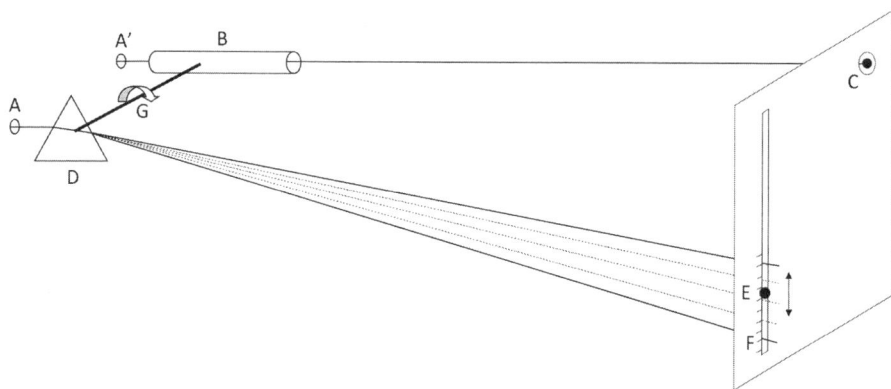

FIGURE 5.2. Daubenton's device for measuring gem color

Daubenton used this device, along with the gem collection at the Jardin du Roi, to turn the color of gems into a number. The idea was to compare a colored gem to a colored ray of light projected onto a colorless crystal.

Two rays of light, A and A', are admitted into a darkened room. One of these rays passes through a tube B onto a colored gem C, which is fixed into an opening in a wooden board. The other ray is refracted through a prism D such that one part of the solar spectrum falls onto the colorless rock crystal E. The operator moves the rock crystal up or down until its hue matches that of the colored gem. A scale F gives the position of the rock crystal and hence a "measure" of the color of the gem C.

The instrument may be adjusted for the motion of the sun and the subtleties of gem optics. The motion of the sun across the sky tended to interfere with measurements by changing the angle of the rays A and A'. To compensate for the sun's motion, the tube B is rotated incrementally around the axis of the rod G. The board is also moved, such that the unrefracted ray always strikes the colored gem C. The rod is attached to the prism as well as the tube, such that the prism turns at the same rate as the tube. In this way, the motions of the tube and board do not change the color of the ray that strikes the gem C. The intensity of the color displayed by the rock crystal can be varied by moving the board closer to, or further from, the prism. A piece of smoked glass placed at the aperture A' can be used to give a blackish tint to the rock crystal and thereby match the color of darker gems. Diagram by the author, based on Daubenton, "Knowledge of Precious Stones" (1750).

glass prisms.[35] Jean-Antoine Nollet, who had learned his physics from Dufay, would describe such experiments a few years later in his popular *Lessons of Experimental Physics*, published in 1753.[36] Without such a guide, it was not obvious what Daubenton meant by "the usual way" or even "equilateral prism." Another challenge was the observed motion of the sun across the sky. This interfered with comparisons between measurements done at different times

35. Newton is named in ibid., 34, 35.

36. Jean-Antoine Nollet, *Leçons de physique expérimentale* (Paris, 1753), vol. 5, 344–47. Note also Daubenton's reference to the *héliostate* (34), a device for compensating for the motion of the sun in optical experiments. On the uptake of Newton's *Opticks* in France, see Alan E. Shapiro, "The Gradual Acceptance of Newton's Theory of Light and Color, 1672–1727," *Perspectives on Science* 4 (1996): 97–101.

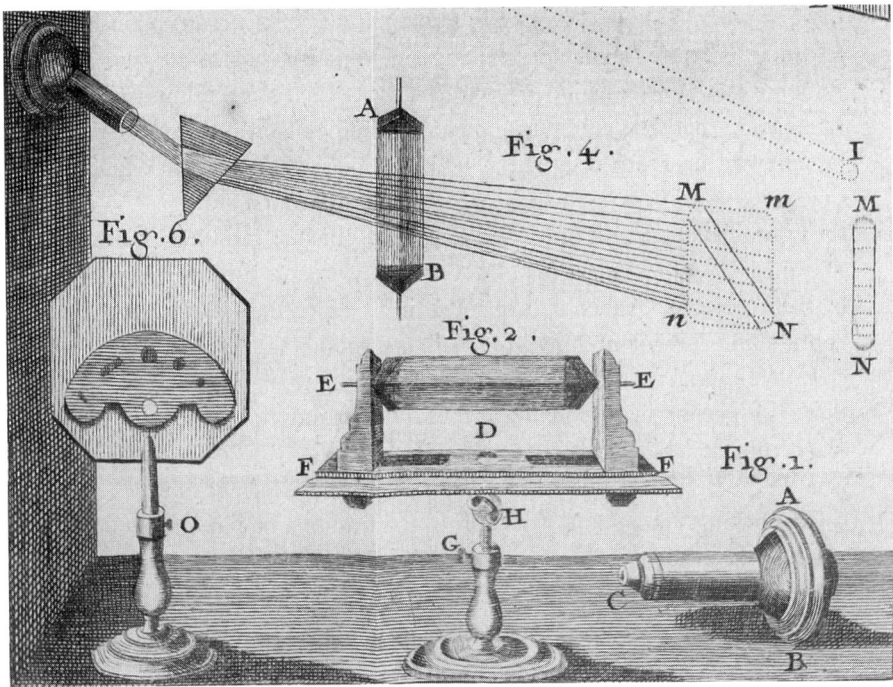

FIGURE 5.3. Nollet's instruments for optical experiments
Instruments such as these were used by Daubenton in his study of gem color. *Figure 1* shows a tube attached to a swivel, *figure 2* a mounted glass prism, and *figure 4* the use of both objects to project a solar spectrum onto a prism. Compare the mounted tube in this image to the rotating tube G shown in figure 5.2. From Nollet, *Lessons in Experimental Physics* (1753), vol. 5, lesson 17, plate 1. Courtesy of the Wellcome Collection.

of the day. Daubenton's solution was to channel the sun's rays through a tube, one that resembled a device used by Nollet for a similar purpose (fig. 5.3). Daubenton's procedure for quantifying colors was part of the tradition of experimental physics established in France by Dufay and Nollet.

At the same time, this procedure embodied a theory of classification developed by Buffon. The famous naturalist was a student of gems and of Newtonian optics as well as being Daubenton's patron and Dufay's successor as director of the Jardin.[37] He made his theory of classification public in

37. Limoges, "Daubenton," 111; Georges-Louis Leclerc, Comte de Buffon, "Septième mémoire: Observations sur les couleurs accidentelles, & sur les ombres colorées," in Buffon, *Histoire naturelle, générale et particulière, Supplément*, vol. 1 (Paris, 1774), 518 (solar spectrum). Buffon notes (537) that he was working on these experiments in 1743. Cf. Georges-Louis Leclerc, Comte de Buffon, "Sixième mémoire: Expériences sur la lumière, et sur la chaleur qu'elle peut produire," in *Supplément*, vol. 1, 469.

the first volume of his *Natural History, General and Particular*, published a few months before Daubenton read his paper on gem color to the academy.[38] Buffon argued that individual specimens are the fundamental unit in natural history. There is a continual gradation of individuals in nature, he maintained, with no sharp distinctions between species and genera. It followed that species and genera are at best artificial and at worst a distraction from the naturalist's true task of describing nature. It is hard to imagine a better instantiation of these ideas than Daubenton's procedure for identifying gems by their color. As Daubenton explained in his paper, the solar spectrum is a continuum. There is nothing in it that corresponds to the notion of a species.[39] All we can say is that two specimens belong to the same species when there is a perceptible difference in their hues as judged by the human eye. Daubenton concluded that it is impossible to determine how many species of gem there are. Indeed, it is impossible to determine the genus to which some species belong, since some species are a mixture of a hue in one genus and another hue in a neighboring genus. For Daubenton, this was not a drawback but an advantage. It meant that the naturalist could recognize all gem species without having to enumerate species or give precise definitions of genera. The solar spectrum allowed the naturalist to "recognize all possible species independently of their genera, just as nature produces species independently of our methods."[40] These are Daubenton's words, but they would not have been out of place in the first volume of Buffon's *Natural History*. The French word "nuance" captures the connection. Daubenton used the word to refer to the shades of gems in the royal collection; Buffon used it to refer to the continuous gradations between specimens of all kinds.[41]

The two naturalists also shared a pragmatic approach to classification. After all, naturalists do succeed in giving lists of species and genera. If these do not represent natural groupings, what do they represent? Buffon's answer in

38. Georges-Louis Leclerc, Comte de Buffon, *Histoire naturelle des minéraux*, vol. 1, 1–62. A good, short account of Buffon's philosophy of natural history is Phillip R. Sloan, "Natural History, 1670–1802," in *Companion to the History of Modern Science*, ed. Robert C. Olby, Geoffroy Cantor, Jonathan Hodge, and John R. R. Christie (London: Routledge, 1990), 295–313. As Sloan notes, Buffon later changed his view of mineral classification. Daubenton's participation in this philosophy has been noted recently in James Llana, "Natural History and the *Encyclopédie*," *Journal of the History of Biology* 33, no. 1 (2000): 11–14; and Jeff Loveland, "Another Daubenton, Another Histoire Naturelle," *Journal of the History of Biology* 39, no. 3 (2006): 462, though this is qualified on p. 480.

39. Daubenton, "Connoissance des pierres précieuses," 34–36.

40. Ibid., 38.

41. Buffon, *Histoire naturelle des minéraux*, vol. 1, 12. Cf. idem, "Couleurs accidentelles," 520, where Buffon used the same word to refer to the hues in the solar spectrum.

1749 was that they reflect the interests of the classifier, "the order of relations that things seem to have to ourselves."[42] There was a similar pragmatism in Daubenton's writings on gems. He was especially interested in the evaluation of foreign gems, giving a detailed account of Asian diamond mines in an article published in the *Encyclopédie*.[43] This text was copied from two earlier ones, both of which stated that jewelers in Asia tended to examine gems in dark rooms rather than in the light of day.[44] Examining gems in dark rooms was, of course, the essence of Daubenton's solar spectrum method. After describing that method, in his 1750 paper, he immediately explained how to use it to make long-distance evaluations.[45] His explanation took the form of a thought experiment. An "Indian" person has come across a new variety of precious stone somewhere in Pegu; he would like to communicate the discovery to his contacts in Paris. Sending the stones themselves would be risky and expensive; sending a qualitative description of the stones' hardness, transparency, and color would be imprecise. The solution was to use Daubenton's procedure to locate the gem's color on the solar spectrum. In this way, Daubenton wrote, "Naturalists will define its genus and determine its species, jewelers will judge its beauty and determine its price, and thus the stone will be thoroughly known." Daubenton aimed to "reliably judge the nature and quality of a stone one has never seen," as Bomare would later write.[46] Classification, identification and evaluation were one and the same.

Daubenton's own term for his project was *"connoissance,"* a word that appeared in the title of his 1750 paper and that denoted, not just knowledge in general, but also knowledge about how to judge material things, whether paintings, horses and gems.[47] This paper, with its elaborate instrument and mathematical ambitions, may seem like a departure from the rich descriptions of gems in Daubenton's articles for the *Encyclopédie*. But the paper and articles were both part of the same project: to use color to codify the connoisseurship of gems.

42. Buffon, *Histoire naturelle des minéraux*, vol. 1, 37.

43. Daubenton, "Diamant," in Diderot and d'Alembert, *Encyclopédie*, vol. 4 (1754), 938–40.

44. The practice of judging gems at night is noted in Tavernier, *Travels*, vol. 2, 75; and Savary des Brûlons, *Dictionnaire universel de commerce* (1726), vol. 1, 1687. See also the 1748 edition of the latter work, vol. 2, 874.

45. Daubenton, "Connoissance des pierres précieuses," 36–37.

46. Bomare, *Dictionnaire raisonné*, vol. 6, 685.

47. On this term, see Bycroft and Wragge-Morley, "Science and Connoisseurship," esp. 452–53.

Refraction and Structure

Daubenton's method was too clumsy to be used in practice, as other naturalists recognized.[48] But the idea of using quantitative physics to evaluate and classify gems was prescient. Buffon would go on to apply this idea in the mineralogical volumes of his *Natural History*, using double refraction rather than color as the key criterion. The volumes in question were only published in the 1780s, but they had a long backstory. Buffon's interest in double refraction went all the back to 1738, when Dufay read his paper on the topic to the academy.[49] Buffon discussed these experiments with Dufay at the time; three years later, he did his own experiments on the double refraction of rock crystal.[50] He then put these interests aside for two decades, probably to focus on the volumes of *Natural History* dedicated to animals.[51]

When he returned to refraction in the 1770s, Buffon developed the theory of double refraction mentioned at the start of this chapter. This theory is usually treated as a Buffonian oddity, or as a failed attempt at optics or crystallography.[52] It was all those things, but it was also an example of a characteristic form of eighteenth-century quantification. It was an attempt to discover the invisible properties of bodies by making careful measurements of their visible properties. This program had been pursued by Newton himself, who had calculated the sizes of the microscopic pores in bodies on the basis of the color and intensity of the light they reflected or transmitted. Newton had also concluded that diamond is a combustible substance on the grounds that the ratio of density to refractive index is the same in diamond as in other combustible substances.[53] Buffon endorsed this argument about diamond in works published in the 1770s and 1780s, citing Newton's *Opticks* as his source.[54] Light is,

48. Bomare, *Dictionnaire raisonné*, vol. 6, 685; Dutens, *Pierres précieuses*, 1–2; René-Just Haüy, *Traité des caractères physiques des pierres précieuses* (Paris, 1817), 235.

49. Chapter 4, this volume, "Physics as Gem Collecting."

50. Buffon, *Histoire naturelle des minéraux*, vol. 4, 115 (rock crystal), 267 (Dufay on double refraction).

51. See the list of volumes in Émilienne Genet-Varcin and Jacques Roger, "Bibliographie de Buffon," 522–23, in *Buffon: Oeuvres philosophiques*, ed. Jean Piveteau, Maurice Fréchet, and Charles Bruneau (Paris: Presses universitaires de France, 1954), 513–70. Daubenton worked on volumes on animals published from 1749 to 1767: Limoges, "Daubenton," 112.

52. Hélène Metzger, "Théorie curieuse de la double réfraction"; Buchwald, "Experimental Investigations of Double Refraction," 334.

53. The significance of this style of reasoning, and its origin in Newton's *Opticks*, was discussed in Cohen, *Franklin and Newton*, 156–63.

54. Buffon, *Supplément*, vol. 1, 13n and 93n; idem, *Histoire naturelle des minéraux*, vol. 4, 262–64. Buffon's wider debt to Newton's *Opticks* in these late works is noted in Jacques Roger, *Buf-*

Buffon wrote, "the most delicate instrument, the finest scalpel, by which we can probe the interior of substances that receive and transmit it."[55]

Buffon used this "scalpel" to determine the internal structure of gems, using double refraction as a guide.[56] The reason rock crystal has two refractions, Buffon argued, is that it is made up of alternating layers of two substances of different densities. The ordinary ray is the result of refraction through one of these substances, the extraordinary ray the result of refraction through the other substance. He assumed, as most naturalists did at the time, that denser substances are the more refractive ones. His theory resembled earlier ones in that it referred to the internal structure of the crystal. But the idea of differential densities was new. It explains Buffon's interest in the number of refractions in transparent stones. Specimens with three or more rays of refraction could be explained by supposing that they were made up of three or more substances of different densities. Variations in the strength of double refraction—that is, in the size of the angle that separates the ordinary and extraordinary rays—could be dealt with in a similar way. The orientation of the effect was more grist to Buffon's mill. Why can two images be seen in peridot when observed through one face, but only one image when observed through another face of the same specimen? Because, Buffon answered, the layers that are responsible for double refraction in peridot run in only one direction. In other words, double refraction has an orientation because the layers have one.

This theory was not just an explanation of double refraction but also a taxonomic device. Buffon used the theory to define the property of "homogeneity" or "simplicity." This was a property "which we can know with great precision in transparent bodies, by the single or double refraction that light undergoes in traversing them."[57] The same property could be known through the use of fire and acids, but "less exactly" than by refraction. The property was central to Buffon's whole mineralogical scheme: he gave it third place, after density and hardness, on his list of the "natural properties" that defined his classification of the mineral kingdom.[58] Double refraction was especially important for precious stones: "We can be sure that the first character of true precious stones (vraies pierres précieuses) is the simplicity of their essence or

---

fon: A Life in Natural History, trans. Sarah Lucille (Ithaca, NY: Cornell University Press, 1997), 390–91.

55. Buffon, Histoire naturelle des minéraux, vol. 3, 446.

56. This paragraph is based on ibid., vol. 3, 443–46 (rock crystal); vol. 4, 117–18 (Iceland spar), 268–69 (diamond).

57. Ibid., vol. 3, 609.

58. Ibid., vol. 3, 609.

the homogeneity of their substance, which is demonstrated by their refraction, which is always simple."[59] The most spectacular casualty of this definition was the emerald, but amethyst, peridot, garnet and hyacinth shared the same fate. The only precious stones that remained were diamond, oriental ruby, oriental sapphire, and oriental topaz.[60]

Buffon's theory had the added advantage of explaining why these stones are so highly valued by humans. Double refraction divides the rays of light in two and thereby discolors them, he explained. He also associated single refraction with transparency, high refraction, and low dispersion. Here he generalized an idea about the quality of telescope lenses: good diamonds, like good lenses, have a high index of refraction and a low index of dispersion.[61] Much of this was traditional—Boodt, too, had understood preciousness in terms of the diamond, and the diamond in terms of purity, homogeneity, transparency and brilliance. Buffon's innovation was to turn these qualities into numbers. He also gave them a history, building them into his grand vision of the earth's history. The single refraction of diamond suggested to him that the stone was formed from the remains of living organisms; the double refraction of Iceland spar suggested that it was formed from successive deposits of chalk.[62] For Buffon, refraction was a way of seeing into stones and a way of seeing into the past and across the surface of the earth.

The material basis for this theory was a partnership between Buffon and the astronomer and instrument maker Alexis-Marie de Rochon.[63] The latter made a name for himself in the late 1760s, traveling to Morocco and the Indian Ocean as a shipboard astronomer. It was on the latter voyage that Rochon came across a deposit of high-quality rock crystal, on the island of Madagascar. He had a prism and a lens made from one of these crystals, finding

---

59. Ibid., vol. 4, 250–51.

60. Ibid., vol. 3 (list of precious stones in table); vol. 4 (definition of precious stones). Buffon sometimes included girasol and vermilion as true precious stones, but only as varieties of the three oriental stones.

61. Ibid., vol. 4, 253–56. Cf. Buffon, "Couleurs accidentelles," 527 (transparency associated with homogeneity).

62. Iceland spar in Buffon, *Histoire naturelle des minéraux*, vol. 4, 119–21, 441–42. Diamond in Buffon, *Supplément*, vol. 1, 13n, 93n, 46–47; idem, *Histoire naturelle des minéraux*, vol. 1, 262–64, vol. 4, 260–61. Buffon's overall scheme is discussed in Martin Rudwick, *Bursting the Limits of Time: The Reconstruction of Geohistory in the Age of Revolution* (Chicago: University of Chicago Press, 2005), 139–50.

63. On Rochon, see Jean-Baptiste Joseph Delambre, "Notice sur la vie et les ouvrages de M. Rochon," *MAS* 1817 (1819): lxii–lxxxii; and Danielle Fauque, "Alexis-Marie Rochon (1741–1817): Savant astronome et opticien," *Revue d'histoire des sciences* 38, no. 1 (1985): 3–36.

that the optical properties of the stone were ideal for making telescopic lenses. When he reported this finding to the academy in 1770, Buffon wanted to know whether this variety of rock crystal refracted doubly, like the varieties of the stone found in other parts of the world.[64] Rochon went on to use the double refraction of rock crystal to invent the instrument for which he became famous, the prismatic micrometer, a telescope for measuring small celestial angles.[65] He measured the refractive index, dispersive index, and number of refractions of numerous substances, including diamond, peridot, and rock crystal. In an article read to the academy in 1780, he noted that Newton himself had been led astray by measuring the refraction and dispersion of only two substances, water and glass. "Nothing shows better," he wrote, "how necessary it is in physics to multiply, vary, and compare experiments, before trying to establish general laws."[66] Meanwhile, Buffon had secured some of Rochon's specimens of rock crystal for the Jardin du Roi, including a "great needle" that he described in his *Natural History*.[67] The latter work is strewn with references to Rochon's experiments on optical and other properties of gemstones; many of these experiments supported Buffon's theory of double refraction (appendix 5).[68] This was an institutional collaboration as well as an instrumental one, with Buffon continuing as director of the Jardin until his death in 1788 and Rochon becoming the curator of the Royal Cabinet of Physics and Optics in 1775.[69] The leading Parisian institutions of natural history and experimental physics combined to produce Buffon's refraction-based classification of gems.

64. Alexis-Marie de Rochon, *Recueil de mémoires sur la mécanique et la physique* (Paris, 1783), 154–56.

65. Fauque, "Rochon," 25–28 (prismatic micrometer). Cf. Delambre, "Vie de Rochon," who refers (lxxiv) to the micrometer as Rochon's "greatest discovery." Rochon, *Recueil*, 156–66, drew heavily on Giambattista Beccaria, "Observations sur la double réfraction du cristal de roche," *Journal de physique* 2 (1772): 504–10.

66. Alexis-Marie de Rochon, "Mémoire sur la mesure de la dispersion & de la réfraction de différentes substances, & description de l'instrument qui a servi à cette détermination," in Rochon, *Recueil*, 284–85.

67. Alexis-Marie de Rochon, *Voyages à Madagascar, à Maroc, et aux Indes Orientales* (Paris, 1801), vol. 2, 191; Buffon, *Histoire naturelle des minéraux*, vol. 3, 463–64.

68. Note also two experiments that are not in appendix 5 and that Rochon did for Buffon, on the hardness and iron content of gems: vol. 3, 445, 446. On some occasions, Buffon made it clear that he got his data directly from Rochon rather than from a publication: vol. 3, 524–25, 446; vol. 4, 240. Note also the reference to Rochon's "large number of experiments on the refraction of transparent stones": *Histoire naturelle des minéraux*, vol. 3, 524–25.

69. Fauque, "Rochon," 12–15.

Density and Variety

Similar partnerships formed around density, with Mathurin-Jacques Brisson as the central figure. In this case, the key idea was variety—small variations in the density of specimens helped to distinguish between varieties of the same species. The most visible outcome was Brisson's tables of the densities of solids and liquids; these were published in 1787, although Brisson had been working on the project since at least 1783.[70] Density tables were already a well-established genre for European savants, but earlier tables had been incomplete in their coverage of materials and were rarely deployed in mineral classification schemes.[71] This changed quickly as a result of Brisson's data, which was incorporated into French treatises on crystallography and mineralogy in the 1780s and 1790s.[72] Haüy improved on Brisson's numbers in his 1801 treatise on mineralogy, but he did not eclipse them.[73] As late as 1821, Brisson was a point of reference for Cyprien-Prosper Brard, a student of Haüy who made a fresh set of measurements of the density of gems.[74]

Brisson's tables are a good example of quantification in experimental physics, but they went well beyond physics. They were "useful in natural history, physics, the arts and commerce," as Brisson made clear in the subtitle of the book. Brisson was well-placed to combine physics and natural history, since his career had spanned both disciplines.[75] He first worked in

70. Romé de l'Isle, *Cristallographie*, vol. 1, 437; vol. 2, 170–302, passim.

71. No systematic use of density data in Argenville, *Oryctologie*, 151–204; Bomare, *Minéralogie*, 1st ed., vol. 1, 222–65; ibid., 2nd ed. (Paris, 1774), vol. 1, 369–425; Bergman, *Manuel du minéralogiste*, ed. Mongez, 132–38, 145–47, 172–75. Exceptions are Romé de l'Isle's figure for the density of diamond in *Essai de cristallographie*, 210, citing John Ellicott, "A Letter from Mr. John Ellicott, F.R.S. to the President, Concerning the Specific Gravity of Diamonds," *PT* 43, nos. 472–77 (1744): 468–72; and Wallerius's for hyacinth in his *Minéralogie*, vol. 1, 225, cf. 210. Wallerius's number is apparently from John Clayton, "Observations of the Comparative, Intensive or Specific Gravities of Various Bodies," *PT* 17, no. 199 (1693): 694, where the same figure is given for "hyacinth (spurious)." For a review of density tables, see Schuh, *History*, 85–88; and Sally Newcomb, *World in a Crucible: Laboratory Practice and Geological Theory at the Beginning of Geology* (Boulder, CO: Geological Society of America, 2009), 18–20, 80–81. The latter writes that Wallerius "routinely included specific gravity in his mineral descriptions," citing the 1753 French translation of his *Minéralogie*. But in that work, hyacinth is the only gem whose density is given as a number.

72. See the passages cited in note 70; and Torbern Bergman, *Manuel du minéralogiste, ou Sciagraphie du règne minéral distribué d'après l'analyse chimique*, trans. Jean-André Mongez, ed. Jean-Claude Delamétherie (Paris, 1792 [1782]), vol. 1, esp. 253–72.

73. Haüy, *Traité de minéralogie*, vol. 1, 260–68.

74. Cyprien-Prosper Brard, *Minéralogie appliquée aux arts* (Paris, 1821), vol. 3, 397.

75. The following biographical details on Brisson are from René Taton, "Brisson, Mathurin-Jacques," in *DSB*, vol. 2, 473–75. See also Heilbron, *Electricity*, 347n18.

natural history, being hired by Réaumur in 1747 to take care of the aging naturalist's voluminous collections of animals and minerals. When Réaumur died
a decade later, these collections were transferred to the Jardin du Roi and
effectively lost to Brisson. He therefore began a new career in la physique expérimentale, teaching the science to the children of the royal family and publishing books on electricity (1771) and on physics in general (1789), thereby
following in the footsteps of his mentor Jean-Antoine Nollet. Brisson's career-
change was reflected in his official position at the Academy of Sciences: admitted in the class of botany (*botanique*) in 1759, he joined the class of general
physics (*physique générale*) in 1785.

He drew on both traditions to compile his table of gem densities. These
densities were measured with a balance sensitive to weights as small as 1/64 of a
grain, or 0.0008 metric grams, about the weight of three poppy seeds.[76] Brisson
weighed each gem in air and water, taking the kinds of painstaking precautions
that distinguished Brisson's generation of physicists from Nollet's. He did all his
measurements at the same temperature; used rainwater that had fallen directly
into a glass or ceramic vessel; and compared measurements done with rainwater with the same measurements done with distilled water.[77] Brisson took
equal care with his specimens, using large specimens to minimize experimental error and "describing exactly" each specimen so that it could be correctly
identified.[78] Brisson scoured the shops and collections of Paris to find the best
specimens, using gemstones from the Jardin du Roi, from private collections,
from the shops of three different jewelers, and even from the crown jewels
(fig. 5.4).[79] The "exact descriptions" of these stones included data on their crystal
form, the shop or collection where they were located, and (for some specimens)
their hardness as determined by a cutter consulted by Brisson. These descriptions were couched in a classification scheme borrowed from Daubenton. As
curator of the Cabinet du Roi, Daubenton also supplied some of the gems, and
most of the other minerals, that Brisson measured for the purpose of his density tables.[80] Equally important was Romé de l'Isle, a self-educated savant who
was never a member of the academy and who never held a post at the Jardin
du Roi, but who made a living by cataloging natural history collections.[81] Romé

76. Brisson, *Pesanteur spécifique*, iv.

77. Ibid., ii–iii, vii–viii.

78. Ibid., ii–iii.

79. Ibid., 59–80. On the Golden Fleece in fig. 5.4, see ibid., 63–64; and François Farges, Scott
Sucher, Herbert Horovitz, and Jean-Marc Fourcault, "The French Blue and the Hope: New Data
from the Discovery of a Historical Lead Cast," *Gems and Gemology* 45, no. 1 (2009): 4–19.

80. Brisson, *Pesanteur spécifique*, xiv, 59.

81. Reijer Hooykaas, "Romé de l'Isle (or Delisle), Jean-Baptiste Louis," in *DSB*, vol. 11, 520–24.

FIGURE 5.4. Drawing of the Golden Fleece of the Colored Adornment (La Toison d'Or de la Parure de Couleur)

This jewel was commissioned by Louis XV in 1749 and crafted by Pierre-André Jacquemin. The large diamond just above the fleece is the famous French Blue (*Diamant bleu de la Couronne*), later recut to become the Hope Diamond. The stone was temporarily removed from its setting so that Brisson could measure its density for his book *Specific Weights of Bodies* (1787). Courtesy of Collection H. Horovitz.

de l'Isle supplied several of the gems for Brisson's measurements, including the garnets in figure 5.7. He also helped the experimenter to choose specimens with well-defined crystals, so that they could be correctly identified using Romé de l'Isle's ideas about crystal form.[82]

Brisson's density measurements were durable and sophisticated, drawing on the best resources of the Parisian scientific establishment, in physics as well as natural history. They were an embodiment of the quantifying spirit in the 1780s. But they were no less evaluative for all that. Brisson presented his data on gems as a solution to the old problem of distinguishing high-quality oriental stones from low-quality occidental ones. He explained how to use density to distinguish any oriental gem from its occidental variant, drawing up a table to aid the reader. The table showed, for example, that oriental rubies had a specific weight of 4.2833, meaning that an oriental ruby was 4.2833 times heavier than an equal volume of water. By contrast, the figure for occidental rubies (or "false rubies") was only 3.1911. In theory, the two stones could be told apart by anyone with a precise balance.[83] This is not surprising given Brisson's sources. The oriental/occidental distinction still had wide currency among the jewelers and cutters who supplied so many of Brisson's specimens. Daubenton himself had codified the distinction in his 1750 article on gemstones, where he divided gems into three broad classes: diamonds, oriental stones, and occidental stones. These classes were defined by hardness, as inferred from their density and their ability to take a polish.[84]

Romé de l'Isle made similar distinctions in his first major treatise, the *Essay on Crystallography* of 1772. There he identified "oriental" varieties of diamond, ruby, sapphire, topaz, emerald, and chrysolite—all except one of the *cristaux gemmes* he discussed, the exception being hyacinth. Like Daubenton before him, Romé de l'Isle treated these oriental gems as a class and not just as a collection of varieties. Take diamond, for example. This was one species of *cristal gemme*; it was made up of an "oriental" variety and a "Brazilian" variety. Later in the book, however, Romé de l'Isle explained that it made just as much sense to group the oriental diamond with all the other oriental stones, and the Brazilian diamond with all the other Brazilian stones. This was an explicitly evaluative scheme. Oriental diamonds were harder than Brazilian diamonds, and therefore better, in Romé de l'Isle's view. An alternative name for the group of all oriental stones was "precious stones of the first

82. Romé de l'Isle, *Cristallographie*, vol. 1, 437 ("homogeneous"); Brisson, *Pesanteur spécifique*, 66, 67, 70, 73, 76, 79.

83. Brisson, *Pesanteur spécifique*, xvii–xviii.

84. Daubenton, "Connoissance des pierres précieuses," 36.

order."[85] These ideas were rendered in color by one of Romé de l'Isle's collaborators, François-Louis Swebach Desfontaines. The artist made many fine, hand-colored engravings of minerals in the 1780s, using Romé de l'Isle's ideas to organize the images, including the idea that "oriental" stones differ from "Brazilian" or "Peruvian" ones (plate 4).[86]

These geographical terms were not simply a hangover from the seventeenth century. They were updated for the needs of the eighteenth century, especially the discovery of diamonds in Brazil in the 1720s. All of a sudden, the oriental/occidental distinction seemed to make sense for the most important gemstone of all. Diamond had been the odd one out ever since Boodt announced that there was only one kind of diamond and that it came from the Orient.[87] When Brazilian diamonds began to trickle into Europe—and especially when the trickle turned into a torrent—the question arose of how these new diamonds related to the ones that naturalists had long identified as coming from Borneo and the Deccan plateau. Were the new diamonds as good as the old ones? Or did they fit the pattern of other gems, with the variety from the East taking precedence over the variety from the West? This was a vexed question for European jewelers. They worried that the price of Indian diamonds would collapse if they were mixed up with a flood of low-quality diamonds from Brazil. This was a particular concern for English merchants, who worried that their Portuguese counterparts would threaten the dominance of London as a European hub of the diamond trade.[88]

One of those English merchants, David Jeffries, tried to solve the problem by formalizing the rules for determining the price and quality of diamonds.

85. Romé de l'Isle, *Essai de cristallographie*, 194–243. The Brazilian/oriental distinction is explained in ibid., 216–20, 244–45, 270–72; quality of oriental diamond on p. 210; premier ordre on p. 243. Cf. idem, *Cristallographie*, vol. 2, 230–44, plate V fig. 20, plate VI fig. 39; François-Louis Swebach Desfontaines, *Manuel cristallographe, ou Abrégé de la cristallographie de M. Romé de l'Isle* (Paris, 1792), 21–22. Romé de l'Isle's "Brazilian" gems became the species "topaz" in Haüy, *Traité de minéralogie*, vol. 2, 504. It is not clear why Desfontaines referred to the stones in plate 9 as "du Perou" rather than as "du Brésil." The crystal forms for these stones in the plate correspond to the forms for stones "du Brésil" in the other works cited in this note. Desfontaines may have simply confused the two descriptors. On the inferiority of Brazilian diamonds, see also Buffon, *Histoire naturelle des minéraux*, vol. 4, 265–66; Jean Démeste, *Lettres au Dr Bernard sur la chymie, la docimasie, la cristallographie, la lithologie, la minéralogie et la physique en general* (Paris, 1779), vol. 1, 408.

86. On Swebach Desfontaines, see Wendell E. Wilson, "Fabien Gautier d'Agoty and His *Histoire Naturelle Regne Mineral* (1781)," *Mineralogical Record* 26, no. 4 (1995): 65–76.

87. Boodt, *Gemmarum*, 59. See also idem, *Parfait joaillier*, 149–50; Berquen, *Merveilles des Indes*, vol. 10, 15; Rosnel, *Mercure Indien*, 11–12.

88. Vanneste, *Blood, Sweat and Earth*, chap. 2; David Jeffries, *A Treatise on Diamonds and Pearls*, 2nd ed. (London, 1751 [1750]), esp. 65–87.

		In Air	InWater	Specif. Grav.
	Water	.	.	1000
N°.		Grains	Grains	
1	A *Brazil* Diamond, fine Water, rough Coat	92,425	66,16	3518
2	A *Brazil* Diamond, fine Water, rough Coat	88,21	63,16	3521
3	Ditto. fine bright Coat, . . .	10,025	7,170	3511
4	Ditto. fine bright Coat, . . .	9,560	6,830	3501
5	An *Eaſt India* Diamond, pale blue, . .	26,485	18,945	3512
6	Ditto bright yellow	23,33	16,71	3524
7	Ditto. very fine Water, bright Coat, .	20,66	14,8	3525
8	Ditto. very bad Water, honeycomb Coat, .	20,38	14,59	3519
9	Ditto. very hard blewiſh Caſt, . .	22,5	16,1	3515
10	Ditto. very ſoft, good Water, . .	22,615	16,2	3525
11	Ditto. a large red foul in it. . . .	25,48	18,23	3514
12	Ditto. ſoft bad Water	29,525	21,140	3521
13	Ditto. ſoft brown Coat, ...	26,535	18,99	3516
14	Ditto. very deep green Coat, . . .	25,25	18,08	3521

The mean Specific Gravity of the *Brazil* Diamonds appears to be : 3513.
The mean of the *Eaſt-India* Diamonds, - - - 3519.
The mean of Both to be - - - 3517.

FIGURE 5.5. Ellicott's table of diamond densities
Ellicott used precise measurements and select specimens to differentiate Indian and Brazilian diamonds. Note the variety in the size and quality of the diamonds in the table, and the comparison between "Brazil" and "East India" diamonds at the bottom. The columns on the right show the weight of each diamond in air and in water, and an expression of its density based on the first two numbers. For example, the first diamond weighs 92.425 grains in air and 66.16 grains in water. The difference of 26.265 grains is the weight of the water displaced by the submerged diamond. 92.425 divided by 26.265 is 3.5189. Ellicott has multiplied this result by 1,000 and rounded down to give 3,518. From Ellicott, "Diamonds" (1744). Courtesy of the Biodiversity Heritage Library, contributed by Natural History Museum Library, London.

His book on this topic appeared in 1750. A few years earlier, in 1745, a certain John Ellicott read a paper on the density of diamonds to the Royal Society of London. Ellicott's measurements were amazingly precise, partly due to his instrument (a balance three times more sensitive than Brisson's would be) and partly due to his specimens (a wide range of large diamonds). Ellicott worked with London merchants to acquire these specimens. He found the average density of "*Brazil* Diamonds" to be 3.513, and that of "*East India* Diamonds" to be 3.519 (fig. 5.5). These data eventually made their way to France, where Romé de l'Isle used them to establish his distinction between Brazilian and oriental diamonds, a distinction that Brisson then borrowed for his own measurements.[89] The oriental/occidental distinction

89. Romé de l'Isle, *Essai de cristallographie*, 203n1, 210.

gained a new lease of life as merchants, naturalists and experimenters made fine distinctions between different kinds of diamond.

The fineness of these distinctions caused problems for the experimenters. The difference between Brazilian and Indian diamonds was tiny—only 6 parts in a thousand, according to Ellicott's numbers. So, were these really two different varieties, in the naturalist's sense of "variety"? After all, these numbers were only averages. Ellicott's numbers for individual specimens varied from 3.501 to 3.525—by over 20 parts in a thousand. One might wonder whether the alleged difference between Brazilian and Indian diamonds was an artifact of Ellicott's imperfect instruments. Alternatively, perhaps it was due to natural variation in the density of diamonds that had nothing to do with their geographical origin. Indeed, mineralogists have long abandoned the idea that Brazilian diamonds differ systematically in density from other diamonds. Even Haüy was skeptical.[90] Still, it remains the case that the oriental/occidental distinction was a stimulus for precise measurements of gem density. It was also a stimulus for early thinking about experimental error. Ellicott rejected Boyle's numbers on diamond density because they varied so widely, from under 3 to 3.4. This was "a much greater Difference than could be expected in any Bodies of the same Kind."[91] Ellicott believed his own numbers because, although they varied, they varied much more narrowly than Boyle's. Later, when Brisson did his own measurements, the question arose of how to reconcile his numbers with Ellicott's. Brisson gave 3.4444 for Brazilian diamonds, a much lower number than Ellicott had given. Conversely, Brisson's numbers for Indian diamonds were slightly higher than Ellicott's. Buffon noticed this discrepancy and tried to resolve it by appealing to the natural variation in diamond density: perhaps Ellicott had measured an unusually heavy Brazilian diamond and an unusually light oriental one.[92] In this way, measurements of diamonds were bound up with ideas about the uniformity or otherwise of species. The variation of numbers and the varieties of classes were studied together.

Crystals and Correlation

I have treated density, hardness and double refraction separately for the sake of analysis, but in reality they were used together. Naturalists looked for connec-

90. Haüy, *Traité de minéralogie*, vol. 3, 295–96.

91. Ellicott, "Diamonds," 469.

92. Brisson, *Pesanteur spécifique*, 61–64, xvii–xviii, 291; Buffon, *Histoire naturelle des minéraux*, vol. 4, 265–66.

tions between these properties, both to group gems into kinds and to discover physical laws about them. They looked for "correlations," to use a contemporary term.[93] This was especially true of crystal form, a new principle of classification that was defended on the grounds that it gave similar results to other properties. This brings us to the history of crystallography, a well-trodden topic for historians of science but one that has never been studied in terms of the quantification of gem quality. It is well known that French crystallographers such as Romé de l'Isle and René-Just Haüy aimed to reform the classification of minerals by making it more exact. But a sharp distinction has been drawn between Buffon on the one hand and crystallographers on the other, with the former seen as qualitative and speculative and the latter as quantitative and precise.[94] These differences were real, but they have been overdrawn. We have already seen that numbers were the backbone of Buffon's classification of minerals, in the form of measurements of density and refraction. We shall now see that crystallography had much in common with Buffon's project, and with those by Daubenton and Dufay before him. Crystallography was as much about gems as about other minerals; it was as much about physics as about crystals; and it was as much about evaluation as about classification.

Start with gems. They were one of the most abundant kinds of mineral in the collection of the Spanish naturalist Pedro Franco Dávila. Romé de l'Isle cataloged this collection early in his career, sketching out his ideas about crystals in the process. Like Bomare's book on mineralogy, which had appeared five years previously, this catalog gave a disproportionate amount of space to stones and metals. The section on stones was dominated by precious stones, agate, jasper and rock crystal.[95] Passages from this catalog were reproduced in the *Essay on Crystallography* a few years later, including descriptions of each of the seven species of precious stone that Romé de l'Isle listed there.[96] As we have seen, precious stones, or "cristaux gemmes," was one of the categories in

93. Haüy, *Traité élémentaire de physique*, vol. 1 (Paris, 1803), 424; idem, "Observations sur l'électricité des minéraux," *Annales du Muséum national d'histoire naturelle* 15 (1810): 1.

94. Metzger, *Genèse de la science des cristaux*, esp. 65–73, 80–86, 125–26, 189–206, 208–24. This contrast is often reinforced by treating the crystallographers as Linnaeans: Burke, *Origins of the Science of Crystals*, 57–59; Laudan, *Mineralogy to Geology*, 71–78, esp. 76; Seymour Mauskopf, "Crystals and Compounds: Molecular Structure and Composition in Nineteenth-Century French Science," *Transactions of the American Philosophical Society* 66, no. 3 (1976): 15–16.

95. Jean-Baptiste Louis Romé de l'Isle and Abbé Dugaut, *Catalogue systématique et raisonné des curiosités de la nature et de l'art, qui composent le cabinet de Mr. Davila* (Paris, 1767), vol. 1, xix–xxi; vol. 2, 254–79. Romé de l'Isle's classification scheme, like Bomare's, was derived from Wallerius: ibid., vol. 1, xix. Dávila's mineral collection is described in general terms in Wilson, *Mineral Collecting*, 137–40.

96. Romé de l'Isle, *Essai de cristallographie*, 199–243.

Romé de l'Isle's classification scheme. It remained so in the revised edition of the work that appeared a decade later.[97] As late as 1792, Swebach Desfontaines used the category in a summary of Romé de l'Isle's crystallography.[98]

By this time, Romé de l'Isle's ideas had been surpassed by Haüy's. But gems persisted. Haüy's theory of crystals is sometimes traced to an episode in which he dropped a calcite crystal while perusing a private collection.[99] Calcite was certainly important for Haüy, but his first published account of the theory was on garnet, not calcite; he applied it to rock crystal and topaz soon afterward.[100] He may have acquired his interest in gems from Daubenton, whose lectures at the Jardin du Roi introduced Haüy to the science of mineralogy and to the minerals in the Cabinet du Roi. Haüy's own collection of gems can be glimpsed in his 1801 treatise, and in earlier articles on diamond, cymophane, hyacinth, and corundum.[101] By the time of his death in 1822, Haüy owned a collection of over two hundred gems, most of them cut and mounted in a gold setting (plate 5). This was in addition to the gems in his research collection, a much larger collection that contained a disproportionate number of gems. The earliest surviving catalog of the collection lists 265 specimens of rock crystal, 216 agates, and over 100 each of topaz, tourmaline, and garnet, in addition to eighty-six emeralds and seventy-one pieces of corundum.[102] Finally, Haüy devoted more time to what we would now call

97. Romé de l'Isle, *Cristallographie*, vol. 2, 170–303.

98. François-Louis Swebach Desfontaines, *Manuel cristallographe, ou Abrégé de la cristallographie de M. Romé de l'Isle* (Paris, 1792), 20 ("crystaux gemmes, ou pierres fines"); François-Louis Swebach Desfontaines, "Recuille complet de minéralogie rangée par ordre d'individus" (1790), frontispiece for vol. 3 ("gemes" [*sic*]), in the library of the Natural History Museum, London.

99. Burke, *Origins of the Science of Crystals*, 83–84.

100. René-Just Haüy, *Essai d'une théorie sur la structure des crystaux* (Paris, 1784), 10 and 76 (calcite), 169–204 (garnet and topaz); idem, "Extrait d'un mémoire sur la structure des crystaux de grenat," *Journal de physique* 19 (1782): 366–70; idem, "Mémoire sur la structure du cristal de roche," *MAS 1786* (1788): 78–93.

101. See the following works by Haüy: "Structure du cristal de roche," 91; "Sur le diamant," *Journal d'histoire naturelle* 1 (1792): 382; "Description de la Cymophane avec quelques réflexions sur les couleurs de gemmes," *Journal des mines* 4, no. 21 (1796): 13; "Sur les pierres appelées jusqu'ici Hyacinthe et Jargon de Ceylon, leurs différences, leurs caractères physiques et géométriques," *Journal des mines* 5, no. 26 (1796–1797): 93, 94; "Observations sur des cristaux trouvés parmi des pierres de Ceylon et qui paroissent appartenir à l'espèce de Corindon vulgairement spath adamantin," *Mémoires de la Société d'histoire naturelle de Paris* (1799): 55–58, on 55, 56, 58; *Traité de minéralogie*, vol. 2, 471, vol. 3, 7, 38, 43, 205, vol. 4, 361.

102. *Catalogue Haüy III*, a manuscript catalog held at the Muséum national d'histoire naturelle. The catalog was drawn up no earlier than 1848, the year the museum purchased Haüy's collection. On the collection at the time of Haüy's death, see Alfred Lacroix, "La vie et l'oeuvre de l'abbé René-Just Haüy," *Bulletin de la Société française de minéralogie* 67 (1944): 92–93. Haüy's

"gemology" than to any other branches of applied mineralogy.[103] In his book
on this topic, published in 1817, he began by reminding the reader of the role
of gems in the recent development of the science. "Among the many proofs
that mineralogy supplies of the progress of chemical analysis and crystallography in modern times, there is none more striking than . . . the ornamental
objects known as precious stones."[104]

The "progress of crystallography" that Haüy had in mind was closely related to experimental physics. Some context will help to show this.[105] The
study of gem crystals had a long history, one that includes everything from
Pliny's remark on the regular shape of the diamond (*adamas*) to the analysis of garnet by the Swedish naturalist Torbern Bergman in 1773.[106] But even
Bergman had analyzed only two minerals, garnet and calcspar. It fell to Romé
de l'Isle and Haüy to develop the analysis of crystals in such a way that it
could be applied to many minerals, not just one or two. Romé de l'Isle showed
how this might be done in his *Essay on Crystallography* of 1772. Like Bergman,
he thought of each crystal as a variant of a small number of simple solids, or
"primitive forms." He generated variants of these primitive forms by mentally
slicing through their points and edges, a process he called "truncation." Starting with a regular octahedron, for example, he generated twenty-one other
forms, as shown in figure 5.6. The primitive forms and the variants could then
be used to define mineral species—see figure 5.7 for the example of garnet,
and plate 1 for diamond and spinel. This was a good start, but many puzzles
remained. If the oriental diamond has one primitive form, and the Brazilian
diamond another, as Romé de l'Isle maintained, why should we call them
both "diamonds"? If oriental diamond and oriental ruby have the same form,
as he also maintained, why did he treat them as different species?[107]

Puzzles such as these were addressed in the three decades after the publication of the *Essay on Crystallography*. Simply put, the study of crystals became more numerical and more experimental. The numerical breakthrough

---

collection of cut gems is noted in Lacroix, "René-Just Haüy," 94. 219 specimens of this collection
are described in *Catalogue Haüy III*, specimens 6973 to 7192, under the heading *Pierres fines
taillées.*

103. On the origins of gemology, see chap. 7, this volume.

104. Haüy, *Pierres précieuses*, i.

105. This paragraph and the next are based on Burke, *Origins of the Science of Crystals*,
esp. 62–106. The examples of diamond and ruby are my own.

106. Pliny, *Natural History*, vol. 10, bk. 37, p. 207; Torbern Bergman, "Variae crystallarum
formae, e spatho ortae," *Nova Acta Regiae Societatis Scientiarum Upsaliensis* 1 (1773): 150–55.

107. Romé de l'Isle, *Essai de cristallographie*, 199, 203, 213–14, "Suite du tableau cristal-
lographique," p. VI (oriental diamond), plate VIII (Brazilian diamond).

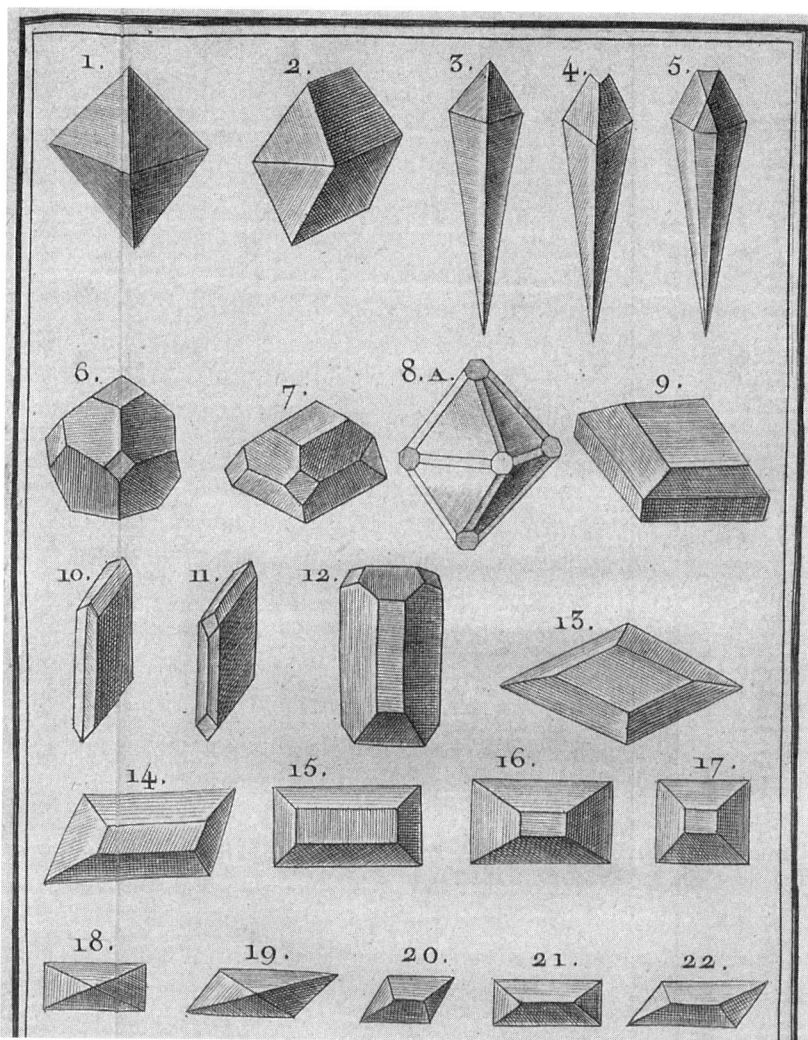

FIGURE 5.6. Romé de l'Isle's truncations of the octahedron

Figure 1 shows the octahedron, one of the crystallographer's "primitive forms." *Figures 2* through *22* show variants of this form, generated by mentally truncating the primitive form. Romé de l'Isle held that *figure 1* is the form of oriental diamond, oriental ruby, and oriental sapphire, as well as of very different substances such as alum and gold. According to his theory, oriental diamond can also take the forms of *figures 19, 20,* and *22.* From Romé de l'Isle, *Essay on Crystallography* (1772), plate 6. Courtesy of ETH-Bibliothek Zürich, Rar 2708, https://doi.org/10.3931/e-rara-16480.

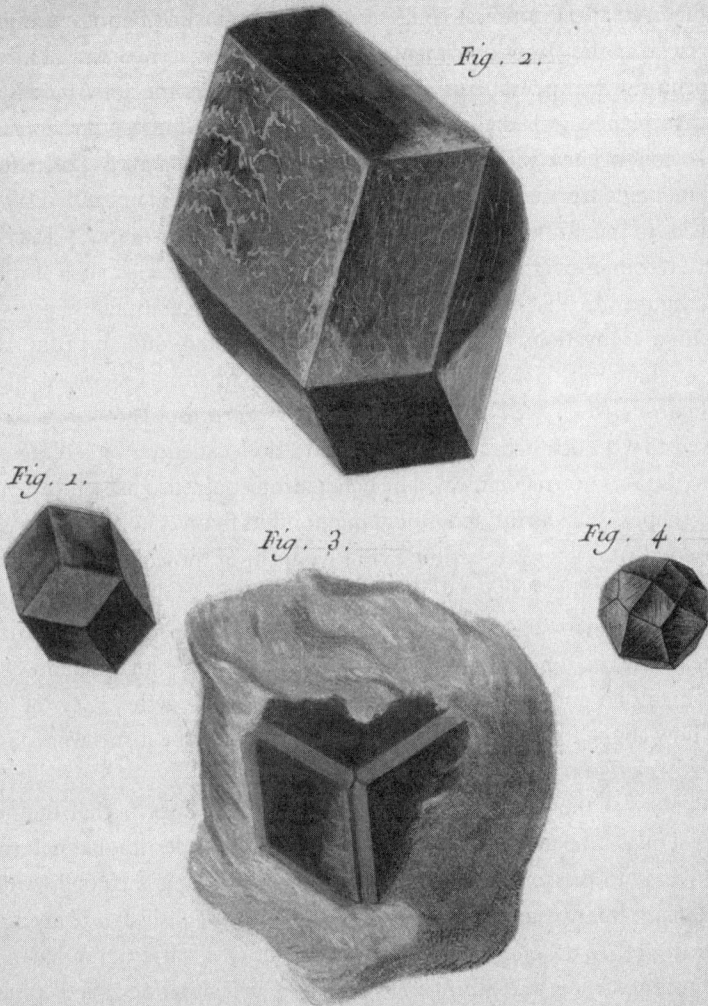

FIGURE 5.7. Romé de l'Isle's crystallography of garnet
The primitive form of garnet was a dodecahedron (*fig. 1*). Variants included the irregular dodecahedron (*fig. 2*), the truncated dodecahedron (*fig. 3*), and a trapezoid (*fig. 4*). The illustrations were based on specimens in Romé de l'Isle's collection, two of which (*figs. 2* and *4*) were probably used by Brisson to measure the density of garnet. From Romé de l'Isle and Gautier d'Agoty, *Natural History* (1781), plate 26. Courtesy of MINES ParisTech.

was the invention of the contact goniometer, a device for measuring the angles between two faces of a crystal. Stones that had the same shape—two dodecahedrons, for example—could now be distinguished by measuring their interfacial angles. The experimental breakthrough was a new way of identifying primitive forms. Haüy used a sharp tool to divide specimens along their cleavage planes until their shape was unaffected by further divisions. This was linked to a new way of generating variants from a primitive form. Rather than mentally truncating the primitive forms, Haüy mentally added layers of particles to them. This layering followed strict rules. For example, each layer differed from its neighbors by an integral number of particles, such as the layers in figure 5.8A that decrease in size by one particle at a time.

These innovations may again be illustrated by diamond and ruby. In the case of diamond, Haüy noticed that the dodecahedral specimens yielded octahedrons when they were divided with a sharp tool (fig. 5.8B). He also showed that an octahedron could become a dodecahedron by adding particles to its faces in a regular way. The octahedron could also become a range of other shapes, such as the spheroidal diamond in figure 5.8C. A similar study of oriental ruby showed that the primitive form of these stones was not the octahedron but the hexagonal prism (fig. 5.8D).[108] Crystallography now combined, rather than separating, the octahedral or "Indian" diamond and the dodecahedral or "Brazilian" diamond. Conversely, crystallography now separated diamond and ruby rather than lumping them together. Crystal forms were now aligned with mineral species. That is to say, the forms were aligned with other properties that defined mineral species.

What were those other properties? The short answer is that they were physical ones. In particular, they were the physical properties that naturalists from Dufay to Brisson had been trying to quantify. This included hardness and density, but not color and transparency. Romé de l'Isle and Haüy agreed on this point. To hardness and density, Haüy added electricity, double refraction, luminescence, and a range of other properties that he called "physical characters."[109] These characters, along with crystal form, were "the most constant, the most general, and the most closely tied to the constitution" of each species.[110] Physical characters have been overlooked by most historians of mineralogy, who have instead focused on the distinction between "internal"

108. The figures relate to the discussions of diamond and telesia in Haüy, *Traité de minéralogie*, vol. 2, 480, 482–83; vol. 3, 288–93.

109. Romé de l'Isle, *Caractères extérieurs*, 5–6, 9–12, 31–39; Haüy, *Traité de minéralogie*, vol. 1, xl–xlii, 210–73, vol. 5, "Système de caractères relatifs aux minéraux"; idem, *Pierres précieuses*, 64–185.

110. Haüy, *Traité de minéralogie*, vol. 1, xxxix.

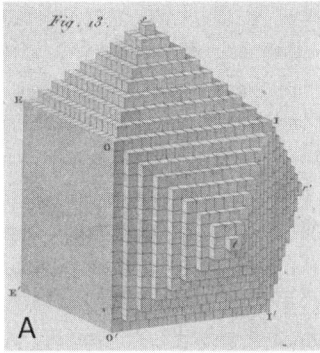

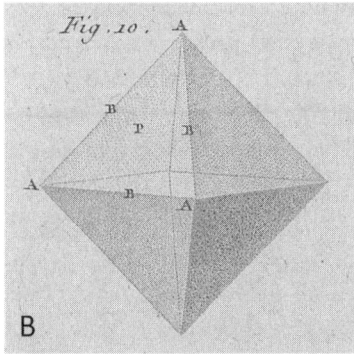

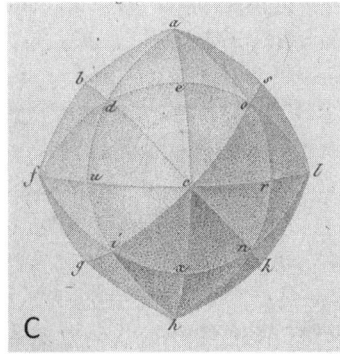

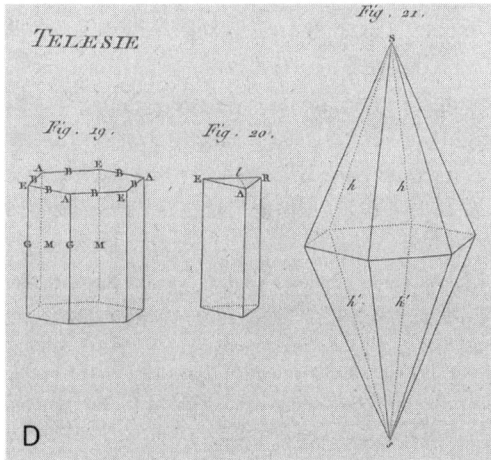

FIGURE 5.8. Haüy's crystallography applied to gems
(A) A cube can be turned into a dodecahedron by adding layers of particles or "constituent molecules." Each new layer retreats from the previous layer by a distance of one particle. (B) An octahedron, the "primitive form" of diamond. (C) The "spheroidal" variety of diamond, formed by adding particles to an octahedron. (D) Forms of oriental ruby, sapphire, and topaz, all of which Haüy called "télésie." This image shows, from left to right, the primitive form of telesia, a constituent molecule, and a variety. From Haüy, *Treatise on Mineralogy* (1801), vol. 5, plates 2, 42, and 62. Courtesy of gallica.bnf.fr / Bibliothèque nationale de France.

properties such as chemical composition and "external" properties such as color and hardness.[111] This distinction made little sense to Haüy, who thought that hardness was a guide to the internal composition of bodies and that color was not. For Haüy, the key division was not between "internal" and "external" characters but between "accidental" and "essential" ones, with physical characters prominent among the latter.[112] For both men, crystal form was first among equals in the classification of minerals. Romé de l'Isle put it like this: "It is not from the consideration of one or other of their external characters, taken in isolation, but from the comparative study of all these characters, that the most natural and illuminating distribution [of minerals] must be established."[113]

Gems illustrate how this "comparative study" worked in practice. We have already seen that Romé de l'Isle used density, not just crystal form, to establish the distinction between oriental and Brazilian diamonds. More generally, he used density and hardness to transform the classification of gems between the first and second editions of his major work on crystallography. In the first edition, oriental ruby, oriental sapphire, and oriental topaz were three different species; in the second, they were varieties of the same species.[114] The first edition had spinel ruby as a variety of ruby; the second had spinel ruby as a distinct species.[115] The first placed Peruvian emerald and Siberian aquamarine in two different species; the second, in the same species.[116] In each case, Romé de l'Isle took a large step away from traditional species of gem and a large step toward modern species, especially the species now known as corundum, spinel and beryl. In each case, he used density or hardness as well as crystal form to justify the change. In fact, he used density or hardness in this way to defend *each* of the eight species of gems he recognized in the second edition.[117] For Romé de l'Isle, hardness and density were not just a pragmatic way of recognizing minerals in the field, nor a post hoc rationalization of

111. Examples are Spencer St. Clair, "Classification of Minerals," passim; Rudwick, *Bursting the Limits of Time*, 61; Simon, "Mineralogy and Mineral Collections," 135; and Laudan, *Mineralogy to Geology*, 78–83.

112. "Essential" versus "accidental" in Haüy, *Traité de minéralogie*, vol. 1, 279. Elsewhere he contrasted "accidental" properties to "specific" ones: ibid., vol. 1, xl–xli. Note also the references to "essential" characters and "accidents of light" in his descriptions of individual species, such as that of telesia: vol. 2, 480, 483.

113. Romé de l'Isle, *Caractères extérieurs*, 40–41.

114. Romé de l'Isle, *Essai de cristallographie*, 213–30; idem, *Cristallographie*, vol. 2, 212–15.

115. Romé de l'Isle, *Essai de cristallographie*, 216–20; idem, *Cristallographie*, vol. 2, 224–25. See also Content, *Ruby, Sapphire & Spinel*, vol. 1, 67.

116. Romé de l'Isle, *Essai de cristallographie*, 235–42, 243; idem, *Cristallographie*, vol. 2, 245–46.

117. The remaining species in the second edition are (to use Romé de l'Isle's names) topaze, rubis, saphir du Brésil, topaze de Saxe, chrysolite ordinaire, and hyacinte. The arguments for

categories established by other means. They were an essential part of the classification of minerals, including new and contested classifications. The same goes for Haüy. His contributions to experimental physics—articles on double refraction, a book on electricity and magnetism, a textbook on physics, the discovery of electrification by pressure—have been noted by historians.[118] But the tightness of the fit between his physics and his mineralogy has been underestimated. Haüy himself underestimated it. In 1817, he gave four examples of gem species that had been established by crystallography, without noting that they had been established with the help of double refraction and electricity as well, as his earlier works show.[119] Cymophane was one such species. It was usually confused with corundum, as Haüy noted in his 1801 treatise, since the two species had a similar hardness and density and often took the same hexahedral form. "There remained only mechanical division *and double refraction* to clear up the ambiguity," he wrote.[120] Refraction was even more important in the union of emerald and aquamarine (plate 7). Like Romé de l'Isle before him, Haüy was struck by the affinity between these two stones with respect to hardness, density and crystal form. But he kept them separate in an early draft of his treatise. It was only when he found a way to observe double refraction in aquamarine that he revised his draft and turned the two species into one.[121] Electricity was similarly decisive in establishing the species of tourmaline, since it suggested that a Norwegian "aphrazite" was really a black tourmaline and that a French "beryl" was really a white tourmaline.[122] Later

---

these species are in *Cristallographie*, vol. 2, 230–31, 261, 271–72, 282–84, respectively. Cf. *Essai de cristallographie*, 216–20, 221–23, 224–26, 226–30, 230–32, 232–35, respectively.

118. The key physical works by Haüy are the following: "Mémoire sur la double réfraction du spath d'Islande," *MAS* 1788 (1791): 34–61; "Sur la double réfraction du cristal de roche," *Journal d'histoire naturelle* 1 (1792): 406–8; "Sur la double réfraction du spath calcaire transparent," *Journal d'histoire naturelle* 1 (1792): 63–80; "Sur la double réfraction de plusieurs substances minérales," *Annales de chimie* 17 (1793): 140–55; "Sur l'électricité produite dans les minéraux à l'aide de la pression," *Mémoires du Muséum national d'histoire naturelle* 3 (1817): 223–28; *Exposition raisonnée de la théorie de l'électricité et du magnétisme* (Paris, 1787); *Traité élémentaire de physique* (Paris, 1803). Aspects of Haüy's physics are covered in Burke, *Origins of the Science of Crystals*, 138–40; Buchwald, "Experimental Investigations of Double Refraction," 335–42; and Christine Blondel, "Haüy et l'électricité: De la démonstration-spectacle à la diffusion d'une science newtonienne," *Revue d'histoire des sciences* 50, no. 3 (1997): 265–82.

119. Haüy, *Pierres précieuses*, iii–v.

120. Haüy, *Traité de minéralogie*, vol. 2, 495; italics added.

121. Ibid., vol. 2, 527–28; René-Just Haüy, "Suite de l'extrait du traité de minéralogie du Citoyen Haüy," *Journal des mines* 6, no. 33 (1797): 686–88.

122. Haüy, *Traité de minéralogie*, vol. 3, 38–39, 43. The fourth example was the union of Saxon and Brazilian topaz: *Traité de minéralogie*, vol. 2, 514.

in his career, Haüy used physical characters to defend crystallography against chemistry. The crucial case involved calcite and aragonite, two substances that had different crystal forms but that appeared to have the same chemical composition. To some, this was a serious blow to Haüy's program of classifying minerals by their crystal form. Haüy's response was to point out that calcite and aragonite differed in their hardness, density, and double refraction. In this case, physical characters tipped the balance in favor of crystallography.[123]

The use of physical characters to classify minerals occurred alongside the study of the physical characters themselves, to the extent that the two activities are often hard to distinguish in Haüy's research. Consider his observation of double refraction in aquamarine, an essential step in uniting aquamarine and emerald in the same species. This was not a trivial observation in 1797. It meant making use of "recent research on double refraction in general," as Haüy put it in an article published in that year.[124] In particular, it meant knowing that the observation of double refraction in a crystal depends on the orientation of the crystal with respect to the observer. The same crystal could show two refractions when viewed through one face but only one refraction when viewed through another face. This effect was already known in rock crystal; Haüy found it in Iceland spar, drawing on the mathematical analysis of double refraction in the stone he had published in 1788.[125] Studying other species, he found that double refraction vanished when viewed through a face perpendicular to the axis of the crystal.[126] Here he used the notion of an "axis" from the crystallographic theory he was developing at the same time. Haüy concluded that he had been looking for the double refraction of aquamarine in the wrong place—through a face that was perpendicular to the axis of the specimen. He found what he was looking for by having the specimen cut so that he could view a ray of light through faces that were *not* perpendicular to the axis. Similar points can be made about electricity, a phenomenon that Haüy used to classify minerals while developing new instruments for measuring electricity and new insights about electrical phenomena in general.[127] Haüy continued to look for correlations until the end of his life, when

123. René-Just Haüy, *Traité de minéralogie*, 2nd ed. (Paris, 1822 [1801]), vol. 1, 482, 484. Cf. idem, "Sur l'arragonite," *Annales du Muséum national d'histoire naturelle* 11 (1808): 237–42; and Burke, *Origins of the Science of Crystals*, 127, 130.

124. Haüy, "Suite de l'extrait du traité de minéralogie," 687. Cf. idem, *Traité de minéralogie*, vol. 2, 527–28.

125. Haüy, "Double réfraction du spath d'Islande."

126. Haüy, "Double réfraction de plusieurs substances minérales."

127. Blondel, "Haüy et l'électricité," 271–73; Burke, *Origins of the Science of Crystals*, 138.

he published a remarkable summary of the "most secret analogies" between electrical, optical and crystallographic phenomena.[128]

These correlations helped to evaluate gems as well as to classify and quantify them. Evaluation was central to Haüy's *Treatise on Precious Stones*, with its proposal to use an array of physical characters as a nondestructive test of the authenticity of cut gems. This book appeared late in Haüy's career, in 1817, which may give the impression that gem appraisal was a topic he took up only after he had mastered the science of mineralogy. But nothing could be further from the truth. As early as 1784, in his first book on crystallography, Haüy mentioned Daubenton's device for evaluating gems by measuring their colors.[129] A decade later he engaged with another technology of gem appraisal, Brisson's table of the densities and refractive indices of gems; this was the inspiration for his technique of cutting gems to better observe their refractions.[130] In the same period, he looked for ways to use physical and crystallographic characters to correct perceived errors in the gem trade. "Skilled cutters in Paris confuse cymophane and chrysolite," he wrote in one paper, before explaining how to tell the two stones apart using crystal form in tandem with density, refraction, and hardness.[131] These tests were codified in the *Treatise on Mineralogy*, where they took the form of "distinctive characters," properties that were characteristic of a given species and that served as convenient tests for distinguishing that species from similar ones.[132] These tests reflected a commercial context that Haüy knew well. Zircon was often cut to resemble diamond, for example, so it is no surprise to find the following in the list of distinctive characters of zircon: "Between cut zircon and the other stones called gems: it differs markedly by the force of its double refraction."[133] It was also in the 1790s that Haüy began to explain the attractive properties

128. René-Just Haüy, "Mémoire sur l'électricité des minéraux," *Annales de chimie et de physique* 8 (1818): 378.

129. Haüy, *Structure des crystaux*, 188–89. Cf. Farges and Kjellman, "Bicentenaire du décès de René-Just Haüy," 36–37.

130. Haüy, "Double réfraction de plusieurs substances minérales," 155; idem, "Description de la Cymophane," 13; idem, "Les pierres appelées hyacinthe," 89–90, 96n1.

131. Haüy, "Description de la Cymophane," 15n1. Cf. the tests for distinguishing cymophane and telesia in the same article (16), and for distinguishing zircon and topaz in Haüy, "Les pierres appelées hyacinthe," 96.

132. Haüy did not mention "distinctive characteristics" when explaining the format of his descriptions in *Traité de minéralogie*, vol. 1, xl, 279–82. But he did use this notion in the descriptions themselves, under the headings "caractères distinctifs," "caractères distinct.," "caractères dist.," "caract. dist.," etc.: ibid., vol. 2, 407, 467, 492, 481, etc.

133. Ibid., vol. 2, 468, 478–79.

of gemstones in terms of Newtonian optics.[134] These explanations drew on his own crystallographic principles, such as his account of the chatoyance of cymophane in terms of the orientation of the stone's crystals.[135]

Gem appraisal linked Haüy to the wider world of gems. Haüy saw himself as an *amateur*, a lover of fine stones, as well as a *physicien*.[136] The gem trade was the source of some decisive specimens, such as a piece of Siberian aquamarine that Haüy "found in commerce" and that bore a suggestive resemblance to a variety of Peruvian emerald (fig. 5.9).[137] Haüy obtained some of his most valuable gems from Henry-Philip Hope, the Anglo-Dutch banker who would later acquire one of the large diamonds in the crown jewels whose density had been measured by Brisson.[138] The most perfect specimen of euclase, as far as Haüy was concerned, was owned by the aristocrat Etienne-Gilbert, Marquis de Drée, whose vast collection of gems was designed to train the eyes of jewelers.[139] The pyroelectricity of tourmaline was known to jewelers and collectors before it became one of the pillars of Haüy's classification of gems.[140] Tourmaline was worn on rings and used as "a little electrical instrument, which need only be presented to the fire, to be able to attract and repel light bodies," according to Haüy.[141]

134. René-Just Haüy, "Sur les couleurs de l'agathe opaline nommé communément opale," *Journal d'histoire naturelle* 2 (1792): 9–18; idem, "Sur les hydrophanes," *Journal d'histoire naturelle* 1 (1792): 294–99. Cf. *Traité de minéralogie*, vol. 2, 456–59 (opal), 454–56 (hydrophane); *Pierres précieuses*, 70–72 (opal), 206–10 (hydrophane).

135. Haüy, "Description de la Cymophane," 16. See also his later explanation of the optics of garnet: *Pierres précieuses*, 76–77, 79–80.

136. Haüy, *Pierres précieuses*, xx–xxi; idem, *Minéralogie*, vol. 2, 489. See also the discussion of Haüy's *objets d'art* in Farges and Kjellman, "Bicentenaire du décès de René-Just Haüy," 38.

137. Haüy, "Suite de l'extrait du traité de minéralogie," 687n1. Cf. Haüy, *Traité de minéralogie*, vol. 2, 526; idem, "Sur la cristallisation de l'émeraude," *Journal des mines* 2, no. 19 (1795): 72–74.

138. Haüy, *Pierres précieuses*, xxi–xxii. On the Hope Diamond, see Farges et al., "The French Blue and the Hope."

139. Haüy, *Traité de minéralogie*, vol. 2, 535; Etienne-Gilbert, Marquis de Drée, *Catalogue des huit collections qui composent le musée minéralogique de Et. de Drée* (Paris, 1811), 13. On Drée's collection, see chap. 7, this volume.

140. Jewelers in Burke, *Origins of the Science of Crystals*, 137, citing Franz Aepinus, "Mémoire concernant quelques nouvelles expériences électriques remarquables," *Histoire de l'Académie Royale des Sciences et Belles Lettres de Berlin* 12 (1756): 105; and Roderick W. Home, "Aepinus, the Tourmaline Crystal, and the Theory of Electricity and Magnetism," *Isis* 67 no. 1 (1976): 23–24. Collectors in Romé de l'Isle and Dugaut, *Catalogue du cabinet de Mr. Davila*, vol. 2, 279; Jean-Baptiste Louis Romé de l'Isle, *Catalogue raisonné des minéraux, pierres fines et cristallisées, pétrifications, coquilles, madrépores, et autres curiosités de la nature et de l'art: Qui composent le cabinet de M. Galois* (Paris, 1780), 97.

141. Haüy, *Traité de minéralogie*, vol. 3, 58.

PL. XIV.

VARIETE DU BERIL ET DE L'EMERAUDE

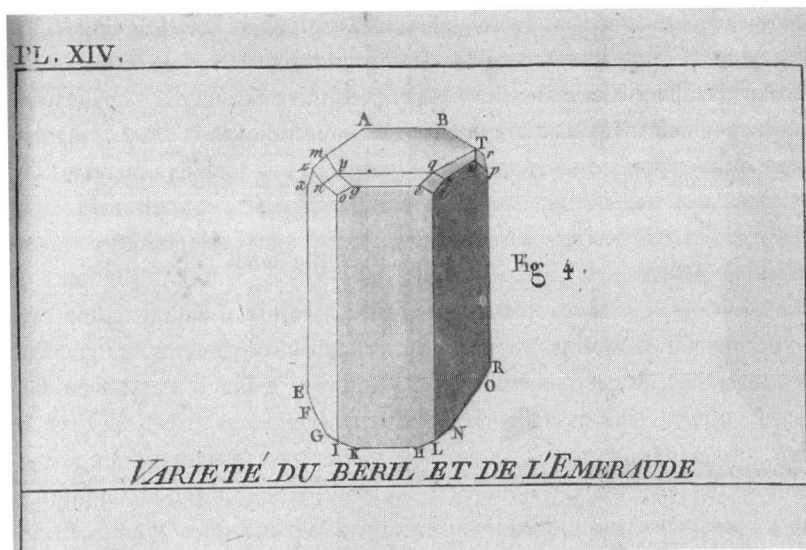

FIGURE 5.9. A variety of beryl and emerald found in commerce
A diagram of a crystal form that Haüy found in specimens of aquamarine (or "beryl") from Siberia, and also in specimens of emerald from the Americas. Note especially the beveled corners and edges. This resemblance was one of Haüy's arguments for treating emerald and aquamarine as varieties of the same species. He observed this form in a piece of aquamarine that he "found in commerce." From Haüy, "On the Crystallization of Emerald" (1795), plate 14. Courtesy of MINES ParisTech.

The names of cutters and jewelers are scattered through Haüy's writings. These men cut gems to facilitate his observations of double refraction; determined the hardness of gems against the grinding wheel; and advised the mineralogist on the commercial names of gems.[142] Cutters were probably the source of the fundamental manual operation in Haüy's crystallography, the mechanical division of crystals. Haüy was coy about the details of this operation, but he did write that "mechanical division" was a synonym for the cutter's term, "cleavage."[143] He might have added that his concept of the "natural joints" of a mineral—the planes through which the mineral can be easily divided—was very similar to the cutter's notion of the "grain" of a diamond. Like d'Augny, the Parisian *amateur* we met earlier, Haüy combined a

142. Haüy, "Double réfraction de plusieurs substances minérales," 151, 153–54 (Carrochés); idem, "Les pierres appelées hyacinthe," 88, 90 (Pichenot); idem, *Pierres précieuses*, 235 (Achard). On Achard, see Farges et al., "The French Blue and the Hope," 11.

143. "Lapidaire," in Diderot and d'Alembert, *Encyclopédie*, vol. 9 (1765), 282; Haüy, *Pierres précieuses*, 3. Haüy's frustratingly brief descriptions of the operation are in "Structure des crystaux de grenat," 367; idem, "Double réfraction de plusieurs substances minérales," 149; and *Traité de minéralogie*, vol. 1, xiv, 20, 243, 255–56.

knowledge of gems with an acquaintance with cutters, jewelers, and collectors. Like his predecessors at the academy and the Jardin du Roi—including Dufay, Daubenton and Brisson—Haüy used these resources to quantify the science of gems. Crystallography was an extension of the tradition that had already been applied to color, density, hardness, and double refraction.

Gems and the Quantifying Spirit

This tradition brought together the material worlds of natural history and experimental physics. It brought together optical instruments and gem collections; the Cabinet of Natural History and the Cabinet of Physics and Optics; naturalists such as Daubenton and experimenters such as Nollet; precise measurements and carefully chosen specimens. It was a form of connoisseurship, in the sense that it was linked to the world of painting and sculpture and to the *amateurs* who judged them. But it was a broad form of connoisseurship, one that included hydrostatic balances and contact goniometers as well as visual judgments. It was linked to public spectacle—recall the electrical tourmaline rings mentioned by Haüy—but it was also linked to trade and empire via the old practice of judging gems by where they came from. Density, refraction, and crystal form were used as proxies for geography, just as hardness had been used for this purpose in earlier centuries. These proxies were used to evaluate gems in new places, such as diamond in Brazil, rock crystal in Madagascar, and the hypothetical gems from Pegu imagined by Daubenton. Natural history was part and parcel of the quantifying spirit.

Moreover, classification evolved in the same way as quantification. Quantification was not new in the physical sciences in the eighteenth century, but an extension to new phenomena of methods that had already been applied to light and motion.[144] Likewise, gem classification was not new in the eighteenth century, having been done with color, transparency, and locality since at least the Renaissance. The novelty in the later eighteenth century was the extension of the old methods to new phenomena, especially density, double refraction, and crystal form. This change in subject matter was accompanied by a change in method, with gem classification becoming quantitative for the first time. The two-stage narrative of quantification has a parallel in the history of classification.

The irony is that color, transparency and locality, which had done so much to nurture the quantitative approach, were eventually consumed by their own children. Diamonds illustrate the point. The distinction between

---

144. Kuhn, "Mathematical Versus Experimental Traditions."

octahedral and dodecahedral diamonds emerged from the distinction between oriental and Brazilian diamonds, but the crystallographic distinction
ended up superseding the geographical one. More generally, there was little
room for the category of precious stones in Haüy's 1801 treatise on mineralogy. Gemstones were scattered through the four broad classes of mineral
(acidic, earthy, combustible and metallic) in the treatise.[145] Most gemstones
were classified as earthy substances. But some important gems, such as diamond and turquoise, were classified elsewhere.[146] And those listed as earthy
substances rubbed shoulders with apparently unrelated minerals, such as talc
and asbestos. All these minerals were classified on the basis of their chemical
composition, crystal form, and "physical characters" such as hardness, density and double refraction. Color and transparency, by contrast, were too capricious to be a guide to the "nature" of mineral species, in Haüy's view. And
the geographical origins of minerals were discussed only in the "annotations"
that followed the characterization of each species.[147] The transparency-color-
locality scheme had become the crystal-chemistry-physics scheme. Physics
and crystallography had been introduced to refine the traditional taxonomy
of gems, but they ended up undermining that taxonomy. They did so in tandem with chemistry, another field in which material evaluation brought together different material worlds.

145. The classification is summarized in Haüy, *Traité de minéralogie*, vol. 2, xxii–xxxiii, vol. 5,
1–10; and Jean-André-Henri Lucas, *Tableau méthodique des espèces minérales* (Paris, 1806–1813),
vol. 1, xxv–xxxviii.

146. Haüy, *Minéralogie*, vol. 2, 164 (marble); vol. 3, 287 (diamond).

147. Haüy, *Traité de minéralogie*, vol. 1, vi–xvii, xl–xlii, xlvii.

6

Gems, the Crafts, and Chemical Composition

Gems were a hard case for chemistry. For a long time, they resisted all efforts to discover what they were made of. As late as 1758, the leading mineral chemist in Europe, Axel Fredrik Cronstedt, had to apologize to his readers for failing to identify the constituents of any of the precious stones he described in his major work on mineralogy.[1] Four decades later, the situation had changed beyond recognition. In a book published in 1801, René-Just Haüy could confidently declare that emeralds are made up of 64.5 parts quartz, 16 clay, 13 beryllia, 3.25 chromium oxide, 1.6 lime, and 2 volatile matter.[2] Two of these substances (beryllia and chromium oxide) had been isolated for the first time in the 1790s. The same substances had been found in aquamarine in the same proportions, a strong hint that emerald and aquamarine belonged to the same species. Haüy gave a similar breakdown of the composition of all the other gems in his treatise. He used these data to reach surprising taxonomic conclusions, such as that the very idea of a gemstone has no place in mineralogy. The chemical analysis of gems is one of the untold stories of eighteenth-century chemistry.[3]

1. Axel Fredrik Cronstedt, *Essai d'une nouvelle minéralogie*, trans. Dreux fils (Paris, 1771 [1758]), xxx.

2. Haüy, *Traité de minéralogie*, vol. 2, 518 and 528–29. Haüy's terms were (respectively) "silice," "alumine," "glucyne," "oxyde de chrome," "chaux," and "matières volatile." On beryllia and chromium oxide, see notes 107, 108, and 122.

3. "Untold" as a general phenomenon. I am indebted to existing studies of the chemistry of individual gems, especially emerald, diamond and corundum. Michel Spiesser, "Nicolas Louis Vauquelin—La découverte de deux nouveaux éléments: Le chrome (1797) et le glucinium (béryllium 1798)," *Bulletin de l'Union des physiciens* 10, no. 807 (1998): 1403–16; Lehman, "Nature of Diamond"; Irish, "The Corundum Stone and Crystallographic Chemistry"; Farges and Kjellman, "Bicentenaire du décès de René-Just Haüy," 31–32.

This story is a good test of transmaterialism. Chemistry lends itself to a materialistic approach, with its focus on material substances and its role in material life outside the laboratory. Eighteenth-century chemistry is no exception to the rule. We have many studies of the role of mining, pharmacy, dyeing, glassmaking, porcelain making, and other crafts in the emergence of chemistry as a distinct science in this period.[4] State-sponsored mining schools, the beginnings of industrial chemistry, and the effort to replicate Chinese porcelain with European ingredients, have all been linked to the science of chemistry. Yet the crafts are usually studied one at a time, with the goal of showing that the science of chemistry was inseparable from its practical applications—that "mind" and "hand" went together, to use the standard shorthand. My goal is different. It is to show that hand-hand coordination mattered as much as hand-mind coordination in the history of chemistry. Metal mining and porcelain making came together in early attempts to classify gems by their composition. Porcelain making and diamond polishing were both needed to show that diamonds are a combustible substance. Drugs and glass were both involved in the first true decomposition of gems, done by the Swedish chemist Torbern Bergman. Ultimately, these chemical analyses of gems were combined with the crystallographic analyses we met in the previous chapter. In France, this meant combining a chemical culture associated with metals and a crystallographic culture associated with gems. These interactions between crafts were not just a matter of generalization or juxtaposition. New kinds of analysis emerged when two or more crafts interacted. Chemistry was greater than the sum of its crafts.[5]

4. The following is a recent, representative sample of a large literature. Mining: Porter, "Promotion of Mining"; Fors, *Limits of Matter*, chap. 5; Ursula Klein, *Technoscience in History: Prussia, 1750–1850* (Cambridge, MA: MIT Press, 2020), pt. 2. Pharmacy: Klein and Lefèvre, *Materials in Eighteenth-Century Science*, pt. 3; Jonathan Simon, *Chemistry, Pharmacy and Revolution in France, 1777–1945* (Aldershot: Ashgate, 2005); Spary, *Eating the Enlightenment*; idem, *Feeding France: New Sciences of Food, 1760–1815* (Cambridge, UK: Cambridge University Press, 2020); Klein, *Technoscience in History*, chap. 3. Porcelain: Mark Pollard, "Letters from China: A History of the Origins of the Chemical Analysis of Ceramics," *Ambix* 62, no. 1 (2015): 50–71; Klein, *Technoscience in History*, chap. 4. For other examples, see Klein and Spary, *Materials and Expertise*; Ursula Klein, "Artisanal-Scientific Experts in Eighteenth-Century France and Germany," *Annals of Science* 69, no. 3 (2012): 303–6; John Perkins, ed., "Sites of Chemistry in the Eighteenth Century," special issue, *Ambix* 60, no. 2 (2013); Roberts and Werrett, *Compound Histories*; and Leslie Tomory, "Trade and Industry: An Era of New Chemical Industries," in *A Cultural History of Chemistry in the Eighteenth Century*, ed. Matthew Eddy and Ursula Klein (London: Bloomsbury, 2022), 137–56.

5. I've not found a clear and general statement of this point in the literature cited in the previous note, though there are suggestive examples in Szabadváry, *History of Analytical Chemistry*,

The story of gem analysis is significant for another reason. The story un-
folds in the latter half of the eighteenth century, a crucial period in the history
of chemical composition. This is the period of the Chemical Revolution, an
event centered on the rejection of the phlogiston theory of combustion and
its replacement by the oxygen theory. It has long been known that the Chemical Revolution was linked to changes in the study of chemical composition.
The anti-phlogistonists are celebrated as much for their compositionism as
for their skepticism about phlogiston.[6] They divided substances into their
components and they used their results to name, define, and measure the
substances so divided. The debate about phlogiston was "a ripple riding on a
large wave, which was the very gradual establishment of compositionism," in
the words of Hasok Chang.[7] Gems rode the same wave. They also help to map
the contours the wave. I return to compositionism in the conclusion, after
explaining how the crafts came together to take gems apart.

Metals and Porcelain

To tell this story, we need to leave Paris for a moment and travel to the mining towns of Germany in the middle decades of the century. Johann Friedrich
Henckel, Christlieb Ehregott Gellert, and Johann Gottlob Lehmann exem-
plify the link between mining and chemical composition in this period.[8] The

43; Klein and Lefèvre, *Materials in Eighteenth-Century Science*, 153, 182; Simon, *Chemistry, Pharmacy and Revolution*, 131–39; Klein, *Technoscience in History*, 35, 52; and elsewhere.

6. A still-current textbook account is William H. Brock, *The Fontana History of Chemistry*
(London: Fontana, 1992), esp. 84–85, 125–26. Current thinking on the topic is summarized in
Matthew Eddy and Ursula Klein, "Introduction: The Core Concepts and Cultural Context of
Eighteenth-Century Chemistry," in Eddy and Klein, *Cultural History of Chemistry*, 13–14.

7. Hasok Chang, *Is Water H2O? Evidence, Realism and Pluralism* (Springer: Dordrecht,
2012), 42. Chang draws on several earlier works (ibid., 37, 39), especially Robert Siegfried, "From
Elements to Atoms: A History of Chemical Composition," *Transactions of the American Philosophical Society* 92, no. 4 (2002): 1–278; and Klein and Lefèvre, *Materials in Eighteenth-Century
Science*. Note also older works such as Szabadváry, *History of Analytical Chemistry*, chaps., 4, 5,
6, 8, esp. p. 71; and Maurice Crosland, *Historical Studies in the Language of Chemistry* (London:
Heineman, 1962), 126–30, and pt. 3, chaps. 3–6. Like any consensus, this one leaves room for
debate: William R. Newman, "Mercury and Sulphur Among the High Medieval Alchemists:
From Rāzī and Avicenna to Albertus Magnus and Pseudo-Roger Bacon," *Ambix* 61, no. 4 (2014):
327–44; Ursula Klein, "A Revolution That Never Happened," *Studies in History and Philosophy
of Science* 49 (2015): 80–90; and Hasok Chang, "The Chemical Revolution Revisited," *Studies in
History and Philosophy of Science* 49 (2015): 91–98.

8. Porter, "Promotion of Mining," 549 (Freiberg Academy); David Oldroyd, "Some Phlogistic Mineralogical Schemes, Illustrative of the Evolution of the Concept of 'Earth' in the 17th and
18th Centuries," *Annals of Science* 31, no. 4 (1974): 278–83 (Henckel), 293–96 (Lehmann); Spencer

first two were involved in the Freiberg Mining Academy. Founded by the king of Saxony in 1765, the academy is now remembered as a precocious example of state-sponsored technical education. Henckel's annual course on metallurgical chemistry, begun in the late 1730s in Freiberg, is usually seen as a precursor to the mining academy; Gellert was a founding professor at the academy. Lehmann followed a similar route, working as a mining adviser in Berlin before becoming a chemistry teacher in St. Petersburg. In the years around 1750, this trio published books that were milestones in the chemical classification of minerals. The titles of the books reflected their origins in metallic mining: *Introduction to Mineralogy, Metallurgical Chemistry*, and *Introduction to the Knowledge Necessary for the Exploitation of Metallic Mines*. Even Henckel's book, ostensibly an introduction to minerals in general, had twice as many pages on metals and metallic ores as it did on all the other classes of mineral combined.[9] Each author sought to classify minerals according to their chemical composition—their "chemical anatomy," "interior," and "essential combination," to use Henckel's terms.[10] Each author was wary of external properties such as color, with Henckel writing that color should be "banned from philosophy" in the study of stones.[11] Each author combined the classification of minerals with detailed descriptions of assaying, the art of separating metallic ores into their components with a view to determining their value. In this way, metal mining was one important source of what we might call analysis "by decomposition"—the practice of determining the composition of substances by extracting their components in the laboratory.

But metallic ores were not gemstones. Methods that worked for the former did not necessarily work for the latter. Henckel's discussion of stones in his *Introduction to Mineralogy* illustrates the point. There he classified stones according to their response to fire and acid.[12] He grouped precious stones

St. Clair, "Classification of Minerals," 127–31 (Henckel). Cf. Karl Hufbauer, *The Formation of the German Chemical Community (1720–1795)* (Berkeley: University of California Press, 1982), chaps. 2, 3, 4, esp. 42–45 and 58 (mining).

9. Johann Friedrich Henckel, *Introduction à la minéralogie*, trans. Paul-Henri Thiry, Baron d'Holbach (Paris, 1756 [1747]), vol. 1, xv–xvi. Cf. Johann Gottlob Lehmann, *L'art des mines*, trans. Paul-Henri Thiry, Baron d'Holbach (Paris, 1759 [1750–1758]), vol. 1, xxii–xv, where metals dominate as well. The same goes for Christlieb Ehregott Gellert, *Chimie métallurgique*, trans. Paul-Henri Thiry, Baron d'Holbach (Paris, 1758 [1751]), vol. 1, viii–xii.

10. Johann Friedrich Henckel, *Idée générale de l'origine des pierres* [1734], trans. Paul-Henri Thiry, Baron d'Holbach, in Henckel, *Pyritologie, ou Histoire naturelle de la pyrite* (Paris, 1760 [1725]), 420 ("anatomy"); idem, *Introduction*, vol. 1, 55–56 and 63 ("interior," "essential combination"). I have relied on the French translations of these German works.

11. Henckel, *Origine des pierres*, 427.

12. Henckel, *Introduction*, vol. 1, 40–55. Cf. vol. 2, 330–34.

with flints and sandstone on the grounds that they do not effervesce with acids and are not calcined by strong fire. Henckel called this group "silicei" and "noncalcareous stones," and he wrote they are "of the nature of flint."[13] These terms suggest that precious stones contain flint. But Henckel did not reach this conclusion by separating precious stones into their components and identifying one of those components as flint. He simply showed that precious stones respond to fire and acids in the same way as flint. As he put it, he found an "analogy" between the "properties" of these substances.[14] His unstated assumption was that stones that behave in the same way as flint are made of flint. He made the same assumption in an earlier work in which he divided gemstones into classes. There he argued, for example, that rock crystal probably contains marl because, like marl, it melts easily in a furnace.[15] Henckel based his analogies on the physical properties of gems as well as on their chemical properties. He placed gems among the silicei because they were hard and gave a high polish, not only because of their response to fire and acid.[16] In short, he did not analyze gems in the same way he analyzed metallic ores. He analyzed gems by analogy rather than by decomposition, and the analogies drew on physical as well as chemical properties. This was a compromise between the ideal of decomposition and the realities of gemstones.

It was also a synthesis of two crafts, metal mining and jewelry. Henckel was familiar with jewelers' taxonomies of gems, describing them in some detail in his *Introduction to Mineralogy*. There he explained how jewelers divide gems into groups according to their hardness, transparency, price and color.[17] He presented these as pragmatic divisions that were inferior to the division by "essential combination," meaning their composition. Yet he used hardness and polish even when he classified gems by their composition, as we have seen. He also referred to the "first order" and "second order" of gems, a distinction based on their price.[18] This was not just a pedagogical device but also a matter of practical significance, as Henckel's study of Saxon topaz shows (plate 6, bottom left). Henckel described what he called "true Saxon topaz" in a short article published in 1737.[19] He maintained that his compatriots had undervalued "our topaz," selling it at low prices to foreign merchants who sold it back to

13. Ibid., vol. 1, 65.
14. Ibid., vol. 1, 64.
15. Henckel, *Origine des pierres*, 420–31.
16. Henckel, *Introduction*, vol. 1, 59–60, 65. Cf. Gellert, *Chimie métallurgique*, 15–18; Lehman, *L'art des mines*, vol. 1, 143–44.
17. Henckel, *Introduction*, vol. 1, 40–55.
18. Ibid., vol. 1, 48–49, 65.
19. Henckel, "Dissertation sur une véritable topase [1737]," in Henckel, *Pyritologie*, 500–503.

Saxons as an "oriental" topaz. Henckel traced this commercial error to a classificatory one, namely the confusion of Saxon topaz with yellow fluorspar and yellow rock crystal. He used external properties to clear up the confusion: the brilliance, hardness, and layered texture of the Saxon topaz convinced him that it was a distinct species. Tellingly, he christened this new species "the precious stone of our Northern Indies."[20] Saxon gem cutters helped with the identification, since they cut the stone into the form of diamond-like brilliants, using a rock in which the stone was found. Henckel invoked this practice to show the hardness of the stone and thereby distinguish it from yellow rock crystal. He also pointed out that topaz was much harder to fuse than rock crystal. The chemical test (fusibility) corroborated the external properties (hardness, brilliance, texture). In short, Henckel identified the new stone by combining the methods of metal miners and gem cutters.

The analysis of Saxon topaz did not end there. Another craft, porcelain making, brought another kind of analysis into play. The key chemist here was Johann Heinrich Pott, a follower of Henckel. Some time in the 1740s, with Henckel's help, Pott was hired by Frederick the Great of Prussia. Pott's task was to discover the recipe for hard-paste porcelain—already known to the king of Saxony—and to replicate the recipe with minerals available in Prussia. The result was what one historian has called a "massive empirical onslaught against the problem of porcelain."[21] Pott carried out thousands of trials on many different minerals using a purpose-built furnace.[22] The format of most of these trials was the same: mix several known minerals; heat them to high temperatures in a crucible; observe the results. Pott was not able to make porcelain, but he did distinguish between four kinds of mineral, namely those that are like chalk, gypsum, clay, and flints.[23] Like Henckel, Pott analyzed minerals by analogy. He assumed that minerals that respond to fire in the same way are made of the same stuff, an assumption he made explicit.[24] But

20. Henckel, *Pyritologie*, 415.

21. Oldroyd, "Phlogistic Mineralogical Schemes," 286. On Pott's career, see Partington, *History of Chemistry*, vol. 2, 717–18.

22. Partington reports that Pott performed thirty thousand trials, but the source of this datum is unclear. About one thousand trials are recorded in Johann Heinrich Pott, *Lithogéognosie*, trans. Didier d'Arclais de Montamy (Paris, 1753 [1746]), vol. 1, 431 onward.

23. Pott, *Lithogéognosie*, vol. 1, chaps. 1–4.

24. Ibid., vol. 2, 7–8, 110–18. Note also his rejection of external properties in ibid., vol. 2, 14–19. Cf. Porter, "Promotion of Mining," 557: "Pott, more than anyone else, deserves to be credited with the analytical definition of simple substance." The quotes from Pott's *Lithogéognosie* that Porter reproduces show only that Pott classified minerals by composition, not that he did this by decomposing minerals.

he gave two new twists to this mode of analysis, twists that reflected his aim of making porcelain. He varied the materials with which he fired the mineral under test; and he paid close attention to the visual properties of the resulting fusion.

Pott's study of Saxon topaz illustrates his technique.[25] He heated samples of the stone with seven other substances, from alkaline salts to borax, trying a range of combinations and proportions of each of these substances. In each case he recorded the visual qualities of the resulting fusions, sometimes using porcelain or gems as an analogy for these qualities. One mixture gave an opaque mass that resembled the white color of porcelain; another a "beautiful blue color"; another a white substance with the consistency of agate. Taken together, these trials helped to distinguish Saxon topaz from other gems, which Pott had studied in the same way. His work on Saxon topaz is of particular interest because he presented it as a model for the analysis of gems in general. Moreover, he explicitly rejected the traditional approach of using solvents ("menstruums") to extract the components of gems that gave them their color. He associated this approach with physicians who hoped to use the extracts of gems for medical purposes. His own approach, which he called "the resolution of precious stones," went in the opposite direction. He inferred the natures of gems from what they made, not from what they were made of. Gems that made white porcelain with borax were of one nature; gems that made something else with borax were another; and so on for additives other than borax. "Resolution" was a form of analysis that made sense to a porcelain maker. At the same time, it was an extension of Henckel's method of analogy, which was in turn derived from assaying and gem cutting.[26] Taking Pott and Henckel together, we must conclude that there was more to German chemical mineralogy than one craft (metallic mining) and one mode of analysis (decomposition). Gem cutting and porcelain making suggested forms of analysis that did not involve decomposition and that were sometimes directly opposed to decomposition. Different crafts lent themselves to different techniques; these techniques intermingled in the work of scholar-artisans such as Pott and Henckel.

Diamond and Porcelain

The ideas of Pott and Henckel quickly migrated to France, where they were involved in perhaps the most spectacular experiments on gems in the eighteenth

25. Pott, *Lithogéognosie*, vol. 1, 254–77, and tables LX to LXXI.
26. Pott acknowledged his debt to Henckel in *Lithogéognosie*, vol. 1, 255.

century. In a flurry of experiments done in 1771 and 1772, a group of Parisian chemists showed that diamonds were as combustible as charcoal. This was a remarkable result given the fabled indestructibility of diamonds. It was also a technically challenging set of experiments, with diamonds being heated to high temperatures in the presence and absence of air. The experiments have been recounted several times before, but they are worth revisiting from the point of view of the crafts involved.[27] In these experiments, jewelers and porcelain makers combined to do something that neither could do on their own.

Both crafts were part of the background to these experiments. Pott's *Lithogeognosy*, translated into French in 1753, introduced Parisian chemists to Pott's systematic search for the ingredients of hard-paste porcelain. Henckel's *Introduction to Mineralogy*, translated in 1756, included a long footnote on the response of diamonds and other gemstones to fire, based on experiments done several decades earlier at the courts of the Grand Duke of Tuscany and the Holy Roman Emperor.[28] These translations fell on fertile ground. The French attack on hard-paste porcelain, begun by Réaumur in the 1710s, was in full swing when the translations appeared. Jean-Etienne Guettard, a student of Réaumur, had revived the project in 1750. Another chemist, Pierre-Joseph Macquer, led a similar effort at the Porcelain Manufactory of Sèvres from 1757. A third team of porcelain hunters was established the following year, with the chemists Jean Darcet and Augustin Roux in charge.[29] There was also a French precedent for heating diamonds. Around 1730, Charles Dufay had heated ten diamonds in a crucible for two hours, such that the crucible was entirely fused or "vitrified" on its external surface.[30] Dufay also reported in unpublished notes that he had placed a diamond "in the focus of

27. The following account expands on Pierre-Joseph Macquer, *Dictionnaire de chimie*, 2nd ed. (Paris, 1778 [1766]), vol. 1, 467–92; Partington, *History of Chemistry*, vol. 3, 381–83; Henry Guerlac, *Lavoisier, the Crucial Year: The Background and Origin of His First Experiments on Combustion in 1772* (Ithaca, NY: Cornell University Press, 1961), 78–90; and Lehman, "Nature of Diamond."

28. Henckel, *Introduction*, vol. 2, 413. Cf. Paul-Henri Thiry, Baron d'Holbach, "Pierres précieuses," in Diderot and d'Alembert, *Encyclopédie*, vol. 12 (1765), 593–95, on 594–95. D'Holbach's sources were Anon. "Esperienze fatte con lo specchio ustorio di Firenze sopra le gemme, e le pietre dure," *Giornale de' letterati d'Italia* 8 (1711): 221–309; and Anon., "Bersuchet, welche mit einigen Edelgesteinen, sowol im feuer, als auch vermittelst eines Tschirnhausischen Brennglases angestellet worden," *Hamburgisches Magazin* 18 (1757): 164–80. Cf. Henry Guerlac, "Some French Antecedents of the Chemical Revolution," *Chymia* 5 (1959): 100–104.

29. Guerlac, "French Antecedents," 83–86; Christine Lehman, "Pierre-Joseph Macquer: An Eighteenth-Century Artisanal-Scientific Expert," *Annals of Science* 69, no. 3 (2012): 322–32.

30. Dufay, "Lumière des diamants," 362–64.

a burning mirror for as long as the hand could tolerate the heat."[31] As we have
seen, Dufay's work of gems eventually led to Louis-Jean-Marie Daubenton's
article on gem color, published in the academy's journal in 1750.[32] Gems and
porcelain came together in a paper read to the academy by Darcet in 1768.
Darcet's porcelain trials, like the trials by Pott on which they were modeled,
had prompted the Frenchman to heat a large number of mineral substances
in a purpose-built furnace. When he heated diamonds in this way, Darcet
found that they vanished like so many drops of water.[33] The cause of this phenomenon was explored in a flurry of experiments in 1771 and 1772 in which
Darcet, Roux, Macquer, Hilaire-Marin Rouelle, and a young Antoine-Laurent
Lavoisier, were key players.

The main result of these experiments was not simply to confirm that diamonds can be destroyed by fire. Nor was it that diamonds are made of the
same stuff as charcoal, a fact that was only demonstrated some thirty years
later. The conclusion was instead that diamond was a combustible substance
rather than a vitrifiable stone. In other words, diamond was reclassified. It
migrated from one broad class of minerals to another. This was the main
conclusion of the two key articles written by Lavoisier in 1772. Lavoisier did
mention the surprising resemblance between charcoal and diamond, but he
immediately qualified this: the resemblance "exists only because they both
seem to belong to the class of combustible bodies, and both may be regarded
as the most fixed substances of this class when they are secured against contact with air."[34] Macquer drew the same lesson when he surveyed the whole
episode in a dictionary published in 1778.[35] This concern for classification explains the continued interest the experimenters showed in stones other than
diamond, including other precious stones. Four of the key experimenters,

31. Dufay, "Notes sur l'électricité." The mirror in question was very likely the one that the
German mathematician Ehrenfried Walther von Tschirnhaus had sold to Philippe II, Duc
d'Orléans, in 1702. After being used for some years by Wilhelm Homberg, this enormous and
expensive mirror was installed at the Bercy laboratory of one of Dufay's associates, Louis-Léon
Pajot, Comte d'Onsenbray, where it remained until the Comte's death in 1754. See Principe,
Transmutations of Chymistry, 185–93, 387–89; Lehman, "Nature of Diamond," 375–82; Bycroft,
"Physics and Natural History," 113–14.

32. Daubenton, "Connoissance des pierres précieuses."

33. Jean Darcet, *Mémoire sur l'action d'un feu egal, violent et continué pendant plusieurs jours
sur un grand nombre de terres, de pierres [et] de chaux métalliques* (Paris, 1766), "Avertissement"
(Pott), 10–11 (furnace); idem, *Second mémoire sur l'action d'un feu égal, violent, et continué pendant plusieurs jours* (Paris, 1771); Guerlac, *Crucial Year*, 79, citing AS PV 87 (1768), f. 72v.

34. Antoine-Laurent Lavoisier, "Second mémoire sur la destruction du diamant," *MAS* 1772
2e partie (1776): 616, cf. 613–14.

35. Macquer, *Dictionnaire de chimie*, vol. 1, 484, 489.

including Lavoisier, combined their study of diamond with studies of ruby, emerald, hyacinth, and other gems.[36] They were interested in precious stones as a category, not just in the nature of diamond. This also helps to account for the paradox that the category persisted even as its most important item disappeared in a puff of smoke. Buffon was confident that all the oriental stones would combust in the same way as diamond, given a strong enough furnace.[37] Lavoisier did not go that far, but in 1782 he still thought of diamond and the other precious stones as a coherent category. In an article published in that year, he distinguished five classes of "precious stone" based on their response to fire, with diamond as one of those classes.[38] Lavoisier was untroubled by the idea that diamond was *both* a precious stone *and* a combustible substance. The experiments on diamonds were a way of studying gems as a whole.

Jewelers played a decisive role in this work. Lavoisier referred to a "famous jeweler" by the name of Leblanc and a "skilled jeweler" by the name of Maillard.[39] Other jewelers were involved as well, but these two men took the initiative at a crucial moment.[40] The destruction of diamond by fire, reported by d'Holbach in his footnote in 1756, had been replicated in Paris on three occasions between 1768 and 1770.[41] But doubt remained about the mode of the destruction. Did the diamond evaporate like boiling water, burn like wood in a flame, or shatter into tiny particles due to the temperature difference between the hot diamond and the cool air that surrounded it? Leblanc and Maillard took three steps toward answering this question. First, they insisted—in the face of the incredulity of other participants, including Macquer—that diamonds could withstand high temperatures if suitably prepared. Second, they

36. Antoine-Laurent Lavoisier, "Premier mémoire sur la destruction de diamant," *MAS* 1772 2e partie (1776): 572 (Darcet on precious stones), 582–83 (Mitouard on precious stones); Macquer, *Dictionnaire de chimie*, vol. 1, 473 (Rouelle on precious stones); Jean Darcet, *Second mémoire sur l'action du feu*; idem, "Expériences faites au feu, sur un Diamant, des Pierres précieuses & des Métaux: Par M. d'Arcet," mentioned in *HAS* 1770 (1773), 119. Lavoisier's efforts to synthesize ruby are vividly recounted in James Evans, "Rediscovering Manufactured Ruby, Part 1," *Gemmology Bulletin* (2020): 1–10.

37. On Buffon's views on gems, see chap. 5, this volume.

38. Antoine-Laurent Lavoisier, "Mémoire sur l'effet que produit sur les pierres précieuses un degré de feu très-violent," *MAS* 1783 (1785): 476–85, esp. 485.

39. Lavoisier, "Premier mémoire sur la destruction du diamant," 573–74 (Leblanc), 576–77 (Maillard).

40. Macquer, *Dictionnaire de chimie*, vol. 1, 474 ("jewelers, lapidaries, and diamond cutters"); Lehman, "Nature of Diamond," 365 ("the jeweler Cordier"), 365 (unnamed jewelers). Darcet worked with "M. Carnay, lapidaire de Paris très-expérimenté": Romé de l'Isle, *Cristallographie*, vol. 2, 190n34.

41. The following account is based on Macquer, *Dictionnaire de chimie*, vol. 1, 474–81.

supplied diamonds to test this hypothesis. Leblanc gave one stone for an experiment in August 1771; Maillard gave three for an experiment in 1772. Most importantly, they supplied a procedure for shielding the diamonds from the effect of extreme heat.

This procedure used a complex device made up of three vessels, each of them carefully packed or sealed. The three diamonds rested in a tobacco pipe filled with crushed charcoal. The pipe was sealed with a mixture of sand and salty water and placed in a crucible filled with chalk. Finally, the crucible was housed in a sphere made of two crucibles sealed in the same way as the pipe. This was not a new device, but one that Maillard and other jewelers commonly used when they heated diamonds to remove imperfections, such as stains due to the oil that was mixed with diamond powder in the polishing process.[42] In "Maillard's experiment," as Macquer called it, the jeweler arranged the apparatus himself. After all, it was his conviction that was under test and his diamonds that were at stake.[43] The experiment was done in the laboratory of Hilaire-Marin Rouelle, the younger brother of the popular chemistry teacher Guillaume-François Rouelle. The apparatus was placed in Rouelle's furnace for several hours until the furnace itself started to melt. To the astonishment of the chemists present—Macquer, Lavoisier, and Louis-Claude Cadet de Gassicourt—the diamonds were intact. The obvious conclusion to draw, at least for Macquer, was that the jewelers' apparatus worked by shielding the diamond from air. "This fact," he later wrote, "seems to me to prove completely that this substance is combustible."[44] In other words, diamonds burn in air, just like wood or coal.

Jewelers evaluated gems in these experiments as well as preparing them, an unspectacular function but a crucial one for the chemists. Leblanc was "a great connoisseur of diamonds," Macquer wrote, and connoisseurship mattered.[45] It is easy to forget that the diamonds had to be identified as such, something that the chemists usually passed over in silence but which came to the surface when things went wrong. In one trial, a stone melted rather than dissipating, and the chemists concluded that it was a peridot rather than a diamond. The chemists explored the possibility that different sorts of diamond behave differently in fire, varying their experiments carefully (and expensively) to rule out this possibility. This meant using diamonds

---

42. Other accounts of this practice are in Dufay, "Lumière des diamants," 362 (heating to remove oil); and Argenville, *Oryctologie*, 77 (heating to remove red points).

43. Macquer, *Dictionnaire de chimie*, vol. 1, 481.

44. Ibid., vol. 1, 478.

45. Ibid., vol. 1, 474.

of different colors. It also meant using jewelers' terms such as "Brazilian diamond," "flat diamond," and "rose-cut diamond."[46] Similar language was used to register the changes wrought by fire. The three diamonds in Maillard's experiment emerged with their facets and polish intact, the only difference being a faint shade of black on their surface. When Maillard had them polished on a wheel, they were as brilliant as ever. Lavoisier and Macquer used these details as evidence that the diamonds were unchanged by fire.[47] When the diamonds did change, the chemists paid close attention to their transparency, color, texture, and the sharpness of their edges, both before and after the trials.[48] Transparency, texture, and color turned out to be important in the comparison that Lavoisier went on to make between diamond and charcoal. Most tellingly, the blackish substance on the surface of the diamonds turned out to resemble soot.

These visual clues were used alongside the meticulous measurement of weight changes for which Lavoisier would become famous. In his very first experiment on diamonds, Lavoisier determined that his specimens had lost $2\frac{22}{32}$ grains of their initial weight, not including the ¾ grains that he recovered from the bottom of the vessel in which they were heated.[49] This experiment was done in the presence of Maillard, using what Macquer called "a very exact assaying balance."[50] Lavoisier's measurements had the same precision (down to ⅛ of a grain) as the data in jewelers' account books of the period.[51] Jewelers had a professional interest in weight loss, since diamonds were priced by weight. As a result, jewelers were hypersensitive to the weight lost during cutting. Maillard was probably a diamond cutter: Macquer referred to him as a "lapidary," and Dufay tells us that diamond cutters heated their own diamonds to remove stains.[52] Lavoisier's precise accounting of the weight of diamonds had much in common with the practice of his cutter-collaborator.

46. Lavoisier, "Premier mémoire sur la destruction de diamant," 585–88 (jewelers' language); Lehman, "Nature of Diamond," 369 (variation in diamond color).

47. Lavoisier, "Premier mémoire sur la destruction du diamant," 577–78, cf. 570–71; Macquer, *Dictionnaire de chimie*, 478.

48. Lavoisier, "Second mémoire sur la destruction du diamant," 614 (transparency), 599–600 and 608 (texture), 615 (soot), 614 (transparency), 597, 601, 602–3, 606.

49. Lavoisier, "Premier mémoire sur la destruction du diamant," 576.

50. Macquer, *Dictionnaire de chimie*, vol. 1, 478. Only later, in a 1782 paper, did Lavoisier report the use of a custom-made balance for his gem experiments: "Pierres précieuses," 479.

51. E.g., Jean-Baptiste Caumon, "marchand joyaillier," 1748–1751, in Archives de la Ville de Paris, D5B6, reg. 1183. The fraction $\frac{22}{32}$ might suggest a precision greater than ⅛ of a grain, but the former fraction was an artifact of Lavoisier's calculation.

52. Macquer, *Dictionnaire de chimie*, vol. 1, 476; Dufay, "Lumière des diamants," 362.

Maillard's experiment was not the last word on the combustion of diamond, however. The experiment raised the question of how to reconcile the findings of jewelers with those of porcelain makers. Diamonds may have been protected by Maillard's pipe-and-crucible arrangement, but they had not survived in Darcet's original experiment in 1768, where they had been surrounded by a sphere made of porcelain.[53] Darcet had used a porcelain furnace, more powerful than the one in Maillard's experiment. So, it made sense to repeat Maillard's experiment using such a furnace. Macquer soon did exactly that at the Porcelain Manufactory at Sèvres. This was in 1772, three years after the large-scale production of hard-paste porcelain had begun at the manufactory, following Macquer's discovery of the key ingredients in 1665.[54] The replication was successful, with Maillard's arrangement protecting his diamonds from "the hottest fire known," as Lavoisier put it in his published account of the experiments.[55]

Why then had Darcet's porcelain sphere failed? If the furnace was not the problem, what was? Macquer went on to show that porcelain shields worked as long as they were fired in advance. By contrast, Darcet had used shields made of porcelain paste that hardened as the diamonds heated, no doubt because this was a natural procedure for a porcelain maker.[56] Further experiments showed the importance of charcoal, since pipes filled with chalk or air did not preserve the diamonds placed inside them. Finally, in a separate investigation, Darcet and Rouelle showed that fired porcelain packed with charcoal was an excellent shield for diamonds.[57] When the diamond did dissipate in the heat, so did the charcoal—a crucial detail for Lavoisier, who inferred that diamond and charcoal are equally combustible.[58] Porcelain making and jewelry were still present in these later experiments, but they had been changed in crucial ways. The pipe had been replaced with porcelain, porcelain paste with fired porcelain, and the sealed hemispheres with a continuous sphere equipped with a self-sealing stopper.

Is this an example of scholars interacting with artisans? Yes, but it is also an example of different sorts of artisan interacting with each other. Diamond cutters and porcelain makers contributed to something that went beyond both groups.

53. Lavoisier, "Premier mémoire sur la destruction du diamant," 571.

54. Guerlac, "French Antecedents," 85; Lehman, "Macquer: An Artisanal-Scientific Expert," 329; idem, "Nature of Diamond," 370.

55. Lavoisier, "Premier mémoire sur la destruction du diamant," 578.

56. Ibid., 580.

57. Ibid., 585–89.

58. Ibid., 589.

Drugs and Glass

Diamonds are not rubies, however. High temperatures worked well for diamonds, but not so well for other gems. Lavoisier is remembered for showing that diamond is combustible, not for his finding that emerald, chrysolite, and garnet melt easily in the focus of a large mirror. The five-fold division of precious stones that he presented to the academy in 1782 was quickly superseded by a new approach to gem chemistry. Like Pott's analyses of minerals, the novelty came from the mining centers of northern Europe, this time from Sweden. Much has been made of the Swedish Bureau of Mines (Bergskollegium) as a catalyst of chemical mineralogy.[59] The bureau was set up in the 1630s to oversee the mining industry in the kingdom of Sweden, which was home to rich deposits of iron and copper. By the middle of the eighteenth century, the bureau was a major employer of university-trained chemists and the site of two well-equipped laboratories. More than any other institution, the bureau explains the disproportionate number of elements on the periodic table that have been discovered in Sweden.[60]

Yet there was more to Swedish chemistry than metals and mining. Drugs and pharmacy were also important, whether in the University of Uppsala, the Royal Academy of Sciences, or in the Bureau of Mines itself, where one of the laboratories was once dedicated to the distillation of medical substances.[61] Pharmacy was central to the career of the doyen of Swedish chemists in the eighteenth century, Torbern Bergman.[62] The son of a tax collector, Bergman learned chemistry under Johan Gotschalk Wallerius at the University of Uppsala. Wallerius taught pharmacy as well as metallurgy.[63] Bergman succeeded Wallerius as professor of chemistry at the university in 1767. His most skilled collaborator was an apothecary, Carl Wilhelm Scheele, whose professional interest in plant and animal substances was the basis for his pioneering work in organic chemistry.[64] Pharmacy lay behind two of Bergman's major publications, a table of chemical affinities (1775) and a general method of analyzing

59. Porter, "Promotion of Mining," 549–51; Anders Lungren, "The New Chemistry in Sweden: The Debate That Wasn't," *Osiris* 4 (1988): 146–68, 147–49; Fors, *Limits of Matter*.

60. Fors, *Limits of Matter*, 100.

61. Lundgren, "The New Chemistry in Sweden," 150–51.

62. Partington, *History of Chemistry*, vol. 3, 179–82; Joseph A. Schufle, "Torbern Bergman, Earth Scientist," *Chymia* 12 (1967): 58–97.

63. Boklund, "Wallerius," 144.

64. Lundgren, "The New Chemistry in Sweden," 156, 158–60; Szabadváry, *History of Analytical Chemistry*, 58–59.

mineral waters (1778).[65] The former was modeled on the table published by
the French chemist Etienne-François Geoffroy in 1715, which was in turn
based on procedures used by apothecaries.[66]

It was probably through the study of mineral waters that Bergman became
interested in what he called "wet" analysis. This meant mixing the substance
under analysis with other liquids to produce tell-tale effects, such as color
changes or the production of solid substances or "precipitates." Solids were
analyzed in the same way, by first converting them into liquids or "bringing them into solution." Bergman contrasted this with "dry" analysis, where
fire rather than liquid was the main agent that altered the substance under
analysis. Although Bergman used both forms of analysis, he tended to favor
the wet over the dry. In doing so, he self-consciously followed the German
chemist Sigismund Andreas Marggraf, who trained as a pharmacist before
studying mining with Henckel in Freiberg.[67] Bergman used the wet way to
examine substances that were associated with mining and that were usually
examined in the dry way. A striking example is a treatise called "The Assaying of Minerals by the Wet Way."[68] There Bergman showed how to measure
the metallic content of ores by dissolving them in acids and precipitating out
the metal, all without having to melt any of the solids involved. The mining
industry in Sweden not only fused the theory of mining with the practice of
mining. It also fused mining with pharmacy.

This merging of crafts was the precondition for Bergman's groundbreak-
ing analysis of gems. In an article published in 1777 and entitled "The Earth of
Gems," Bergman described a procedure for isolating the components of any
gem.[69] The procedure gave a true decomposition, unlike the methods of analogy and resolution discussed in this chapter. This has been called "one of the

65. *De analysi aquarum* (1778); *De attractionibus electivis* (1775). Dates of these works are
from Partington, *History of Chemistry*, vol. 3, 182–83. Bergman's work on mineral waters is analyzed in Szabadváry, *History of Analytical Chemistry*, 73–76; and Schufle, "Torbern Bergman,
Earth Scientist," 60–71.

66. Schufle, "Torbern Bergman, Earth Scientist," 61. On the pharmaceutical origins of Geoffroy's table, see Ursula Klein, "Origin of the Concept Chemical Compound," *Science in Context*
7, no. 2 (2008): 163–204, esp. 175–93.

67. Bergman, *Manuel du minéralogiste*, ed. Mongez, 13. Marggraf's life and work, including
pharmacy and the wet way, are summarized in Szabadváry, *History of Analytical Chemistry*, 55–
59. His link to mining is noted in Porter, "Promotion of Mining," 555–56.

68. *De minerarum docimasia humida* (1780). Cf. Bergman's *De praecipitatis metallicis* (1780);
Szabadváry, *History of Analytical Chemistry*, 78–79.

69. Torbern Bergman, "La terre des gemmes," in Bergman, *Opuscules chimiques et physiques*,
vol. 2, 78–124.

most important steps in the whole history of chemistry."[70] The difficulty of the task is shown by the failure of Bergman's compatriot, Cronstedt, to do the same in a book published twenty years previously. Cronstedt, another student of Wallerius, was an inspector at the Swedish Bureau of Mines. He used his knowledge of assaying techniques, especially the blowpipe, to continue Pott's project of classifying minerals based on their chemical composition, a project that led him to discover the element nickel.[71] But the techniques that worked so well on metallic ores were unsuited to gems. Cronstedt frankly admitted that he was unable to distinguish these stones by chemical tests, as noted at the start of this chapter.[72] He classified gems by color, hardness, and transparency, precisely the sorts of property that he elsewhere derided as "superficial."[73] Even Bergman made little progress on gems with the blowpipe.[74] A new approach was needed to prize apart the components of gems.

The new approach was an amalgam of techniques drawn from Bergman's various lines of chemical inquiry, glassmaking as well as pharmacy and mining. The main challenge was to bring gems into solution so that they could be analyzed in the wet way. Bergman's usual procedure was to grind up his samples and dissolve them in acids. But this did not work on gems.[75] Only iron and calcareous earth could be extracted in this way; the remaining components seemed to be insoluble in all known acids. Fortunately, Bergman had been in this position before.[76] In his extensive experience with the wet way, he had learned that substances that appear to be insoluble in one liquid can become soluble in that liquid if they are first dissolved in another liquid. One solvent "softens up" the substance so that another solvent can act on it.

What could be used to soften up gems? This is where Bergan's knowledge of glassmaking came in handy.[77] Briefly, glassmaking is the art of melting

70. Oldroyd, "Phlogistic Mineralogical Schemes," 303.

71. Cronstedt, *Essai d'une nouvelle minéralogie*, xviii–xx (debt to Pott). Cf. Spencer St. Clair, "Classification of Minerals," 131–61; David Oldroyd, "A Note on the Status of A. F. Cronstedt's Simple Earths and His Analytical Methods," *Isis* 65, no. 4 (1974): 506–12; and Staffan Müller-Wille, "Eighteenth-Century Classifications of Non-Living Nature," in *Spaces of Classification*, ed. Ursula Klein (Berlin: Max Planck Institute for the History of Science, 2003), secs. 3 and 4.

72. Cronstedt, *Essai d'une nouvelle minéralogie*, 66–67.

73. Ibid., xxiii, xxvi, xxxvi (against external properties), 67–76 (division of gems). Garnet was the only exception: 102–5, 108.

74. Bergman, "Terre des gemmes," 83–86.

75. Ibid., 93–96.

76. Ibid., 96–98.

77. As noted in David Oldroyd, "Some Eighteenth-Century Methods for the Chemical Analysis of Minerals," *Journal of Chemical Education* 50, no. 5 (1973): 338. Cf. Partington, *History of Chemistry*, vol. 3, 186.

sand so that it becomes transparent and malleable. But sand does not melt on its own, at least not at the temperatures attained by early modern furnaces; it had to be mixed with substances known as "fluxes." The ratio of flux to sand was crucial. Too little flux and the sand would not melt; too much and the resulting glass would dissolve in acids or even in water. Soluble glass was a bane for glassmakers but a boon for Bergman. Why not treat gems in the way that glassmakers treat sand, softening it up with a flux so that it becomes susceptible to acids? Bergman tried this, mixing his ground-up gems with a common flux (soda) and heating them in a crucible until they formed a glassy mass. This was a success—the glassy mass dissolved in acids that the ground gems had resisted. Bergman then proceeded with the wet way, using a sequence of acids to precipitate out the components of gems before identifying these components by their color, taste, and solubility.[78] Multiple crafts entered here as well. Painting was the origin of Prussian blue, the substance Bergman used to test for iron.[79] Mineral waters had been the occasion for the discovery of one of the components he identified (magnesia[80]) and of the substance he used in the identification (Epsom salt[81]). Even the blowpipe made an appearance: Bergman used it to identify the one component (quartz) that resisted all known acids, even after the gem was heated with a flux.[82] Once again, multiple crafts achieved what no single craft could achieve on its own.

Gems and Metals

Bergman's study of gems was quickly translated into the French language, much like the analyses by Pott and Henckel two decades earlier. The 1777 article was available in French within two years of its publication in Sweden.[83] Bergman's *Sciagraphia*, his mature classification of minerals, was also available in French within two years of its first publication in 1782.[84] This was part

---

78. His general analytical method is in Bergman, "Terre des gemmes," 98–101.

79. Ibid., 99; Szabadváry, *History of Analytical Chemistry*, 56–57.

80. Bergman, "Terre des gemmes"; Szabadváry, *History of Analytical Chemistry*, 31–32.

81. Bergman, "Terre des gemmes," 100; Oldroyd, "Eighteenth-Century Methods," 337; Noel Coley, " 'Cures Without Care': 'Chemical Physicians' and Mineral Waters in 17th Century English Medicine," *Medical History* 23 (1979): 210–12.

82. Bergman, "Terre des gemmes," 100–101.

83. Torbern Bergman, "Recherches chimiques sur la terre des Pierres précieuses ou gemmes," *Journal de physique* 14 (Oct. 1779): 257–80.

84. Torbern Bergman, *Sciagraphia regni mineralis* (Leipzig, 1782); idem, *Manuel du minéralogiste*, ed. Mongez; idem, *Manuel du minéralogiste*, ed. Delamétherie. Mongez's first name does not appear on the second of these works, but the identity of the translator is clear in "Jean-

of the wider uptake of Bergman's ideas in France, including an early attempt at a chemical nomenclature based on composition.[85]

But it would be too hasty to see these translations as the victory of analysis by decomposition in the study of gems in France. Bergman's approach to mineral classification was rather different from the one that Daubenton and others had been developing at the Jardin du Roi. As we saw in the previous chapter, the French tradition was weighted toward gemstones, gem collections, and gem appraisal. This led to a classification of gems based on quantifiable physical properties, such as color, density, double refraction, and, increasingly, crystal form. The contrast between the French approach and the German and Swedish one was already apparent in the *Encyclopédie* in the 1750s and 1760s. There, half the articles on gems were written by Daubenton and the other half by Paul-Henri Thiry, Baron d'Holbach.[86] The articles by Daubenton emphasized physical properties such as color. By contrast, the articles by d'Holbach contained much on the response of gems to fire, the metals that color them, and the manner in which they are counterfeited. This was a reflection of d'Holbach's interest in the chemical approach to minerals coming from the German mining community.[87] In the language of the *Encyclopédie*, Daubenton was doing "histoire naturelle des minéraux" whereas d'Holbach was doing "minéralogie."[88] The challenge was to reconcile mineralogy with the natural history of minerals.[89]

André Mongez," *Biographie universelle ancienne et moderne*, 2nd ed., vol. 29, ed. Louis-Gabriel Michaud (Paris, 1843 [1811]), 622.

85. On the wider uptake, see Crosland, *Language of Chemistry*, pt. 3, chaps. 3 and 4; David Oldroyd, "Mineralogy and the 'Chemical Revolution,'" *Centaurus* 19, no. 1 (1975): 54–71; Porter, "Promotion of Mining," 565–69; and James Llana, "A Contribution of Natural History to the Chemical Revolution in France," *Ambix* 32, no. 2 (1985): 80.

86. Articles by Daubenton: émeraude, diamant, amethyste, aigue-marine, chrysolite. Articles by d'Holbach: péridot, grenat, saphir, rubis, topase, hyacinthe. The article béril was jointly written by both. These articles were published from 1751 to 1765 inclusive. Articles consulted on https://enccre.academie-sciences.fr/encyclopedie/, accessed Mar. 23, 2015.

87. Guerlac, "French Antecedents," 100–101; Loveland, "Louis-Jean-Marie Daubenton and the *Encyclopédie*," 186.

88. D'Holbach, "Minéralogie," in Diderot and d'Alembert, *Encyclopédie*, vol. 10 (1765), 541–43, esp. 542; "Histoire naturelle," in Diderot and d'Alembert, *Encyclopédie*, vol. 8 (1765): 225–30, esp. 228. The latter article was unsigned but is usually attributed to Daubenton, perhaps with input from Denis Diderot: Loveland, "Louis-Jean-Marie Daubenton and the *Encyclopédie*," 212–14; Llana, "Natural History and the *Encyclopédie*," 11.

89. The paragraphs that follow expand on earlier accounts of the chemistry/crystallography relationship around 1800, especially Mauskopf, "Crystals and Compounds," 7–20; Oldroyd, "Mineralogy and the Chemical Revolution," 64–65, citing Burke, *Origins of the Science of Crystals*, 126–32; and Irish, "The Corundum Stone and Crystallographic Chemistry."

One approach was to nest one inside the other. The idea was to use mineralogy to define classes and genera, then to populate these classes and genera with species defined by natural history. That way, there was no conflict between the two because they were never used to define the same thing. This approach was taken by Jean-André Mongez, the French naturalist who edited the first translation of Bergman's mineralogical treatise. Mongez used "internal characters" to define classes and genera and "external characters" to define species, to use the terms that Mongez did much to disseminate.[90] But this approach became increasingly untenable as the followers of Bergman began to use chemistry to define species, not just classes and genera. A key figure here was the Prussian chemist Martin Heinrich Klaproth, whose collected works contained a disproportionate number of articles on gems (*Edelstein*) when they were published in 1795 and 1797.[91] Meanwhile, crystallographers such as Romé de l'Isle and Haüy had published descriptions of several gem species based on their crystal form, corroborated by hardness, refraction and density. By the time the second French edition of Bergman's book came out, in 1792, chemistry and crystallography were on a collision course over the definition of mineral species. The editor of that translation, a partisan of Daubenton, used his extensive annotations to champion external properties over internal ones and thereby undermine the text he was annotating. But even here, there were signs of a rapprochement. Take the stone then known as "Ceylonese jargon." Romé de l'Isle had shown in 1783 that this had a distinct crystal form. A few years later, Klaproth took the hint and analyzed the stone chemically. He thereby discovered a previously unknown substance, one that was eventually identified as the element zirconium.[92] Chemistry and crystallography could work together, at least for some species.

So far this agreement was mainly textual, however. It was one thing to translate works by Bergman and Klaproth into French. It was another thing

90. Bergman, *Manuel du minéralogiste*, ed. Mongez, xi–xii, lxxix–lxxx, lx, 10, 12–13; the gems are on pp. 132–38.

91. Martin Heinrich Klaproth, *Beiträge zur chemischen Kenntnis der Mineralkörper* (Posen and Berlin, 1795–1815), vol. 1, ix ("Edelstein"), xv–xvi (contents); vol. 2, ix–xii (contents). Cf. Martin Heinrich Klaproth, *Analytical Essays Towards Promoting the Chemical Knowledge of Mineral Substances*, trans. Gruber, vol. 1 (London, 1801), vii ("gems"). New analyses by Bindhem, Achard, Wieblig, Sage, Heyer, Gmelin, as well as Bergman and Klaproth, are mentioned in Bergman, *Manuel du minéralogiste*, ed. Delamétherie, vol. 1, 253–72. On Klaproth's career, see Klein, *Technoscience in History*, chap. 2.

92. Romé de l'Isle, *Cristallographie*, vol. 2, 229; Martin Heinrich Klaproth, "Chemical Examination of the Circon, or Jargon of Ceylon," in Klaproth, *Analytical Essays*, 175–94; Bergman, *Manuel du minéralogiste*, ed. Delamétherie, vol. 1, 150–54, 270–71.

to import the institution that had made their analyses possible, the state-sponsored mining academy. And it was another thing again to combine this institution with the culture of gem collecting that had nurtured crystallography in France, especially at the Jardin du Roi. The first step in this direction was the creation of the Royal School of Mines (*École royale des mines*) in Paris in 1783. Like the Freiberg Mining Academy on which it was modeled, the School of Mines was a state-sponsored training program with the goal of supplying the national mining industry with experts in the analysis of valuable minerals. It was Balthazar-Georges Sage, the son of a Parisian apothecary, who pushed for the creation of the school, which was established at the Royal Mint with Sage as director of studies and professor of docimastic mineralogy.[93] The room in which Sage gave his lectures contained both a chimney for doing chemical experiments and a set of mahogany cabinets for displaying the mineral collection.[94] The mineral collection was itself a hybrid, including both raw specimens and the results of chemical analyses.[95] Precious stones featured strongly, both in the public collection and in the set of objets d'art that Sage showed only to close acquaintances and prestigious visitors. The public collection was organized like a utilitarian treatise on mining chemistry, with ruby and emerald grouped with granite and schorl.[96] The private collection recalls those of Louis XIV and the Chevalier de la Roque: porphyry vases, rock crystal glasses, rings set with topaz and sapphire, all displayed alongside busts, paintings, and porcelain.[97] Sage's scientific work had a similar duality. His writings on mineralogy covered the response of gems to fire *and* their crystal form.[98] He used his mineral cabinet to observe minerals *and*

93. Birembaut, "Enseignement de la minéralogie," 387–93. On Sage himself, see Doru Todericiu, "Balthasar-Georges Sage (1740–1824), chimiste et minéralogiste français, fondateur de la première École des mines (1783)," *Revue d'histoire des sciences* 37, no. 1 (1984): 29–46.

94. Birembaut, "Enseignement de la minéralogie," 392–93 (chimney); Maddalena Napolitani, "'Born with the Taste for Science and the Arts': The Science and the Aesthetics of Balthazar-Georges Sage's Mineralogy Collections, 1783–1825," *Centaurus* 60, no. 4 (2018): 244 (mahogany cabinets).

95. Napolitani, "Taste for Science and Arts," 247–48.

96. Balthazar-Georges Sage, *Description méthodique du Cabinet de l'École royale des mines* (Paris, 1784), 57–79; idem, *Analyse chimique et concordance des trois règnes* (Paris, 1786), 53–88. Sage notes that this collection was started around 1760: *Supplément à la Description méthodique du Cabinet de l'École royale des mines* (Paris, 1787), 1.

97. Balthazar-Georges Sage, *Description des objets d'art de la collection de B. G. Sage de l'Institut de France* (Paris, 1807), passim, esp. 5–11 (hard-stone vessels), 55–67 (various gems). The location and use of this collection is discussed in Todericiu, "Balthasar-Georges Sage," 250–53.

98. Sage, *Analyse chimique*, 53–88.

to analyze them chemically.[99] The props for his lectures included models of machines used in mines *and* models of the crystal forms of minerals.[100] His students included the chemist Jean-Antoine Chaptal *and* the crystallographer Romé de l'Isle.[101] Sage is usually remembered as a rather backward chemist, an opponent of Lavoisier's anti-phlogistic chemistry.[102] Yet it was Sage who supplied the specimens for Lavoisier's experiments on the fusibility of gems.[103] More broadly, Sage's mining school brought together the sciences of chemistry and crystallography. It did so by bringing together the arts of assaying ores and appraising gems.

The missing ingredient at Sage's school was the decomposition of gems. Sage preferred to classify gems by their response to fire, borrowing Lavoisier's classification scheme for this purpose.[104] It was Sage's successor, Louis-Nicolas Vauquelin, who introduced the German and Swedish art of gem decomposition into France. Vauquelin was the first teacher of docimastics at the House of Mines (*Maison des mines*), an institution created during the French Revolution to supersede the School of Mines.[105] Vauquelin's career had a familiar arc. An apothecary by profession, he caught the attention of the chemistry community for his skill in analyzing plant and animal substances; he then transferred these skills to minerals under the influence of the mining industry.[106] He is remembered primarily for his role in the discovery of the elements now known as chromium and beryllium.[107]

99. Sage, *Description méthodique*, ii.

100. Birembaut, "Enseignement de la minéralogie," 392–94.

101. Todericiu, "Balthasar-Georges Sage," 40.

102. Ibid., 34–36, with a qualification on pp. 33–34; Birembaut, "Enseignement de la minéralogie," 387, 410; Henry Guerlac, "Sage, Balthazar-Georges," in *DSB*, vol. 12, 63–69.

103. Lavoisier, "Pierres précieuses," 480, 481.

104. Sage, *Description méthodique*, 57–58 (Lavoisier). Note, however, the decomposition of aquamarine in Sage, *Analyse chimique*, 69. Cf. Bergman, *Manuel du minéralogiste*, ed. Delamétherie, vol. 1, 267 (hyacinth).

105. Birembaut, "Enseignement de la minéralogie," 402–9; Isabelle Laboulais, *La maison des mines: La genèse révolutionnaire d'un corps d'ingénieurs civils, 1794–1814* (Rennes: Presses universitaires de Rennes, 2012), 12–13.

106. Partington, *History of Chemistry*, vol. 3, 551–52; Guillaume Valette, "La vie de Vauquelin," *Revue d'histoire de la pharmacie* 51, no. 177 (1963): 89–96. Simon, *Chemistry, Pharmacy and Revolution*, 151–59, emphasizes his link to pharmacy, whereas I wish to emphasize the links he forged between pharmacy to mining.

107. Partington, *History of Chemistry*, vol. 3, 553 (chromium and beryllium discoveries). Cf. Marcel Delépine, "Ses oeuvres chimiques [de Vauquelin]," *Revue d'histoire de la pharmacie* 51, no. 177 (1963): 83–85. Vauquelin did not name the earth he found in beryls; it was briefly called "glucine" by French writers, then "beryllia" by German ones.

He ought also to be remembered as a gem chemist. Between 1796 and 1799 he decomposed topaz, peridot, hyacinth, garnet, chrysolite, spinel, emerald, aquamarine, and tourmaline, often doing analyses on two or more varieties of these species. His analyses of aquamarine, spinel, and emerald were linked to the discoveries of chromium and beryllium.[108] In 1799, he described a general procedure for dividing stones into their components. These findings were widely circulated, many of them published both in the *Journal des mines* (the journal of the House of Mines) and the *Annales de chimie* (the journal of Lavoisier and his followers).[109] Vauquelin was familiar with analyses done by Bergman and by the German chemist Franz Carl Achard.[110] He saw himself as continuing the tradition of "lithogeognosy" that had been begun by Bergman and perfected by "the celebrated Klaproth," to use Vauquelin's terms. It is no coincidence that Vauquelin's gem analyses appeared immediately after the publication, in 1795 and 1797, of the first two volumes of Klaproth's collected works. These volumes were published in German, with the French translation of the work appearing only in 1807.[111] Vauquelin lost no time in

108. Louis-Nicolas Vauquelin, "Analyse du rubis spinelle" and "Analyse de l'émeraude du Pérou," *Journal des mines* 8, no. 37 (1797): 81–92, 93–97.

109. Vauquelin's articles on gems in the *Journal des mines* from 1796–1799 are "De la topase blanche de Saxe," vol. 4, no. 24 (1796): 1–3 (progress since Bergman, "lithogeognosy"); "Du péridot du commerce," vol. 4, no. 37 (1796): 37–44; "Analyses comparées des hyacinthes de Ceylan et d'Expailly," vol. 5, no. 26 (1796): 97 (Klaproth analysis, Bergman errors); "Expériences sur les grenats blancs ou leucite des volcans," vol. 5, no. 27 (1796): 201 (Klaproth analysis; "celebrated"); "De la chrysolite des jouailliers," vol. 7, no. 37 (1797): 20 (Achard analysis); "Du rubis spinelle," 82 (Klaproth and Achard analyses); "De l'émeraude du Pérou," 93 (Klaproth analysis, praise for Klaproth); "Analyse de l'aigue-marine ou beril, et découverte d'une terre nouvelle dans cette pierre," vol. 8, no. 43 (1798): 553–54 (progress since Bergman); "Analyse des grenats noirs du pic d'Erès-Lids," vol. 8, no. 44 (1798): 573 (Achard analysis); "Des grenats rouges du même pic," vol. 8, no. 44 (1798): 574–75; "Analyse de la tourmaline de Ceylan," vol. 9, no. 54 (1799): 479 (Bergman analysis). Later Vauquelin articles on gems in the same journal: "Du Béril de Saxe, dans lequel M. Tromsdorf a annoncé l'existence d'une terre nouvelle qu'il a nommée Agustine," vol. 15, no. 86 (1803): 81–87; "Expériences sur les topases," vol. 16, no. 96 (1804): 469 (Klaproth analysis). Some of these articles were published in the *Annales de chimie*. See Anon., *Table générale raisonnée des matières contenues dans les . . . volumes des Annales de chimie* (Paris), vol. 1 (1801): 424–26; vol. 2 (1807): 330–32; vol. 3 (1821): 413–17. Note also Vauquelin, "Réflexions sur l'analyse des pierres, et résultats de plusieurs de ces analyses," *Annales de chimie* 30 (1799): 66–106, esp. 66–67 ("marvelous"). The 1804 article on topaz was published in revised form as "Analyse des topazes de Saxe, de Sibérie et du Brésil," *Annales du Muséum national d'histoire naturelle* 6 (1805): 21–25.

110. On Achard, see Partington, *History of Chemistry*, vol. 3, 592–93.

111. House of Mines made its own translations of French and German works: Partington, *History of Chemistry*, vol. 3, 654; Birembaut, "Enseignement de la minéralogie," 380, 396, 407.

replicating and extending Klaproth's results. Chemical mineralogy had well and truly arrived in France.

At the same time, chemical mineralogy merged with the natural history of minerals. The key partnership was between Vauquelin and Haüy. Both men taught at the House of Mines from 1795, with Haüy's courses on mineralogy coexisting with Vauquelin's on docimastics.[112] Haüy used these courses to develop the material that would eventually be published in 1801 as his *Treatise on Mineralogy*. Like Vauquelin, he used the house's mineral collection and correspondence network. Extracts of Haüy's treatise appeared in the *Journal des mines*, sometimes invoking analyses by Vauquelin.[113] Haüy paid tribute to his colleagues, including Vauquelin, in the finished treatise: "several important points were soberly and peacefully discussed in private conversations."[114] The collaboration continued at the Museum of Natural History, with Haüy taking up the chair of mineralogy there in 1802 and Vauquelin joining him two years later as the professor of practical chemistry (*chimie des arts*).[115] Haüy continued to rely on Vauquelin's analyses of gems in his publications, notably in an article on tourmaline published in 1804.[116] Vauquelin stopped working on gems around this time, but he continued to work on minerals and Haüy continued to cite him. Vauquelin's name came up again and again in the 1809 book in which Haüy compared the results of chemistry with those of crystallography. By this time, Vauquelin had analyzed twenty-six out of the forty-three species that Haüy listed in the class of earthy substances.[117]

A study of chrysolite set the pattern for this partnership. The stone was usually classed as a precious stone of the second rank, as Vauquelin noted in an article published in 1797.[118] Achard had decomposed a stone of this name that seemed to confirm this classification, but Vauquelin's results differed dramatically from those of the German chemist. Whereas Achard had found lime, clay, quartz, and iron, Vauquelin found only lime and phosphoric acid.

112. Reijer Hooykaas, "Haüy, René-Just," in *DSB*, vol. 6, 178.

113. Extracts of Haüy's treatise in *Journal des mines* 5, no. 27 (1796): 209–30; 5, no. 28 (1796–1797): 249–334, esp. 250–51 (Haüy's use of collections, correspondents, and lectures at House of Mines), 288 (Vauquelin on topaz); 6, no. 33 (1797): 655–92, esp. 688–91 (Vauquelin's yet-unpublished finding on chrysolite).

114. Haüy, *Traité de minéralogie*, iii.

115. Caroline Kaspar, "L'oeuvre minéralogique et pétrographique des pharmaciens du Muséum," *Revue d'histoire de la pharmacie* 93, no. 347 (2005): 404–5.

116. René-Just Haüy, "Mémoire sur les tourmalines de Sibérie," *Annales du Muséum national d'histoire naturelle* 3 (1804): 243–44. Cf. Haüy, *Tableau comparatif des résultats de la cristallographie et de l'analyse chimique, relativement à la classification des minéraux* (Paris, 1809), 146.

117. Haüy, *Tableau comparatif*, 151–212.

118. Vauquelin, "Chrysolite des jouailliers."

Moreover, these two substances occurred in proportions that closely matched the composition of apatite, a substance Klaproth had recently analyzed. Klaproth had found 55 parts of lime and 45 parts of phosphoric acid in apatite; Vauquelin's numbers for chrysolite were 54.28 and 45.72. This suggested that chrysolite and apatite were varieties of the same species. It followed that chrysolite was not a precious stone, or even a stone, but a salt. Looking for confirmation, Vauquelin asked Haüy whether he had compared the crystal forms of apatite and chrysolite. Haüy had not done the comparison, but he had given the forms of each mineral separately, one in a just-published extract of his treatise and the other in the course on mineralogy he had given at the School of Mines. Comparing these two results showed that the two minerals indeed had the same primitive form.[119] Crystallography chimed with chemistry, to the credit of both. "This satisfying agreement between two apparently distant sciences," Vauquelin explained, "shows at the same time that they are founded on sound principles."[120]

More such agreements followed in the next few years. Saxon beryl was grouped with apatite, Brazilian topaz with pycnite, and aquamarine with emerald.[121] In each case, Haüy noticed a geometric resemblance and asked Vauquelin to find a chemical one, or vice versa. The collaboration was especially rich in the case of aquamarine and emerald (plate 6, bottom right; plate 7). In this case, Haüy noticed the geometric resemblance between the two stones. He asked Vauquelin for an analysis of aquamarine to supplement Klaproth's of emerald. In doing so, Vauquelin found a new earth, one that was briefly called "glucyne" and is now known as "beryllia." He then analyzed emerald, encouraged by the geometric resemblance. He thereby found beryllia in emerald as well. He later found the same substance in a French stone that was soon classed as an emerald.[122] An attempt to corroborate crystallography led to a triple discovery in chemistry.[123]

Vauquelin's work with Haüy was one example of his engagement with the natural history of minerals. It was not the only example. Vauquelin's articles bear witness to the continued importance of color, commerce, collections

119. René-Just Haüy, "Analyse de la chrysolite de M. Romé de l'Isle," *Journal des mines* 6, no. 33 (1797): 688–91.

120. Vauquelin, "Chrysolite des jouailliers," 21.

121. Vauquelin, "Du Béril de Saxe," 86; Haüy, "Chrysolite de M. Romé de l'Isle," 41–42.

122. René-Just Haüy, "Émeraude et béril," *Journal des mines* 6, no. 33 (1797): 686–88; Vauquelin, "De l'émeraude du Pérou," 97; idem, "De l'aigue-marine," 554; Haüy, *Traité de minéralogie*, vol. 4, 316, 512. Cf. Farges and Kjellman, "Bicentenaire du décès de René-Just Haüy," 31–32.

123. Only later, in the case of tourmaline, did Haüy favor crystallography at the expense of chemistry in the classification of a gem: Haüy, "Les tourmalines de Sibérie," 244.

and appraisal in chemical mineralogy. The gem trade was present in the titles of his articles, in phrases such as "commercial peridot" and "jewelers' chrysolite." Like so many naturalists before him, Vauquelin relied on merchants for specimens and for information about their identity. The question of identity became pressing when multiple analyses gave different results, as in the case of chrysolite, where Vauquelin explained away the discrepancy between his analysis and Achard's by saying that the German chemist had not used "true chrysolite."[124] Commerce also informed his study of hyacinth, where he confirmed that the stone by that name found in France was a "true hyacinth," one with the same composition as the hyacinth found in Ceylon.[125] Collections were never far from the surface, whether institutional collections such as those at the École Polytechnique or the many private collections in Paris.[126] Vauquelin echoed the call of his mentor, Antoine-François, comte de Fourcroy, for a chemical study of "the objects of natural history conserved in cabinets."[127] Several of Vauquelin's specimens came from afar: chrysolite purchased in Spain, aquamarine from a collector in Berlin, rubies seen in "lithological collections" in London.[128] The latter specimens were a clue to the presence of chrome in spinel, since the range of colors in them (blue, green, white) seemed to Vauquelin to be inconsistent with the presence of iron.[129] Vauquelin dismissed color as a guide to the classification of minerals but continued to use it indirectly for this purpose. His articles are dotted with phrases such as "pale lilac," "handsome emerald-green," and "red color perfectly similar to a ruby," not in reference to gems themselves but to the substances he extracted from them.[130] The language of gem appraisal had not disappeared from chemistry but been transferred from gems to their components. With Vauquelin, then, mineralogy is hard to distinguish from the natural history of minerals. The world of gem appraisal—with its private

124. Vauquelin, "Chrysolite des jouailliers," 19.

125. Louis-Bernard Guyton de Morveau, "Mémoire sur l'Hyacinthe de France, congénère à celle de Ceylan, et sur la nouvelle terre simple qui entre dans sa composition," *Annales de chimie* 22 (1797): 73 ("vraies hyacintes"); Vauquelin, "Analyses comparées des hyacinthes de Ceylan et d'Expailly," 97–98.

126. Guyton de Morveau, "Hyacinthe de France," 73.

127. Vauquelin, "Chrysolite des jouailliers," 19.

128. Vauquelin, "Chrysolite des jouailliers" (Spain); idem, "Du rubis spinelle," 82 ("collections lithologiques" of "Greville" and "Hawkins"); idem, "Du Béril de Saxe," 82 (beryl from "Karsten"). On these collectors, see Wilson, *Mineral Collecting*, 174 (John Hawkins), 173 (Charles Francis Greville), 178 (Dietrich Ludwig Gustav Karsten).

129. Vauquelin, "Du rubis spinelle," 85.

130. Examples of such terms in ibid., 85; Vauquelin, "De l'émeraude du Pérou," 95. Dismissal of color in idem, "Réflexions sur l'analyse des pierres," 67.

collections, visual judgments, and its links to merchants and cutters—had merged with the world of mining and assaying.

The end result of this merging was the collapse of gems as a scientific category. This may come as a surprise, given the importance of that category in eighteenth-century chemistry. Indeed, one aim of this chapter has been to show that gemstones were a major topic in the work of Henckel, Pott, Lavoisier, Bergman, and other leading chemists. But the cumulative effect of this work was to undermine the category on which it was based. Lavoisier's reclassification of diamond as a combustible substance was a sign of things to come. Soon afterward, Bergman reclassified garnet and tourmaline as schorls, rather than as precious stones, on the basis of their composition. In Bergman's eyes, garnet and tourmaline were more like granite than ruby.[131] In the 1790s, Vauquelin argued that chrysolite, Brazilian topaz, and Saxon topaz were not stones at all but salts and acids. His analyses also showed the difficulty of classifying gems by traditional criteria. Aquamarine and emerald were different colors but belonged to the same species; ruby and emerald were different species both colored by the same substance, namely chromium. Neither the colors of stones, nor their colorants, were a reliable guide to their identity. These results were corroborated by German and Irish chemists, who showed that ruby and sapphire had the same composition as the coarse, opaque stone known as adamantine spar.[132] The upshot was the decline of gems as a category in mineral chemistry. This was reflected in the titles of Bergman and Vauquelin's general analytical procedures for minerals: "the earth of gems" of 1777 had become "the analysis of stones" in 1799.[133] The category of precious stones is nowhere to be found in Haüy's 1801 textbook on mineralogy, largely due to the classification of minerals by their chemical composition. By decomposing gems, chemists had decomposed the very idea of a gem.

Compositionism About What?

This process was driven by hand-hand coordination, by the fusion of different crafts. Early work on the chemistry of gems, by Pott and Henckel, was not only informed by mining but also by gem cutting and porcelain making. The result was Henckel's method of analogy and Pott's method of resolution.

131. Bergman, *Manuel du minéralogiste*, ed. Mongez, 135–38.

132. Irish, "The Corundum Stone and Crystallographic Chemistry," 307–11, 321; Content, *Ruby, Sapphire & Spinel*, vol. 1, 66–74.

133. Vauquelin, "Réflexions sur l'analyse des pierres," with scattered references to gems on pp. 85, 105–6.

Later, jewelers and porcelain makers were both involved in the Parisian experiments on the combustion in diamonds. These experiments took Pott's method in a new direction, using fine judgments about the value of diamonds to measure their response to heat. Other gems required a different approach. The breakthrough came from a Swedish chemist with a background in glassmaking and pharmacy as well as metal mining. By combining techniques from these crafts, Bergman came up with a general method for separating a gem into its components. This method, like those of Pott and Henckel, was quickly adopted in France and altered in the process. The decomposition of gems was used in tandem with the study of their crystal forms to define new mineral species. This was done at institutions—the School of Mines and House of Mines—that were dedicated to the art of mining but that were open to the arts of gem appraisal. The interaction between different crafts was at least as important in the chemistry of gems as the interaction between this or that craft and science.

What does this mean for the Chemical Revolution and compositionism? The basic point is that compositionism worked out differently for different materials. This point, implicit in much literature on eighteenth-century chemistry, is worth bringing into the foreground.[134] For one thing, it helps to bring some clarity to the question of when compositionism first arose. The answer is that it arose at different times for different materials, with neutral salts being treated in a modern way by 1750 and airs, earths, and gems following in the next half-century.[135] For gold, silver, and other metals, aspects of the modern understanding of composition may have emerged as early as the thirteenth century.[136] Different materials also lent themselves to different kinds of compositionism, or at least to different ways of identifying the components of substances. Decomposition worked well for metallic ores but was much harder to apply to gems; intense heat worked well for diamonds but not for rubies or emeralds. These differences were partly due to the material properties of substances—the fact that diamond is made of carbon, for example—but they were also due to the way the materials were valued by humans. After all, the aim of chemical analysis was very often to evaluate a material. The point was to determine the medical virtues of a spring, the type

134. These remarks build on Klein and Lefèvre, *Materials in Eighteenth-Century Science*, a major theme of which is the difference between plants and minerals as subjects of compositional chemistry.

135. Support for this claim can be found in Siegfried, "Elements to Atoms," chap. 4 (neutral salts), chap. 9 (airs); Oldroyd, "Mineralogy and the Chemical Revolution," 55–60 (earths); Klein and Lefèvre, *Materials in Eighteenth-Century Science*, 171n57 (earths), chap. 8 (neutral salts).

136. Newman, "Mercury and Sulphur Among the High Medieval Alchemists."

of metal in an ore, the proportion of gold in an alloy, and the potential of a clay as an ingredient in porcelain, to use examples from this chapter. Other examples abound in the history of analytical chemistry.[137] This mattered for the type of analysis carried out. For example, one reason external properties were so important in the history of gem chemistry is that gems were valued for these properties, especially transparency, hardness, and color.

Moreover, this mattered for the geography of chemistry, since different places specialized in different materials and their associated crafts. Most obviously, Sweden and Saxony were major centers for the mining of metallic ores in eighteenth-century Europe. This was significant, not just because it brought scholars and artisans together, but also because it brought the analysis of metallic ores into contact with the analysis of other materials, especially medical drugs. It is striking how many eighteenth-century chemists were trained as apothecaries before becoming involved in the mining industry— Marggraf, Klaproth, Sage, and Vauquelin are examples from this chapter. It is equally striking how many innovations in French chemistry can be traced, at least in part, to an influx of mining expertise from elsewhere in Europe. The combustion of diamonds and the decomposition of gems are two examples from this chapter. To these we might add the mineralogical turn at the academy earlier in the century, which had its origins in a tour of the mining regions of eastern Europe and Scandinavia undertaken by Wilhelm Homberg in the 1680s.[138] This transnational movement has been noted before by historians, but usually in a way that separates mining from other crafts.[139] In fact, mining made a difference to French chemistry precisely because it was combined with other crafts that were already well-developed in France, such as pharmacy and gem cutting. The Chemical Revolution was a meeting of different material worlds.

137. This statement is based on a reading of Szabadváry, *History of Analytical Chemistry*.

138. Principe, *Transmutations of Chymistry*, 14, 34–37.

139. For example, Guerlac, "French Antecedents," 100–104.

7

The End of Gems and the Origins of Gemology

The category of gems was ejected from mineralogy around 1800, a casualty of the rise of physics, chemistry, and crystallography as a basis for the classification of minerals. But the science of gems persisted. It persists to this day in the form of gemology, a field that makes judgments about gems with the help of mineralogy, geology, physics, chemistry, and other disciplines that are firmly established in science faculties in Western universities. There are no gems in the *Strunz Mineralogical Tables*, but there is plenty of mineralogy in *Gemmology*, a current textbook in the field.[1] This is a puzzle for anyone interested in science as a form of material evaluation. On the one hand, gems were ejected from science because they were seen as an aesthetic and commercial category rather than a natural one. On the other hand, science came to be seen as a reliable judge of the aesthetic and commercial value of gems. Science seems to be both indifferent to human values and a source of special insight about them. How is this possible? How can there be such a thing as value-free evaluation?

We can see how something is possible by looking at how it came to be. Postrevolutionary France was one context in which something like gemology came to be. The first three decades of the nineteenth century saw the invention of a new field of study that, though not identical to modern gemology, did share the broad project of bringing the natural sciences to bear on the practical question of the value of gems. The most obvious manifestation of this new field was a series of books dedicated to the "science of precious

---

1. Strunz and Nickel, *Strunz Mineralogical Tables*; Read, *Gemmology*. On the history of gemological institutes in the twentieth century, see introduction, note 20.

.

stones applied to the arts," to quote the title of one contribution to the genre.[2] The applied science of gems had other manifestations as well, including a widely publicized trial that involved René-Just Haüy, an enterprising merchant from Rouen, and 70 kilograms of stones that may or may not have been diamonds. Gem appraisal was a major application of the earth sciences in this period, alongside such things as mining and porcelain production.[3]

How did this field emerge? The existing literature on the history of material evaluation suggests three answers, none of them complete. One answer is that there was a shift from connoisseurship to the analysis of production in writing about craft and commerce. The old practice of documenting the qualities of existing materials gave way to a new practice of analyzing the manufacture of new materials.[4] A second answer is that a gulf opened up between science and values, between description and judgment, such that whole idea of reading values off the natural world came to seem wrong-headed.[5] This chimes with the widespread thesis that science became more esoteric around 1800, with its own journals, societies, instruments, and languages, each adapted to specific disciplines.[6] A third answer goes in the opposite direction, noting that science became increasingly embedded in everyday life in this period, whether through industry, government, advertising, education, or popular culture.[7] Arguably, scientists had a greater stake than ever before in the material world outside the laboratory and the lecture hall. This fits a wider thesis in the history and sociology of science, which says that the much-vaunted objectivity of science is in need of deflation.[8]

2. Caire-Morant, *Science des pierres précieuses*. Read dates the "science of gemology" to the second half of the nineteenth century, and the institutionalization of this science to the early twentieth century: Read, *Gemmology*, 1, 2–3.

3. Mining and porcelain making in Laboulais, *Maison des mines*; Derek E. Ostergard, ed., *The Sèvres Porcelain Manufactory: Alexandre Brongniart and the Triumph of Art and Industry, 1800–1847* (New Haven, CT: Yale University Press, 1997).

4. Pickstone, "Thinking over Wine and Blood."

5. Lorraine Daston, "The Naturalistic Fallacy Is Modern," *Isis* 105, no. 3 (2014): 579–87.

6. This is sometimes called "the second scientific revolution" or "the transition to modern science." Recent surveys are Heilbron, "History of Science"; Knight, *Voyaging in Strange Seas*, 277–78, 282–88.

7. This is the spirit of Charles C. Gillispie, *Science and Polity in France: The Revolutionary and Napoleonic Years* (Princeton, NJ: Princeton University Press, 2004), with its emphasis on "action"; and Spary, *Feeding France*, esp. 15–18. Both the gaps and the bridges are covered in Nicole Dhombres and Jean Dhombres, *Naissance d'un nouveau pouvoir: Sciences et savants en France 1793–1824* (Paris: Payot, 1989), esp. chap. 5, though without reference to material evaluation.

8. A clear statement of this view is in Steven Shapin, "The Sciences of Subjectivity," *Social Studies of Science* 42, no. 2 (2012): 170–71.

My own view is that these answers work best together. They each describe one side of a three-sided process. There was a process of bifurcation, with gaps opening up between science and society, or at least with a widening of existing gaps. These gaps were not just rhetorical, at least as far as gems were concerned. They were materialized in collections, institutions, and literary genres, and they had an intellectual basis in new ways of classifying minerals. At the same time, there was an effort to build bridges across the widening gaps. Nature and value were not two sides of an impassable chasm but two languages in need of translators, to borrow the linguistic metaphor often used at the time. The result of these two processes—the opening of gaps and the building of bridges—was the emergence of new forms of material evaluation. These included instruments such as new kinds of balance, literary devices such as tables of gem color, and social arrangements such as committees of scientific experts. One might think of this as a shift away from "connoisseur-ship" and toward the "analysis of production," but these terms are mislead-ing insofar as they suggest that evaluation was superseded by production. Naturalists still aimed to evaluate gems after 1800, no less so than before 1800. The difference was that they now drew a sharper distinction between tech-niques drawn from mineralogy and techniques drawn from cutters, jewelers and connoisseurs.

The whole process may be illustrated by the careers of the people who did the most to create the applied science of gems. Haüy was a research sci-entist in all but name.[9] He was a member of the Academy of Sciences and the House of Mines before becoming the professor of mineralogy at the National Museum of Natural History in 1802, a post he held until his death in 1822. Haüy worked on the practical problem of gem appraisal; in this sense he was a "scientific-technological expert" of the kind that Ursula Klein has found in Prussia in the same period.[10] But Haüy was a different sort of expert to Antoine Caire-Morant, Jean-Baptiste Pujoulx, and Cyprien-Prosper Brard, three understudied proponents of the applied science of gems.[11] All three were shaped by Parisian scientific institutions, but none had a permanent place in those institutions. Caire-Morant, the oldest of the three, was trained

9. Haüy's posts in Hooykaas, "Haüy."

10. Klein, *Technoscience in History*, esp. chap. 13.

11. These three men are absent in relevant surveys, such as Charles C. Gillispie, *Science and Polity in France at the End of the Old Regime* (Princeton, NJ: Princeton University Press, 1980); idem, *Science and Polity in France: The Revolutionary and Napoleonic Years*; Dhombres and Dhombres, *Naissance d'un nouveau pouvoir*; and Guy Vautrin, *Histoire de la vulgarisation scientifique avant 1900* (Les Ulis, France: EDP Sciences, 2018). Pujoulx is mentioned in passing in Dominique Poulot, *Musée, nation, patrimoine, 1789–1815* (Paris: Gallimard, 1997), 225.

as a gem cutter in Turin. He found his way to Paris in the 1770s, when he presented at least two inventions to the academy, took courses in the natural sciences, and persuaded the French state to support a rock crystal manufactory in his hometown of Briançon.[12] Pujoulx had a very different background, working as a playwright before turning his talents to science between 1800 and his death in 1821. He attended lecture courses on the sciences in Paris, including Haüy's at the museum, while working for three of the four most important Parisian newspapers in the period.[13] Brard, the youngest of the three men, was educated at the House of Mines and the museum. He then left the museum in 1813 to supervise a series of industrial projects in the French provinces, including a decade as the director of a charcoal mine in Dordogne.[14] These men bridged the gap that had opened up between the esoteric new mineralogy and its users. The same pattern can be found in books, collections, tests, and legal expertise associated with the applied science of gems. I take these topics in turn, before returning to the paradox of value-free evaluation at the end of the chapter.

12. Biographical details from Paul Guillaume, "Autobiographie de Caire-Morand, fondateur de la manufacture de cristal de roche de Briançon, en 1778," *Bulletin de la Société d'études des Hautes-Alpes* (1883): 142–70; Jean-Armand Chabrand, *Antoine Cayre-Morand: Fondateur de la Manufacture de Cristal de Roche de Briançon* (Grenoble, 1874). See also the invention in "Mémoire qui a été lu à l'Académie Royale des sciences le 22e janvier 1777," AN F/12/2274, dossier 7. The details of Caire-Morant's state funds are in AN F/12/2274, dossier 7; AN F/12/1323, dossiers II.1, II.2, and II.3; and Pierre Bonnassieux and Eugène Lelong, *Conseil de commerce et Bureau du commerce 1700–1791: Inventaire analytique des procès-verbaux* (Paris: Imprimerie nationale, 1900), 472b and 483a.

13. Biographical details and bibliography in "Pujoulx, Jean-Baptiste," in *Annuaire nécrologique [for 1821]*, ed. Alphonse-Jacques Mahul (Paris, 1822), 265–68; Jean-Baptiste Pujoulx, *Minéralogie à l'usage des gens du monde* (Paris, 1813), 10 (Haüy's lectures), 11n1 (physics lectures), 99 (chemistry lectures); idem, *Promenades au Jardin des plantes, à la ménagerie et dans les galeries du Muséum d'histoire naturelle* (Paris, 1803), 287n1 (Haüy's lectures). Pujoulx worked for the *Gazette de France*, *Journal de l'Empire*, and *Journal de Paris*; the fourth daily was the *Moniteur universel*. On the "big four" in this period, see Gilles Feyel, *La presse en France des origines à 1944: Histoire politique et matérielle* (Paris: Ellipses, 2007), chap. 4, esp. 56, 59–60, and 62. The details of Pujoulx's employment at the *Journal de Paris*, where he rose to the position of editor-in-chief, are in AN F/7/3452/1, f. 404, f. 413, f. 415, f. 416.

14. Biographical details from François-René-Bénit Vatar-Jouannet, *Notice historique sur Cyprien-Prosper Brard, ingénieur civil des mines* (Périgueux: Dupont, 1839); Michel Combet and Anne-Sylvie Moretti, *La Dordogne de Cyprien Brard* (Périgueux: Archives départmentales de la Dordogne, 1995). Brard's major works on minerals were *Manuel du minéralogiste et du géologue voyageur* (Paris, 1808); *Traité des pierres précieuses* (Paris, 1808); *Minéralogie appliquée aux arts* (Paris, 1821); *Minéralogie populaire* (Paris, 1826); *Description historique d'une collection de minéralogie appliquée aux arts* (Paris, 1833).

Books

Haüy's five-volume *Treatise on Mineralogy*, published in 1801, became the standard work on the topic in France for the next two decades. The book played much the same role in mineralogy as Pierre-Simon Laplace's *Treatise on Celestial Mechanics* (1798–1825) and Georges Cuvier's *Lessons in Comparative Anatomy* (1800–1805) did in their respective fields.[15] The book cemented the author's reputation, defined the scope of the field, and served as a pedagogical aid for teachers. It was summarized, translated, and tabulated by one of Haüy's assistants at the museum, Jean-André-Henri Lucas, who wrote that the work was "now regarded as the manual for mineralogists everywhere."[16] All this came at a cost, however. It meant that "everyone who cultivated this science was concerned solely with the great discoveries made by Haüy," as Brard later recalled. The result was that "the most useful substances, and therefore the most precious ones, were looked upon with a kind of disdain."[17] Brard had no doubts about the intellectual value of Haüy's work, which amounted to a "revolution" in the study of minerals.[18] But he worried that this work was known only to savants (*savants*) and not to the artisans (*artistes*) who worked with gems on a daily basis.[19] Hence his *Treatise on Precious Stones*, published in 1808, which aimed to bring the new mineralogy to bear upon the lapidary arts, especially gem cutting, gem engraving, and marble cutting.

Brard believed, correctly, that he had written a new kind of book. It is true that earlier works had described the lapidary arts in a systematic way. Brard acknowledged his debts to earlier writers such as David Jeffries, Jean-Baptiste Tavernier, and even to Pliny and Theophrastus.[20] Brard was also aware of earlier authors who had combined a knowledge of natural history with a knowledge of the practical arts, especially Anselmus Boethius Boodt, Robert de Berquen, Pierre Rosnel, and Louis Dutens. These authors had all noticed differences between the naturalist's taxonomy and that of the jeweler or cutter. But these differences had been minor—small enough that authors had a realistic chance of ironing them out to create a single taxonomy that combined the strengths of both. Even Dutens, the most recent precursor to Brard,

---

15. Haüy's other major works after 1800 were *Tableau comparatif* and *Traité de cristallographie*. The importance of such field-defining books is noted in Gillispie, *Science and Polity in France at the End of the Old Regime*, 519–20.

16. Lucas, *Tableau méthodique*, vol. 1, iii.

17. Brard, *Minéralogie appliquée aux arts*, i–ii.

18. Brard, *Traité des pierres précieuses*, xx.

19. Ibid., xxxi.

20. Ibid., xvii–xx.

believed that he was improving upon the taxonomies given by naturalists like
Daubenton and jewelers like Berquen. For Brard, there was no question of
improving on the taxonomy that Haüy had laid down in his 1801 treatise. Nor
did he wish to impose Haüy's taxonomy wholesale upon artisans. His aim was
not to collapse the two taxonomies into one but to connect them. As he put it,
he hoped to "establish a certain concordance between the language of artisans
and that of savants."[21]

Haüy and Caire-Morant had the same goal in their books on precious
stones, published in 1817 and 1826 respectively. To these we can add books
on applied mineralogy published by Pujoulx in 1813 and Brard in 1821, both
of which gave more space to gems than their titles would suggest. Pujoulx
dedicated a disproportionate number of pages to gemstones, as figure 7.1
shows.[22] These works had much in common with a published catalog of the
"mineralogical museum" of the Parisian aristocrat Etienne-Gilbert, Marquis
de Drée, a catalog that contained a thirty-page guide to the classification,
cutting, mounting, and pricing of gems.[23] These books were designed for different audiences, with Pujoulx writing for curious members of the middle
classes (*gens du monde*, in his phrase) rather than for jewelers or gem cutters.
It is tempting to distinguish between applied science and popular science,
but the distinction was less clear-cut at the time than it is now. Pujoulx aimed
both to popularize (*populariser*) mineralogy and to make it useful (*utile*), as
did Brard.[24] All these authors aimed to bring the latest findings of mineralogy
to a wider audience.

The first challenge in the neo-lapidaries was to define their subject matter.
What was a precious stone, now that mineralogists had ejected this category
from their taxonomies? Brard's treatise shows the difficulty of giving a clear
answer to this question. He used three words interchangeably for his subject matter: gemmes, pierres fines, and pierres précieuses. He also used these
words to refer to three different groups of objects.[25] The broadest group was
made up of all the mineral substances cut or polished by artisans, including

21. Ibid., xxvi.

22. Haüy, *Pierres précieuses*; Caire-Morant, *Science des pierres précieuses*; Pujoulx, *Minéralogie*; Brard, *Minéralogie appliquée aux arts*, vol. 3, 145–431.

23. Drée, *Catalogue des huit collections*, 82–115; idem, *Description des objets composant les 4
collections du Marquis de Drée* (Paris, 1816).

24. Vatar-Jouannet, *Notice historique*, 11 (populariser); Pujoulx, *Minéralogie*, 1 (utile), 4
(populariser). Note also the title of Brard's *Minéralogie populaire*. On the history of French popular science in this period, see Vautrin, *Vulgarisation scientifique*, 83–100.

25. Brard, *Pierres précieuses*, xvi–xviii (gemmes, pierres fines), xxii (cut by lapidaries), ii and
xxiii (stones that cut quartz), 78 (pierres fines proprement dites). Note also the interchangeable

0   5   10   15   20

hyalin-quartz*
agate-quartz*
corundum*
feldspar
jasper-quartz*
topaz*
resinite-quartz*
emerald*
diamond*
garnet*
talc
spinel*
tourmaline*
mica
asbestos
zircon*
lazulite*
cymophane
jade*
disthene
amber*
peridot*
epidote
idocrase
jet*
amphibole
diallage
analcime
axinite
pyroxene
stilbite
meionite
turquoise*
mesotype
prehnite
staurolite
macle
paranthine
pinite
harmotome
leucite
nepheline
laumontite
apophyllite
hypersthene
haüyne
triphane
dypire
gadolinite
wernerite
andradite
anthophyllite

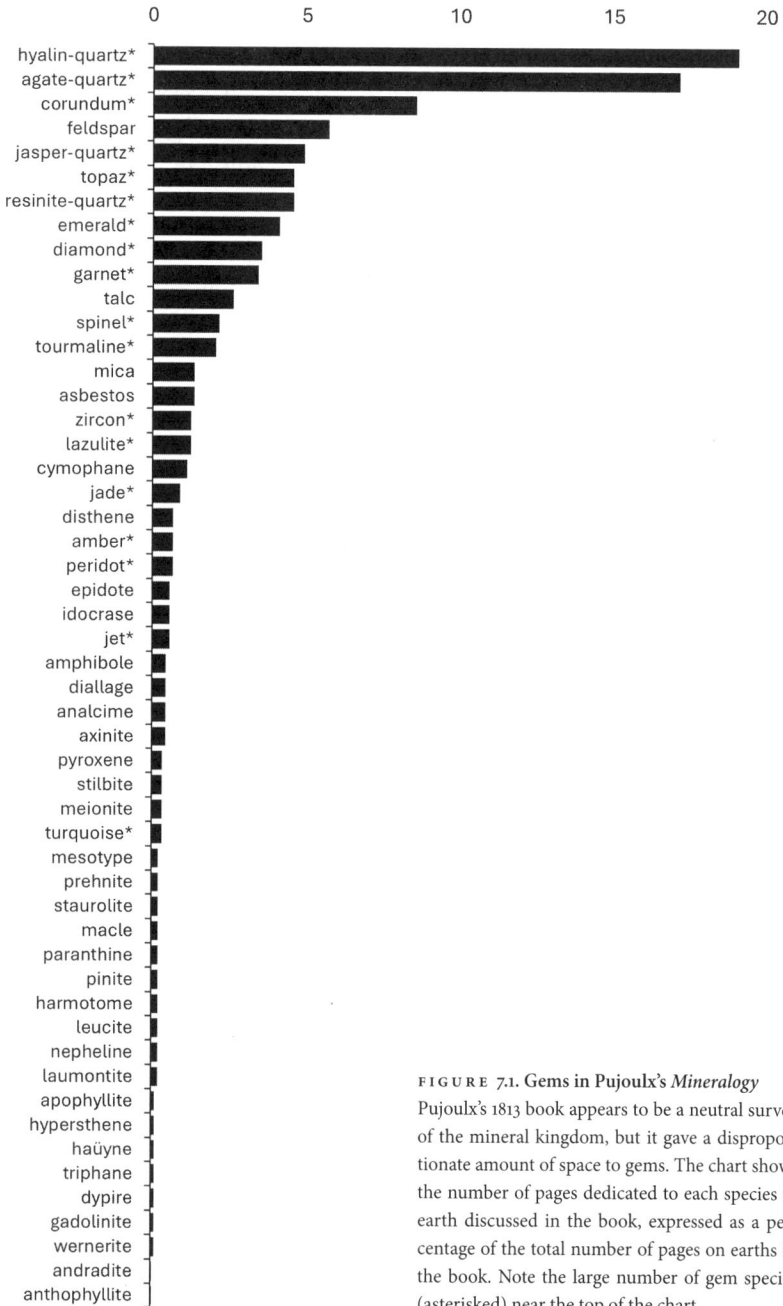

FIGURE 7.1. Gems in Pujoulx's *Mineralogy*
Pujoulx's 1813 book appears to be a neutral survey
of the mineral kingdom, but it gave a disproportionate amount of space to gems. The chart shows
the number of pages dedicated to each species of
earth discussed in the book, expressed as a percentage of the total number of pages on earths in
the book. Note the large number of gem species
(asterisked) near the top of the chart.

everything from diamond to talc. Then there were stones that fitted into this category but were hard enough to scratch quartz, a group that included diamond and rock crystal but not amber or turquoise. Within this group, some were harder, rarer or more expensive than the rest. Brard called these "pierres fines proprement dites," a category that included garnet and topaz but not rock crystal or cymophane. Even Haüy had trouble defining his subject matter. With the help of a jeweler, he gave a list of fourteen species of mineral that he had discussed in his 1801 treatise and that furnished all the known pierres précieuses.[26] But he also discussed forty-three other substances that were somewhat less precious, and his distinction between the two groups hinged on the vague phrase "favored as ornamental objects."[27] Moreover, his list of fourteen pierres précieuses did not agree with the seven listed by Brard, nor with the eighteen described by Caire-Morant. Haüy's list did not even agree with itself. Haüy could not make up his mind whether topaz was a pierre précieuse or merely a pierre fine, so he included it in both groups.[28] Coming up with a hybrid definition of precious stones—one that both naturalists and jewelers could relate to—was harder than it looked.

Coming up with a hybrid taxonomy was no easier. There were a range of solutions, each of them characteristic of the author's approach to the problem of reconciling mineralogy with its public. Brard hoped to find a true hybrid, a distinct taxonomy that would be a middle ground between those of the naturalist and of the gem cutter. Brard organized his gems by the one character they had in common: hardness. "Hard stones" were those that scratch quartz, whereas "soft stones" were those that do not scratch quartz but that do take a polish.[29] This approach was unusual, however. The other authors started with one of the existing taxonomies and adjusted it discreetly to take the other taxonomy into account. Caire-Morant started with the jewelers' taxonomy and incorporated the mineralogists' one when he found this convenient. Pujoulx took the opposite approach, using Haüy's taxonomy as the basis for his treatise and making subtle changes to this taxonomy for the benefit of gens du monde.[30] His book used two font sizes, a small one for dull or obscure

use of pierres fines, pierres précieuses, and gemmes in Pujoulx, *Minéralogie*, 55 and 57; Drée, *Catalogue des huit collections*, 108–10.

26. Haüy, *Pierres précieuses*, 1, 31–63. The jeweler (Archard) is mentioned on p. 235.

27. Ibid., 186–231, esp. 186 ("ornamental objects").

28. Ibid., 32, 190.

29. Brard, *Pierres précieuses*, 15–16. Note also the remarks on hardness at the start of Brard's descriptions of individual species.

30. Pujoulx, *Minéralogie*, 91–93, 122.

species and a large one for the rest, many of which were precious stones.[31] He started with the class of earthy substances, not acidic substances as Haüy had done, reasoning that the most valuable stones should come first. And he largely ignored the crystallographic dimension of Haüy's treatise. Haüy took a different approach again. Rather than glossing over the mathematical side of mineralogy, he tried to make it accessible to cutters and connoisseurs. He referred to the "cleavage" of crystals rather than the "mechanical division" of crystals, thereby playing up the resemblance between the crystallographer's operations and those of the cutter.[32] And he omitted most of the crystal forms that he had described in his 1801 treatise, focusing on the forms he had found in gem-quality specimens and on those that had a straightforward relationship with their primitive form.[33] Each of these authors sought a concordance between scientists and nonscientists, but each did this in his own way. They did not agree on how to make scientists and nonscientists agree.

This diversity manifested itself in everything from images to prose style. There was no standard way of describing the green color of emerald, for example. Descriptions ran the gamut from austerity to lyricism:[34]

Green. (Lucas, 1813)

Very pure green. (Haüy, 1801/1817)

The Peruvian is the most highly valued kind of emerald; in particular, it is notable for its color, which is a deep, velvety green. (Pujoulx, 1803)

The Peruvian emerald, which owes its handsome color to a 1/100th part of chrome, has great commercial value. (Brard, 1808)

The Peruvian emerald is considered the emerald *par excellence* because of its beautiful color that is so much admired. It should be a superb green, like a vivid field of grass, and have a velvety appearance that is clear and resplendent. (Caire-Morant, 1826)

---

31. He explained this method in ibid., 19–20, 125.

32. For details, see chap. 5, this volume.

33. Haüy, *Pierres précieuses*, 55–56, 60.

34. The quotes are drawn from the following works. Lucas, *Tableau méthodique*, vol. 1, 44; Haüy, *Traité de minéralogie*, vol. 2, 521; idem, *Pierres précieuses*, 242; Pujoulx, *Promenades au Jardin des plantes*, 240; Brard, *Pierres précieuses*, 70; Caire-Morant, *Science des pierres précieuses*, 123; Pujoulx, *Minéralogie*, 269; Haüy, *Traité de minéralogie*, vol. 2, 529–30. For more on prose style, see Michael Bycroft, "La fin des pierres précieuses?," in *Avoir une âme pour les pierres: Arts, sciences et minéralité, du tournant des lumières au crépuscule du romantisme*, ed. Pierre Glaudes, Anouchka Vasak, and Baldine Saint Girons (Rennes: Presses universitaires de Rennes, 2024), 189–202.

[The Peruvian emerald] is most valuable when it is a clear, handsome shade of green. I would go so far as to say that this hue, which is very pleasing to the eye, is found only in the emerald; it is all the more valuable for a velvety texture that does nothing to diminish the clarity or brilliance of the stone. (Pujoulx, 1813)

The emerald, inferior in hardness to several other gems, makes up for this by the charm of its color. The glittering red of the ruby, the golden yellow of topaz, the celestial blue of sapphire, are colors that are pleasing to contemplate one after the other, and the beauties of the one distract us from those of the other. But the green of the emerald is a friend to the eye—the eye seems to settle upon it after playing momentarily with the other colors. It is the only stone that satisfies without satiating, as Pliny wrote. It draws the eye to the depths of the happy scene that nature displays to us when spring is in its full splendor. (Haüy, 1801)

Nomenclature was just as varied as prose style, and more contentious. Corundum is a striking example, since it was a large class of very precious stones that seemed to require a complete overhaul in light of the new mineralogy. Haüy made much of this species in his 1801 treatise, holding it up as an example of the dangers of classifying minerals by superficial characters such as color. Oriental ruby, oriental sapphire, and oriental topaz were usually thought of as distinct species, he noted, but they were simply differently colored varieties of the same species. Yet he did not wish to abandon tradition altogether. He named the new species "telesia," by which he meant "perfect body," a nod to the aesthetic qualities of the stone. He also suggested the name "oriental gemstone," for "those of us who wish to bring a little rigor to the study of precious stones, as an object of pleasure." As well as inventing new names, he made it clear how jewelers' names mapped on to mineralogical ones. "Oriental ruby," he explained, is another name for "red telesia"; "water sapphire" another name for "blue transparent quartz." As he later wrote, he was seeking "a concordance between the nomenclature dictated by art and that which is based on scientific principles."[35]

This was far from being the last word on the matter, however (table 7.1). Mineralogists quickly abandoned the term "telesia" in favor of "corundum," when they discovered that ruby, sapphire, and topaz were made of the same substance as the opaque, unattractive minerals known as adamantine spar and emery.[36] Yet one of the key figures in this discovery, Louis-Jacques, Comte de Bournon, continued to use old terms such as "oriental sapphire" and "oriental

35. Haüy, *Traité de minéralogie*, vol. 2, 480, 483, 489. For simplicity, I have translated Haüy's French terms into their nearest English equivalents. Haüy, *Pierres précieuses*, ix (concordance).

36. Haüy was teaching the identity of corundum and telesia as early as 1802: Lucas, *Tableau méthodique*, vol. 1, 40, 257. On the discovery of this identity, see chap. 6, note 132.

TABLE 7.1. Early names for corundum

By 1801 it was well known that oriental ruby, oriental sapphire, and oriental topaz had much in common, despite their different colors. Haüy saw them as varieties of a species he called "telesia" (*télésie*). But the old names persisted, and there was little agreement about how to reconcile the old and the new. The table shows names given to the gems now known by gemologists as "corundum" and "ruby." Original French names are given alongside the nearest English translation. These are the *main* names used by each author; they each used more than one name for these stones.

	French term	Translation
	Corundum	
Haüy 1801	télésie	telesia
Brard 1808	saphir	sapphire
Pujoulx 1813	corindon hyalin	transparent corundum
Bournon 1817	gemme orientale	oriental gem
Caire-Morant 1826	corindon hyalin	transparent corundum
Farges and Segura 2023	corindon	corundum
	Ruby	
Haüy 1801	télésie rouge	red telesia
Brard 1808	saphir rouge	red sapphire
Pujoulx 1813	corindon hyalin rouge	red transparent corundum
Bournon 1817	rubis oriental	oriental ruby
Caire-Morant 1826	rubis oriental	oriental ruby
Farges and Segura 2023	rubis	ruby

Sources: Haüy, *Traité de minéralogie* (1801); Brard, *Pierres précieuses* (1808); Pujoulx, *Minéralogie* (1813); Bournon, *Collection minéralogique* (1817); Caire-Morant, *Science des pierres précieuses* (1826); Farges and Segura, *Pierres précieuses* (2023).

gemstone."[37] Brard, writing in 1808, preferred "sapphire" to "corundum" as the name of the species. In his language, "red sapphire" was the new name for oriental ruby, "yellow sapphire" the new name for oriental topaz, and so on. He argued that the word "sapphire" was the best of both worlds, familiar to artisans but also suggestive of the mineralogical insight that the various oriental stones belonged to the same species.[38] To Pujoulx, however, Brard had achieved the worst of both worlds. Brard's use of the word "sapphire," he argued, was both counterintuitive and inaccurate. He had used an old word to denote a new category, and a common term to refer to a mineralogists' species. Instead, Pujoulx stuck as closely as possible to the new mineralogical terms. "Transparent corundum" was his name for the new species, "red transparent corundum" for the red variety of this species.[39] Finally, Caire-Morant approached the issue from

37. Louis-Jacques, Comte de Bournon, *Collection minéralogique particulière du roi* (Paris, 1817), 25–32, esp. 31; idem, *Traité de minéralogie* (London: 1808), 140.

38. Brard, *Pierres précieuses*, 49, 50–51, 55.

39. Pujoulx, *Minéralogie*, 241, 245–52.

PIERRES VIOLETTES.	PIERRES BLEUES.	PIERRES VERTES.	PIERRES JAUNES.	PIERRES ROUGES.	PIERRES LIMPIDES OU BLANCHES.	PIERRES NOIRES.	PIERRES IRISÉES.
Améthyste 8	Saphir. 49	Emeraude. 65	Topazes (du Brésil.) 62	Saphir rouge. 50	Diamant. 31	Jayet. 183	L'opale. 128
Saphir violet. 56	Emeraude. (Aigue-mar.) 66	Saphir vert. 50	Emeraude miellée.) 66	Rubis. 58	Saphir. 50	Obsidienne. 109	Feld-spath (Pierre de Labrador.) 161
Grenat syrien. 75	Quartz. (Saphir d'eau.) 89	Idocrase. 120	Quartz jaune. 90	Grenat. 74	Topaze (de Saxe.) 62	Canel-coal. 185	Quartz irisé. 93
Spath-fluor. 174	Lapis. 169	Péridot. 151	Spath-fluor. 175	Topaze (brûlée.) 63	Quartz (cristal de roche.) 79		
Lépidolithe. 171	Disthène. 156	Tourmaline verte. 154	Jaspes jaunes.135	Hyacinthe. 71	Feld-spathe. (adulaire.) 159		PIERRES NACRÉES.
	Turquoise. 194	Prase. 104		Cornaline. 101			Cymophane. 56
		Chrysoprase. 104		Aventurine.			Feld-spath. (nacre.) 16?
		Spath-fluor. 174		Tourmaline (apyre.) 154			Quartz na-cré. 93
		Feld-Spath, (pierre des Amazones.) 163					
		Jaspes verts. 139					
		Malachite. 107					

TABLEAU DES PIERRES RANGÉES PAR ORDRE DE COULEURS.

PRÉCIEUSES.

NOTA. Les chiffres qui sont à côté de chaque nom, indiquent la page où l'on a traité de ces pierres.

207

FIGURE 7.2. Brard's table of stones organized by color

Each column lists gems of a different color, from purple to black. Each gem is listed alongside the page number on which it is described in the main text of the book. Readers in possession of a violet stone (for example) are invited to identify the stone by comparing the descriptions of the various violet stones in the main text. From Brard, *Treatise on Precious Stones* (1808), 207. Courtesy of gallica.bnf.fr / Bibliothèque nationale de France.

the other direction. For him, the umbrella category was not the mineralogist's "corundum" but the jeweler's "ruby." He did use the term "red transparent corundum," but he preferred the jeweler's term "oriental ruby." And he treated this as a variety of ruby rather than as a variety of corundum. The other varieties of ruby were the spinel, ballas, and Brazilian rubies, none of which were varieties of corundum. Caire-Morant did not do this out of ignorance, but because of his distinctive approach to concordance, which meant giving priority to jewelers over naturalists rather than the other way round.[40]

Deciding which terms to use was one thing; finding a way to display them on the page was another. Even something as mundane as chapter headings required careful thought. The headings in Brard's *Treatise on Precious Stones* gave multiple names for the same stone, each with a different typography. In the same work he provided a "table of stones arranged in order of their colors," intended for "people who have no knowledge of precious stones" (fig. 7.2).[41] Suppose you are such a person, and you would like to know the technical

40. Caire-Morant, *Pierres précieuses*, 93–104. The names now used by gemologists are "corindon" and "rubis," along with "saphir" for blue corundum gems and "saphir de couleur" for corundum gems that are neither blue nor red: Farges and Segura, *Pierres précieuses*, 167, 169, 170.

41. Brard, *Pierres précieuses*, vol. 1, 206–7.

name for a stone that you possess and about which you know nothing except that it is violet. Go to the column for violet stones, Brard advised, look up the page numbers listed there, and read the descriptions on those pages. The species name you are looking for is the one whose description best matches your stone. This table was distinctive, but the principle was widespread in the period: use a table to guide the layperson from a color-based understanding of gems to a mineralogical understanding, one based on density, double refraction, hardness, electricity, and so on. Brisson, Haüy, and Pujoulx each published one or more such tables; Brard published a more elaborate one in a later work.[42] There was no such table in Haüy's 1801 textbook, nor in the summaries of it made by his assistant Lucas. The color-based table was an innovation of the neo-lapidaries. It was a product, not of the new mineralogy, but of the effort to communicate the new mineralogy to a wide audience.

Collections

The new books went hand in hand with new collections. A gap opened up between what were now known as research collections (*collections d'étude*) and collections designed for beauty or utility.[43] At the same time, there was an effort to bridge this gap by combining the two sorts of collection in the same space. Both trends can be tracked in detail at the National Museum of Natural History in Paris. Haüy was appointed professor of mineralogy there in 1802, a year after the publication of his famous treatise.[44] He lost no time in bringing the collection into line with the treatise. The reorganization was complete in 1803, when Pujoulx published a guide to the museum and its gardens. There is no real catalog of the mineral collection for this period—systematic recording of the museum's acquisitions began in 1822, under Haüy's successor—but

42. Brisson, *Pesanteur spécifique*, xvii–xviii, cf. chap. 5, this volume; Brard, *Minéralogie appliquée aux arts*, 402–12; Pujoulx, *Minéralogie*, 59–63; Haüy, *Pierres précieuses*, 236–53.

43. Examples of this language will be given, but see also Louis-Jacques, Comte de Bournon, *Catalogue de la collection minéralogique du Comte de Bournon* (London, 1813), vi, xxxv, xcvii, lxxxvi. This section may be seen as a hybrid of Dorinda Outram's emphasis on the significance of space in the museum's collections and Emma Spary's emphasis on the separation of natural history into savant and popular varieties. See Dorinda Outram, "New Spaces for Natural History," in Jardine et al., *Cultures of Natural History*, 249–65; and Emma Spary, "Forging Nature at the Republican Muséum," in *The Faces of Nature in Enlightenment Europe*, ed. Lorraine Daston and Gianna Pomata (Berlin: BWV-Berliner Wissenschafts-Verlag, 2003), 163–80.

44. J.-A. Hugard, *Galerie de minéralogie et de géologie* (Paris, 1855), 16–17. On the wider history of the museum in this period, see Gillispie, *Science and Polity in France: The Revolutionary and Napoleonic Years*, chap. 3; and Sue Ann Prince, ed., *Of Elephants and Roses: French Natural History 1790–1830* (Philadelphia: American Philosophical Society, 2013).

a detailed picture of the collection can be built up from successive guides to the museum.[45]

Consider figure 7.3, which shows the collection as it appeared in the first decade of the century.[46] There were sixty numbered cabinets (*armoires*) spread across two rooms. Cabinet 1 contained a goniometer, a Nicholson balance, and wooden models of crystals, all of them familiar to readers of the introduction to Haüy's treatise. The remaining cabinets housed the minerals described in the treatise, in the same order as they appeared there: acidic, earthy, combustible, and metallic substances, with the species in each of these classes appearing in the same sequence in the text and the cabinets. Even the appendixes in the treatise—which covered aggregates, volcanic minerals, and poorly known minerals—had their own cabinets.[47] Within each cabinet, the research collection was placed on a platform made up of four terraces, each of them about 5 centimeters high. Here, too, the order of the specimens matched Haüy's treatise. Each specimen was attached with wax to a wooden stand, with the technical name for the variety written on the stand. Names of species, genera and orders were written on labels of suitable sizes and placed at the head of the specimens they encompassed. Empty stands represented varieties that were known to exist but that the museum did not yet possess. The goal was to present "the uninterrupted series of minerals distributed by classes, orders, genera, species, and varieties."[48]

So much for the research collection. Alongside this collection, but distinct from it, were many other objects that did *not* appear in Haüy's treatise. Cabinets 30 and 31, for example, had no equivalent in the treatise. They

45. Hugard, *Galerie de minéralogie*, esp. 22 (systematic recording); Pujoulx, *Promenades au Jardin des plantes*; Lucas, *Tableau méthodique*, vols. 1–2; Joseph Deleuze, *Histoire et description du Muséum Royal d'histoire naturelle* (Paris, 1823).

46. The numbers of the cabinets, and the names of the gems that appear in them, are based on the index to the cabinets in Lucas, *Tableau méthodique*, vol. 1, 367–72. The distribution of the cabinets around the rooms is based on Pujoulx, *Promenades au Jardin des plantes*, 199–208. The placement of the windows and doors, and the dimensions of the rooms, are based on Bourdier, "Cabinet du Jardin Royal," 45. Cabinet 50 is not shown on the figure because its location is unclear from the sources just cited.

47. Haüy, *Traité de minéralogie*, vol. 5, "Distribution méthodique," for the four classes and three appendixes. The meteorites in the collection correspond to one of the five appendixes in Lucas, *Tableau méthodique*, vol. 1, 329–41. The fifth appendix there (substances modified by the action of nonvolcanic subterranean fire) was not new, since it was part of the appendix of volcanic substances in Haüy's *Traité de minéralogie*. The instruments and crystal models are described in Pujoulx, *Promenades au Jardin des plantes*, 206–8; and in Lucas, *Tableau méthodique*, vol. 1, xvi, and in the index to the cabinets.

48. Lucas, *Tableau méthodique*, vol. 2, x–xiii.

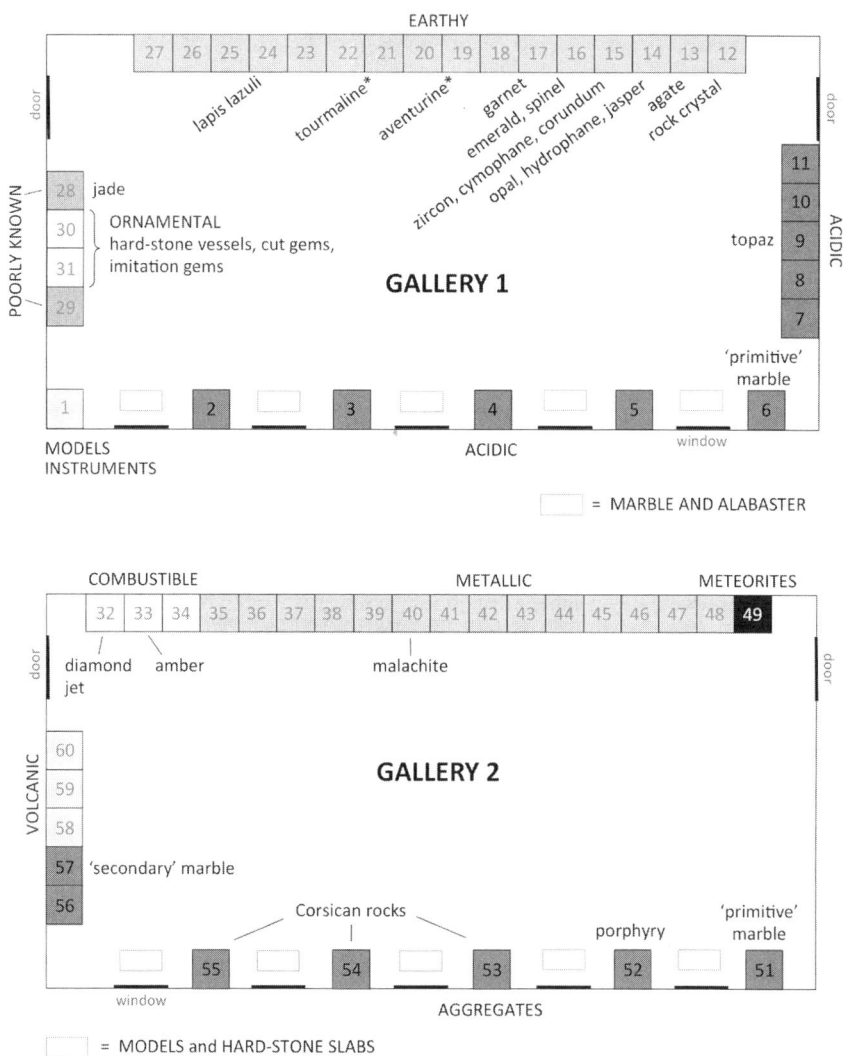

EARTHY

GALLERY 1

27 26 25 24 23 22 21 20 19 18 17 16 15 14 13 12

lapis lazuli
tourmaline*
aventurine*
garnet
emerald, spinel
zircon, cymophane, corundum
opal, hydrophane, jasper
agate
rock crystal

door

door

POORLY KNOWN

28 jade

30 } ORNAMENTAL
31 } hard-stone vessels, cut gems,
29 } imitation gems

11
10
topaz 9
8
7

ACIDIC

'primitive' marble

1   2   3   4   5   6

MODELS
INSTRUMENTS

ACIDIC

window

= MARBLE AND ALABASTER

COMBUSTIBLE   **METALLIC**   **METEORITES**

32 33 34 35 36 37 38 39 40 41 42 43 44 45 46 47 48 49

door

door

diamond   amber
jet

malachite

VOLCANIC

60
59
58
57 'secondary' marble
56

GALLERY 2

Corsican rocks
porphyry
'primitive' marble

55   54   53   52   51

window

AGGREGATES

= MODELS and HARD-STONE SLABS

FIGURE 7.3. **Mineral galleries at the National Museum of Natural History, Paris, ca. 1806**
The research collection was an embodiment of Haüy's 1801 textbook on mineralogy. It was divided into acidic, earthy, metallic, and combustible substances, the four main classes in Haüy's scheme. It was juxtaposed with ornamental stones, including cabinets 30 and 31 and window casements displaying marble, alabaster, and slabs of hard stones such as agate and jasper. The gems in the collection are named and their locations shown. Starred species may be in the cabinet to the right of the one indicated. Diagrams by the author, based on data in Pujoulx, *Promenades* (1803), and Lucas, *Methodical Table* (1806).

contained what later guides to the collection would call ornamental objects (*objets d'ornement*). According to Pujoulx, these were "choice pieces that have been worked and polished, most of which are extremely expensive," including a sphere of rock crystal, twelve columns of amethyst, and two lapis lazuli vases.[49] These two cabinets were strategically located, directly in front of a person entering the room from the main entrance.[50] More ornamental objects were displayed in the window casements. These objects were not in numbered cabinets, but they recalled the research collection in subtle ways—the casements containing polished marble alternated with the cabinets made of the same substance, calcium carbonate.[51] There was a similar mixture in the cabinets themselves, where the research collection occupied the middle section only, with ornamental and technological objects in the upper and lower sections. In the cabinet containing diamonds, for example, the middle section contained raw diamonds, anthracite, coal, jet, and sulfur, in accordance with Haüy's treatise. By contrast, the other sections in the cabinet held diamonds of different colors and with different types of cuts.[52] The correspondence between sections was deliberately loose. "To please the eye," Lucas explained, "the professor [Haüy] thought it best that the order of the varieties should not be strictly the same."[53]

This spirit animated the reorganization of the galleries in 1811. A third room became available for the mineral collection (fig. 7.4), allowing for a sharper distinction between the research collection and other objects.[54] Most of the ornamental objects were transferred to the new room, where they were joined by two collections of faceted gems, one made of genuine stones and the other of glass imitations.[55] Volcanic rocks and aggregates were also transferred, so that the original two rooms were now almost exclusively dedicated to Haüy's four principal classes of mineral. Yet there were still some ornamental objects in these two rooms, both in the window casements and in the

49. Pujoulx, *Promenades au Jardin des plantes*, 246–47.

50. Usual entrance noted in ibid., 175n.

51. Pujoulx, *Promenades au Jardin des plantes*, 214, 247. Note also the casements that combined crystal models and hard-stone slabs, especially marble: Pujoulx, *Promenades au Jardin des plantes*, 249; Lucas, *Tableau méthodique*, vol. 1, xv (models), index to cabinets ("marbres de divers pays I à V," "modèles de cristaux I à VII"); Lucas, *Tableau méthodique*, vol. 2, xii.

52. Pujoulx, *Promenades au Jardin des plantes*, 255 (layout of diamond cabinet); Lucas, *Tableau méthodique*, vol. 1, index on cabinets (species in diamond cabinet).

53. Lucas, *Tableau méthodique*, vol. 2, xi–xii.

54. Deleuze, *Histoire et description du Muséum*, 361 ("salle des roches"); Lucas, *Tableau méthodique*, vol. 2, xii and 183 ("troisième salle," "salle des roches").

55. Deleuze, *Histoire et description du Muséum*, 369–70; Lucas, *Tableau méthodique*, vol. 2, 249.

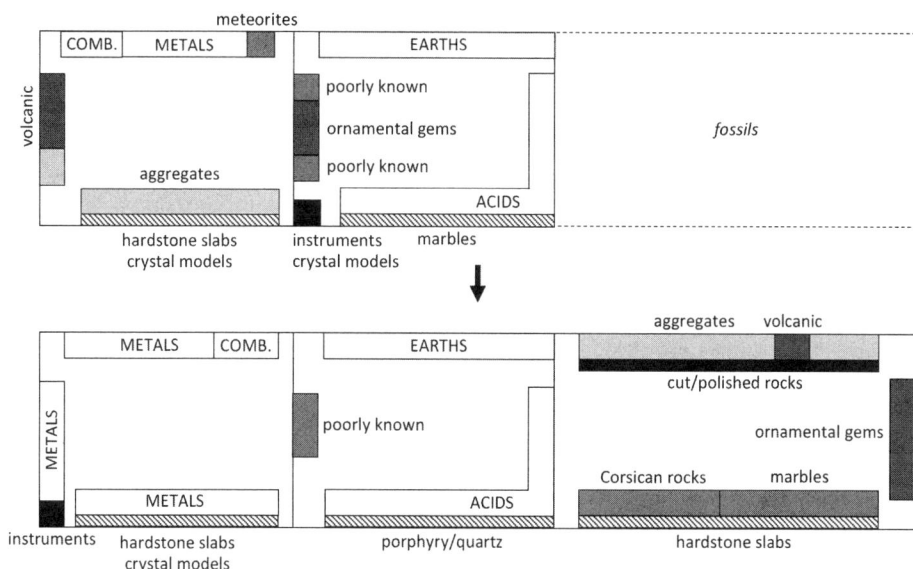

FIGURE 7.4. Evolution of mineral galleries at the National Museum of Natural History, Paris
The top image shows the galleries in 1806, the bottom in 1813. The annexation of a new room by the mineral collection (previously the fossil room) led to a clearer demarcation between the research collection (metals, acids, earths, and combustible substances) and other objects. Some ornamental stones nevertheless remained in the first two rooms, such as hard-stone slabs and quartz/porphyry specimens. Windows and doors are not shown. Hatched areas represent collections placed in window casements. Diagram by the author, based on data in Pujoulx, *Mineralogy* (1813), Lucas, *Methodical Table* (1806), and Lucas, *Methodical Distribution* (1813).

numbered cabinets. The trend was to differentiate research minerals from other minerals without separating them completely.

There was a problem, however. The mineral galleries were designed by Haüy and his assistants. These men had the concerns of researchers and students in mind, especially the students of Haüy's annual course on mineralogy. These were not the concerns of the casual visitor, someone looking for a few hours of pleasant instruction rather than a year of hard study. The gap between the two audiences is measured by the amount of work required to bridge it. Much of this work was done by Pujoulx, the author of the first modern guide to the museum, the *Promenades at the Botanical Garden, at the Menagerie, and in the Galleries of the Museum of Natural History*, published in Paris in 1803. As the title suggests, the guide took the form of a walk through the gardens and galleries, with Pujoulx leading the way. The conceit was based on Pujoulx's experience of physically leading friends along the same route, and of visiting the gardens with his wife—he could write a guide because he *was* a guide.[56] The aim of the book was to make the visit "pleasant and instructive for *gens du monde* of both sexes, and of all ages."[57]

This meant taking a different route through the mineral galleries than the one implied by the physical layout of the collection.[58] Pujoulx quite literally took his own path. Rather than moving from cabinet 1 to cabinet 60 in numerical order, he first went from cabinet 1 to cabinet 31, then from cabinet 60 to cabinet 32.[59] Why? Because that way he went from the earths to the rocks, rather than from earths to metals, thereby respecting the commonsense idea that rocks and earths have more in common with each other than either do with metals. Even when he followed the curator's sequence, Pujoulx was very selective in his attention. He mentioned the attractive pieces of polished marble in one of the window casements, but he virtually ignored the hundreds of crystal models on display in the same place.[60] The average visitor had no time to study the geometric principles behind these models, he explained, and anyway she rarely encountered minerals in crystalline form. The crystallographic names on the labels of the research collection, so carefully prepared by Haüy and his assistants, were virtually meaningless to the average visitor.[61] Not surprisingly, Pujoulx spent a disproportionate amount of time on the gemstones. He had much to say about cabinet 15 (zircon, corundum, cymophane), and almost nothing to say about cabinet 20 (axinite). He ignored half the collection of earthy substances on the grounds that it was "unknown to *gens du monde*, and employed neither in luxury goods nor in the useful arts."[62] He reorganized the collection to suit his audience, just as he would later reorganize Haüy's taxonomy to suit the readers of his popular book on mineralogy.

Pujoulx encouraged readers to build collections of their own, which would themselves be a compromise between savants and gens du monde. His advice was based on his own collection, one that he began as early as 1803 and that he still owned a decade later.[63] He favored well-known species (such as topaz or agate) over those that were of interest only to naturalists (such as euclase). He placed worked gems alongside raw ones, noting that this "rapprochement of

56. Pujoulx, *Promenades au Jardin des plantes*, 4, 13.

57. Ibid., 13.

58. The following expands on Emma Spary's remarks on Pujoulx's selective attention, in "Forging Nature," 178–79.

59. Pujoulx, *Promenades au Jardin des plantes*, 205 and 205n1.

60. Ibid., 249.

61. Ibid., 200–202, 206–7, 208–9.

62. Ibid., 242.

63. Pujoulx, *Minéralogie*, 121.

objects that seem so different" was pleasing to visitors.[64] He had strong opinions about the labels on mineral specimens, preferring labels that showed the "vulgar" names of each specimen as well as the "scientific" ones.[65] Neither the museum (which showed only scientific names) nor the collection at the mint (which showed only vulgar ones) met with his approval.

His aim was not just to mediate between scientists and laypeople, but also to show that the gap was smaller than most people thought. A research collection, he argued, was much easier to form than a collection based on "vanity" or "luxury."[66] He reported that he owned a diamond crystal that cost him 5 francs and was "more valuable for study [*l'étude*] than the regent diamond or the diamond of the king of Morocco."[67] Field trips made collecting more pleasant, not less so: "The piece of metal we value most is the one we detach from the mine ourselves."[68] Scientific knowledge was a form of social mobility, allowing people of modest means to visit first-rate collections as long as they were able to ask pertinent questions about the specimens and give precise descriptions of their own specimens.[69] The overall message was clear: the decision to focus on science, rather than on beauty or prestige, makes it easier rather than harder to collect minerals. Collections were another way to "establish a point of contact" between scientists and the wider public, as Pujoulx described his overall project in his *Mineralogy*.[70]

Other collectors had similar aims, but they realized them in very different ways. Consider the collections of Brard and the Marquis de Drée. The Marquis described his collections in catalogs published in 1811 and 1816; Brard described his in a work published in 1833. Both men were connected to the museum and the House of Mines in the first decade of the nineteenth century.[71] Both believed that "this age must bring about a change in the organization of natural history collections," as Drée put it.[72] Both had one foot in the world of savants and the other in the world of artisans. But the similarities ended there. Whereas Brard's minerals ranged right across the trades, from pottery

64. Ibid., 20n1.
65. Ibid., 33, 35–36.
66. Ibid., 23.
67. Ibid., 26.
68. Ibid., 27.
69. Ibid., 22–23.
70. Ibid., 8.
71. Brard, *Description historique d'une collection*, vi (School of Mines); Drée, *Catalogue des huit collections*, 1 (School of Mines). Drée's collection and catalogs are praised in Lucas, *Tableau méthodique*, vol. 1, 132, 507. The 1816 catalog of Drée's collection is *Description des 4 collections*.
72. Drée, *Catalogue des huit collections*, 1.

to architecture, Drée's were confined to gems. The highlight of Brard's collection was a set of twenty-two porcelain plates, each showing ten of the metal oxides used as pigments by painters.[73] By contrast, the highlight of Drée's collection was a series of two hundred precious stones, each mounted in a gold ring, with each species of stone represented in a range of colors and hues so that jewelers and connoisseurs could educate their eyes by detecting the fine differences between adjacent specimens.[74]

Drée's collection was one of largest in Europe, including at least twelve thousand specimens in numerous drawers and cabinets.[75] By contrast, Brard's 1,500 specimens were contained in a single, purpose-built cabinet.[76] Drée prided himself on acquiring specimens from across the globe; Brard focused on minerals found in France and used in French industry.[77] Drée's collection was one of a kind; Brard hoped that his would be replicated across the country in order to "propagate and popularize [populariser] knowledge about useful mineral substances."[78] The differences were partly institutional, since Drée was an wealthy aristocrat whereas Brard was a mobile state official. The differences were also political. For Brard, a student of the protosocialist Henri de Saint-Simon, the discovery of a seam of coal was far more significant than the discovery of a deposit of rubies. "On the one side there is work, industry and general well-being," he explained, "on the other, elite and ephemeral pleasure."[79] His collection was designed for what he called "the productive class" (classe executante), whereas Drée's was unapologetically aristocratic.[80] Just as research collections diverged from other sorts of collection, so the latter diverged from each other.

Tests

The same pattern that we have seen for books and collections also played out in discussions of tests for distinguishing good gems from bad. Naturalists had long been proposing new tests of this kind, usually with the aim of improving

---

73. Brard, Description historique d'une collection, 5.

74. Drée, Catalogue des huit collections, 80.

75. Wilson, Mineral Collecting, 46, 168.

76. Brard, Description historique d'une collection, 2–3.

77. Drée, Catalogue des huit collections, 1; Brard, Description historique d'une collection, vi, viii.

78. Drée, Catalogue des huit collections, iii.

79. Brard, Description historique d'une collection, 68–69. Brard's politics in Combet and Moretti, Cyprien Brard, 20–23.

80. Brard, Minéralogie populaire, "Avis aux cultivateurs et aux artisans français, sur le but de cet ouvrage."

on tests used by cutters and goldsmiths. Boyle offered electricity as a way
of distinguishing real emeralds from glass ones; Daubenton quantified color
in the hope of rigorously identifying gems; Buffon thought double refraction was the key to gem appraisal. But in each of these cases, the naturalists
drew no distinction between the nature of gems and their qualities. The same
tests were meant to say *what* a gem was and *how good* it was. This changed
with the rise of chemistry and crystallography in mineral classification. These
techniques were now seen as the best guide to the natures of gems. In theory,
they were equally good guides to their qualities. In practice, however, they
were too elaborate and intrusive to serve the latter role. Chemical tests usually meant destroying the gem under test; crystallographic tests could only be
used on raw gems, not on faceted ones. Both called for specialized hardware,
such as goniometers and platinum crucibles, that was not found in a typical
workshop of a jeweler or goldsmith. The tests that mattered most for mineralogists were worse than useless for practitioners. If mineralogy was going to
be useful in the arts, it had to be adapted to the purpose.

This adaptation was the aim of Haüy's *Treatise on Precious Stones* of 1817,
his mature contribution to the applied science of gems. This has been called,
with good reason, the founding text of modern gemology. Haüy began by arguing that existing tests (*épreuves*) are inadequate.[81] He conceded that hardness is a reliable test for cutters but hastened to add that it is nearly as destruc-
tive as chemical analysis. Color and brilliance are misleading, he continued,
because they vary widely within each species. As a result, an atypical member
of one species can easily be mistaken for a typical member of another species.
And they *are* mistaken, Haüy observed, not just by mineralogists but also
by experienced jewelers. To use his examples, jewelers confounded yellow
aquamarine with Brazilian topaz, red spinel with oriental ruby, and both of
the latter with a red variety of Brazilian tourmaline.[82] These mistakes could
be costly—no one wanted to buy a spinel for the price of an oriental ruby or
sell an oriental ruby for the price of a spinel.

Why not use the new mineralogy to clear up these confusions? Chemistry
and crystallography were out of the question, for the reasons given earlier in
this chapter, but there remained the physical characters (*caractères physiques*)

81. Haüy, *Pierres précieuses*, v–vii, xi–xiv. On the book's significance, see Farges and Segura,
Pierres précieuses, 80; and Farges, *Gems*, 243–45. The book covers most of the broad categories
of test (hardness, density, refraction, etc.) that are covered in modern equivalents such as Reed,
Gemmology. The main exception is spectroscopy, first used for gem identification in the 1860s:
Reed, *Gemmology*, 1.

82. Haüy, *Pierres précieuses*, xiv–xvii. Concrete examples of these mistakes are given in ibid.,
xiv (Brazilian tourmaline); Pujoulx, *Minéralogie*, 260 (red spinel), 268 (yellow aquamarine).

that played such an important role in Haüy's mineralogy. Density, double refraction, electricity and magnetism were so many ways of correcting the judgments of the eyes and hands. Haüy described these tests in detail, complete with diagrams of two key instruments, the electroscope and hydrostatic balance. He even named a pair of Parisian artisans who made these instruments.[83] To identify the most decisive tests for a given stone, he supplied one of the color-based tables discussed earlier.[84] Suppose a jeweler owns a red stone that appears to be an oriental ruby but may be a spinel. A glance at the table of red stones shows that oriental ruby can be distinguished from spinel by its greater density and by its double rather than single refraction. The jeweler could then use a hydrostatic balance to measure the density of the stone. If that were inconclusive, he could follow Haüy's instructions for determining the stone's refraction. Either way, the species of the gem would be known. According to one journalist at the time, Haüy had "found the secret of making the deepest truths of natural science accessible to everyone."[85]

But there were problems with this approach, just as there had been problems in Haüy's approach to displaying the museum's collection. In both cases, Haüy's bridge was too short—he underestimated the gap between his own situation and that of his intended audience. The difficulty of putting his tests into practice is shown by Brard's successive works of applied mineralogy, starting with his *Treatise on Precious Stones* of 1808. In that work, Brard was optimistic about the practical value of Haüy's tests. He simply described these tests in commonsense terms, with a few remarks on the circumstances in which they should be used. Crystal structure is very useful, he noted, but only for raw stones; double refraction works on cut stones as well, but only by taking careful precautions; and so on.[86] Thirteen years later, writing his *Mineralogy Applied to the Arts*, after a decade as a director of provincial mines, Brard was much more sensitive to the demands of practical people. He explained that they had a limited amount of time to do the tests and even less time to learn how to do them. He now claimed that crystal form was "of no use whatsoever to lapidaries and jewelers," not only because it was inapplicable to cut stones but also because it required a thorough knowledge of the primary and secondary forms of crystals.[87] No cutter would go to the trouble of acquiring such knowledge when he could identify stones quickly and easily by applying

83. Haüy, *Pierres précieuses*, xxii.
84. Ibid., 238–39.
85. *Moniteur universel*, Dec. 13, 1819, p. 3.
86. Brard, *Pierres précieuses*, 3–30, esp. 3 and 6 (crystal form), 19 (double refraction).
87. Brard, *Minéralogie appliquée aux arts*, vol. 3, 148–50, esp. 148 ("no use whatsoever").

them to the grindstone and thereby judging their hardness. Double refraction and electricity were equally worthless, and for the same reason.[88] As for magnetism, fusibility, phosphorescence, and acidic reactions, these were so seldom of any practical use that Brard ignored them altogether in his 1821 book.[89]

The one technical procedure that Brard retained was the hydrostatic balance, but even here he adapted the procedure for practical use. The instrument that mineralogists usually used for density measurements at the time was the Nicholson balance, named after the British chemist William Nicholson, who had described it in 1785. Haüy gave this instrument his blessing in his 1801 treatise; in the following years it was frequently described in French mineralogical treatises as well as being sold commercially in kits for field naturalists (fig. 7.5).[90] The instrument had the advantage of being cheap, portable, and sufficiently precise for the purposes of the mineralogist, as Haüy explained in 1817. As Brard explained four years later, however, these purposes were not the same as those of the jeweler or cutter. As well as being hard to operate, the instrument required a simple but off-putting calculation, that of dividing the weight of the stone's volume in water by the weight of the stone in air. The "learned terms" in which mineralogists described the instrument were equally unappealing to artisans. A final factor was the jeweler's habit of weighing stones in their hands rather than with a balance, something that they were able to do with a speed and accuracy that greatly impressed naturalists such as Pujoulx.[91] Brard's conclusion was damning: "lapidaries and jewelers have not drawn the least benefit" from the available data on the density of gems.[92]

The problem of the learned terms was easily solved: Brard spent one paragraph explaining Archimedes's principle, rather than the four pages that Haüy devoted to the topic.[93] The unwieldy instrument was another matter. Brard's solution was to do away with Nicholson's balance altogether, replacing it with one that resembled the scales that goldsmiths and jewelers already used to measure small quantities of precious stones and precious metals.[94] The only difference between the existing scales and Brard's balance was the presence of an extra pan on one arm of the scale (fig. 7.6). The extra pan hung below the usual one, such that it was submerged in a container of water when the arm

88. Ibid., vol. 3, 150–54, 164–66.

89. Ibid., vol. 3, 147.

90. Haüy, *Pierres précieuses*, 88–97; Pujoulx, *Minéralogie*, 79–82; Brard, *Manuel du minéralogiste*, 7–9, and note the commercial device on pp. 458–59; cf. Newcomb, *World in a Crucible*, 21–22.

91. Pujoulx, *Minéralogie*, 82.

92. Brard, *Minéralogie appliquée aux arts*, vol. 3, 157.

93. Ibid., vol. 1, 155–56; Haüy, *Pierres précieuses*, 84–87.

94. Brard, *Minéralogie appliquée aux arts*, vol. 3, 160–64, 430–32, plate VIII.

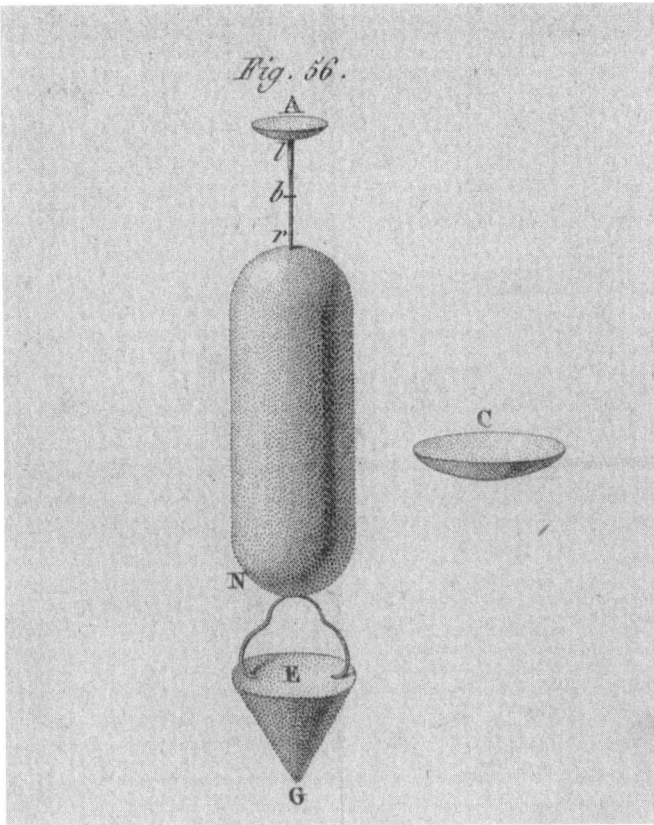

FIGURE 7.5. Nicholson's balance

A balance used by mineralogists in the decades around 1800. The balance allowed the user to weigh a mineral twice, first in air, then in water, and thereby to calculate the density of the mineral. For the first weighing, the whole device was placed in a container of water and bowl C placed on pan A. Weights were added to C until the line *b* was level with the water. The stone was then added to C, and weights removed until *b* was level with the water again. The weight removed was the weight of the stone in air. A similar procedure, with bowl C now placed on E, allowed the stone to be weighed in water. E was lined with weights that kept the instrument vertical when immersed in water. From Haüy, *Treatise on Precious Stones* (1817), plate 2. Courtesy of gallica.bnf.fr / Bibliothèque nationale de France.

of the scale was horizontal. To weigh a stone in air then water, the operator simply weighed it on the top pan and then on the bottom pan. To make the scales "familiar to everyone," by which he meant familiar to artisans, Brard had the weights of all denominations of gold coins used in France engraved on one side of the device.[95] He also tested the instrument in the presence of jewelers, incorporating their suggestions into the final design.

95. Ibid., vol. 3, 162, 430–31.

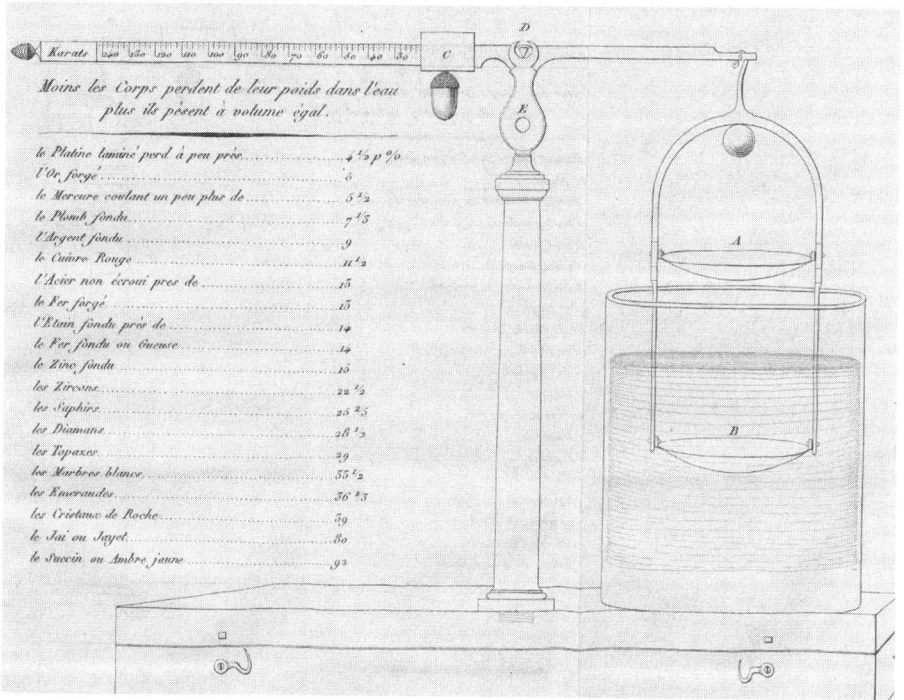

FIGURE 7.6. Brard's hydrostatic balance
A more practicable version of the balances used by mineralogists, such as the Nicholson balance in figure 7.5. Brard modeled this device on balances already used by jewelers and goldsmiths. The extra pan (*B*) on the right allowed the same object to be weighed twice, first in air and then in water. The table on the left shows the percentage of weight that is lost between the first and second weighing, for a range of gems and metals. The numbers range from platinum (4.5 percent lost) to amber (92 percent lost). From Brard, *Mineralogy Applied to the Arts* (1821), vol. 3, plate 8. Courtesy of ETH-Bibliothek Zürich, Rar 8672: 3.

Finally, Brard revised the tables of data on specific weight that Haüy and Brisson had already published. This meant generating new data, both to verify the accuracy of the new instrument and to refine the earlier data. This was done using specimens chosen by the Comte de Bournon from the king's private collection, indicating that the measurements were done after Bournon's return from exile in Britain in 1814. The collection at the museum was used as well, with the help of one of Haüy's assistants.[96] Brard then drew up a new table of gem densities, one that was organized by color (like Haüy's) but that contained data on specific gravity alone (unlike Haüy's).[97] It was also

96. Ibid., vol. 3, 163, 397; Bournon's movements in John G. Burke, "Bournon, Comte de, Louis-Jacques," in *DSB*, vol. 2, 355.
97. Brard, *Minéralogie appliquée aux arts*, vol. 3, 397–401, 402–17.

much longer than the earlier tables, because Brard gave data on a range of carat weights for each gem. On the table of green stones, for example, he gave the weight in water of stones that ranged from one grain to one hundred grains (fig. 7.7). Jewelers could now use the hydrostatic balance without writing down a single number. According to Brard, the whole procedure took less than six minutes per specimen. Haüy had been right to think that balances could identify gems, but wrong to think that mineralogists' balances could be transferred unchanged into the workshop.

Haüy's tests were also challenged from another direction. The problem was not just that his tests were less effective than he thought but also that existing tests were more effective than he thought. This is evident in the case of color, a criterion that Haüy repeatedly dismissed as a poor guide to the nature of gems. "Color," he wrote in 1796, "far from being used in the determination of species, is going to vanish in the eyes of science."[98] Haüy saw the abandonment of color as one of the main achievements of the new mineralogy. But this view was not universally held among people who otherwise adhered to Haüy's system. The Marquis de Drée was such a person. Drée certainly believed that practical knowledge of precious stones was in need of reform. But for him, the solution was not to abandon color but to train the eye more thoroughly in the detection of color. He distinguished the color of stones (*couleur*) from their hues (*teintes*) and their shades (*nuances*). To use his examples, red was a color, poppy-red was a hue, and dark poppy-red was a shade.[99] He acknowledged that color was a misleading and accidental property, as the case of corundum showed. But hues and shades were another matter. His vivid description of sapphire was characteristic:

> The shades of this stone vary from dark Prussian blue to the faint shade of blue that one sees in a large quantity of water. The indigo hue is fairly common. There are some sapphires that, with an admixture of purple, take on a violet hue, especially under light. The most highly prized color is a handsome cornflower blue, which is even more precious when it has a velvety appearance, though this is very rare. Still, the somewhat lighter shades are often rather attractive, dark sapphires having the tendency to develop black spots at their corners.[100]

Here Drée not only described the different hues and shades of sapphire but also said something about the rarity and value of each. Elsewhere he remarked

98. Haüy, "Description de la Cymophane," 11. Cf. Pujoulx, *Minéralogie*, 54; Brard, *Minéralogie populaire*, 74.

99. Drée, *Catalogue des huit collections*, 85.

100. Ibid., 85.

POIDS dans L'AIR. Grammes ou grains.	POIDS DANS L'EAU.					
	Saphirs verts.	Péridot.	Tourmaline verte.	Émeraude.	Aigue-marine.	Chrysoprase.
1	0,766	0,708	0,690	0,633	0,633	0,611
4	3,06	2,83	2,76	2,53	2,53	2,42
8	6,12	5,66	5,52	5,06	5,06	4,86
12	9,18	8,49	8,28	7,59	7,59	7,31
16	12,25	11,32	11,04	10,12	10,12	9,75
20	15,31	14,16	13,80	12,65	12,65	12,19
24	18,37	16,99	16,56	15,19	15,19	14,64
28	21,44	19,82	19,32	17,72	17,72	17,08
32	24,51	22,65	22,08	20,25	20,25	19,53
36	27,57	25,48	24,84	22,77	22,77	21,98
40	30,64	28,32	27,60	25,30	25,30	24,43
44	33,71	31,15	30,36	27,83	27,83	26,88
48	36,76	33,98	33,12	30,36	30,36	29,32
52	39,82	36,81	35,88	32,89	32,89	31,77
56	42,89	39,64	38,64	35,43	35,43	34,21
60	45,95	42,48	41,40	37,94	37,94	36,66
64	49,01	45,31	44,16	40,47	40,47	39,11
68	52,08	48,14	46,92	43,00	43,00	41,56
72	55,14	50,97	49,68	45,53	45,53	44,00
76	58,21	53,80	52,44	48,07	48,07	46,44
80	61,28	56,64	55,20	50,60	50,60	48,88
84	64,34	59,47	57,96	53,13	53,13	51,32
88	67,41	62,30	60,72	55,66	55,66	53,76
92	70,47	65,13	63,48	58,19	58,19	56,21
96	73,54	67,96	66,24	60,72	60,72	58,65
100	76,60	70,80	69,00	63,25	63,25	61,09
Pes. spéc.	4,27	3,42	3,22	2,72	2,72	2,56

FIGURE 7.7. Brard's table of gem densities

This table was designed to be used along with the balance in figure 7.6 to identify a gem by its density, without doing any sums. The left column lists a range of weights that a gem might have when weighed in air. The remaining columns show the weights, in water, of six different species of green gem. Suppose you have a green gem of unknown species. Using the balance in figure 7.6, you find that the stone weighs 36 carats in air and 24.8 carats in water. Turn to the table, find "36" in the first column, then browse the other columns in that row, looking for "24.8" or as close as possible. The best match is green tourmaline. Brard supplied similar tables for gems of other colors. From Brard, *Mineralogy Applied to the Arts* (1821), vol. 3, plate 4. Courtesy of ETH-Bibliothek Zürich, Rar 8672: 3.

that apricot-colored corundum is exceptionally rare, so much so that he only knew of two examples, one in his collection and one that he saw in the hands of a royal jeweler.[101] Some species vary greatly in their color, he observed, whereas others are more uniform. This is true even of varieties that mineralogists place in the same species, such as emerald and aquamarine. The former has only one hue—its value depends instead on its shade, texture, and brilliance—whereas the latter can be sky blue, aquamarine, honey-yellow, pale lemon, or straw-colored.[102] A jeweler or connoisseur who was blind to these nuances would run a constant risk of mistaking valuable stones for mediocre ones and vice versa. A goldsmith who mounted a set of rubies in the same ring, without regard to their hues and shades, would produce an unattractive dappled effect.[103] Hues and shades helped to differentiate stones that belonged to the same species but that differed in their commercial or aesthetic value.

They also helped to distinguish one species from another, at least according to Drée. How can we tell an oriental ruby from a spinel ruby of the same color? Haüy's answer was to abandon color in favor of density and refraction. Drée's answer was to study color more closely. Spinels nearly always have a hint of yellow, he noted, whereas oriental rubies never do.[104] More generally, hues and shades "are often characteristic of a substance, when they are examined closely." The blue of a sapphire is never the same as the blue of a beryl, the green of emerald never that of peridot, and so on. These differences were subtle, and only a "very practiced eye" could detect them.[105] Drée presented his collection as the perfect place to "educate the eye"—he might have added that double refraction and the hydrostatic balance also required extensive training.[106]

There was a similar emphasis on color in books by jewelers or by people who worked closely with jewelers. The English jeweler John Mawe went so far as to publish a color chart showing the characteristic hues of major gem species (plate 8).[107] In France, the cutter Caire-Morant sprinkled his book with remarks on the colors of gems: the hardest diamonds are those that are the color of red vinegar; diamonds that have a milky color when raw will be sky blue when cut; the hue of a piece of rock crystal is a sure guide to the

101. Ibid., 87.

102. Ibid., 92–93.

103. Ibid., 86, 113–14.

104. Ibid., 89.

105. Ibid., 89. Drée also discussed (95) the art of distinguishing garnets from oriental amethysts by color.

106. Brard, *Description des 4 collections*, 92.

107. John Mawe, *A Treatise on Precious Stones* (London, 1813), v–vi.

mine from which it was taken; a certain French diplomat has a collection of aquamarines in a great variety of colors, from sky blue to golden green to rusty white.[108] These observations were all part of Caire-Morant's attempt at a rigorous study of "precious lithology," as he called the field. Mineralogists may have abandoned color, but cutters and jewelers and connoisseurs had not. On the contrary: they gave a more systematic account of gem color than most earlier naturalists had done. For them, color was as much a part of the new science of gems as the hydrostatic balance.

Expertise

All the ingredients of the applied science of gems discussed so far—careers, books, collections, tests—came to a head in the latter part of 1819. The occasion was the trial of Charles-François Legigand, a canvas merchant based in Rouen. Legigand had been accused of an audacious crime: buying a large quantity of nearly worthless topazes and selling them as diamonds for an enormous profit. Legigand had bought the stones in Rouen the previous year, paying 2,000 francs for 150 livres (about 70 kilograms) of stones that the seller identified as "white topazes." After having them cut and polished, Legigand sold them as "Brazilian diamonds" in the capital. Through his business partner, a count, he advertised these diamonds in letters to "the most distinguished persons at court, who themselves recommended Legigand to ambassadors of foreign powers." The count himself bought 13 thousand francs worth of the stones. The market price of 150 livres of diamonds was in the order of 50 million francs, a profit of 3 million percent.

But the count's jeweler smelled a rat. The trial took place in November 1819, with Haüy as the star witness. One of the main Parisian dailies, the *Moniteur universel*, gave a blow-by-blow account of the "verification of the nature of the stones" carried out by the "celebrated mineralogist." All the latest techniques in the science, from double refraction to crystallography, were brought to bear on the case. The tests were done in Haüy's rooms at the museum, using specimens from his own collection. Readers interested in the technical details were directed to Haüy's recently published *Treatise on Precious Stones*. The upshot was that Haüy confirmed the suspicions of the jeweler. Legigand's stones, it seemed, were not diamonds after all. "All the treasure of the mines

108. Caire-Morant, *Pierres précieuses*, 33 (rock crystal and milky diamonds), 34 (red vinegar), 151–52 (aquamarine). Color is still seen as an unreliable guide to the identity of most gem species, although spectroscopy gave color a new lease of life in gem appraisal in the twentieth century: Read, *Gemmology*, 1–9, 78, chap. 11.

of Golconda, which he imagined himself to possess, collapsed before the trials of chemistry," in the words of a journalist.[109]

This trial was a notable event in the history of gemology. It is one of the earliest examples—perhaps *the* earliest example—of scientific expertise being used to appraise a gem in a legal trial. It is also a precocious example of scientific expertise of any kind, medical expertise excepted, being used in the courtroom anywhere in Europe.[110] For this very reason, however, the journalist's account of the trial should be taken with several grains of salt. The practice of using science in courtrooms was itself on trial in this period. Even an apparently straightforward case could be a bone of contention, as the French public discovered two decades later during the famous trial of the alleged murderer Marie Lafarge. The trial was eventually settled with the help of the Marsh test, a sensitive test for arsenic developed by the English chemist James Marsh; but this was achieved only after a long series of mutually contradictory experiments involving multiple chemists, doctors and apothecaries.[111] "Scientists who appeared as witnesses quickly learned that the transfer of knowledge from the laboratory to the courtroom was fraught with difficulties," as the historian Mark Essig has written about the North American context.[112] The situation in France was complicated by the French Revolution and the formalization of medical expertise it entailed. Chairs of legal medicine were created in all French medical schools in 1794. An 1803 law made formal training in this field a requirement for anyone serving as a medical expert in the courtroom.[113] This did not prevent controversies such as the one

109. *Moniteur universel*, Dec. 13, 1819, p. 3 ("verification," "celebrated mineralogist"). See also *Le constitutionnel*, Nov. 26, 1819, p. 2; and *Journal de Paris*, June 16, 1825, pp. 2–3 ("trials of chemistry"). Legigand's first name does not appear in these sources, but a "Charles-François Legigand" turned up two decades later, describing himself as a "traveler trading in Brazilian diamonds": *Journal de la ville de Saint-Quentin et de l'arrondissement*, Dec. 9, 1838, p. 2.

110. This claim is based on Katherine D. Watson, *Forensic Medicine in Western Society: A History* (London: Routledge, 2010), chap. 3; Tal Golan, *Laws of Men and Laws of Nature: The History of Scientific Expert Testimony in England and America* (Cambridge, MA: Harvard University Press, 2004), chaps. 1 and 2; and Déirdre M. Dwyer, "Expert Evidence in English Law Courts, 1559–1800," *Journal of Legal History* 28, no. 1 (2007): 93–118.

111. E. Claire Cage, *The Science of Proof: Forensic Medicine in Modern France* (Cambridge, UK: Cambridge University Press, 2022), chap. 2; José Ramón Bertomeu Sánchez and Agustí Nieto-Galan, eds., *Chemistry, Medicine, and Crime: Mateu J. B. Orfila (1787–1853) and His Times* (Sagamore Beach, MA: Science History Publications, 2006).

112. Mark Essig, "Poison Murder and Expert Testimony: Doubting the Physician in Late Nineteenth-Century America," *Yale Journal of Law and the Humanities* 14, no. 1 (2002): 181–82, cited in Watson, *Forensic Medicine*, 66.

113. Watson, *Forensic Medicine*, 51.

surrounding the Marsh test, but it may have raised expectations about the cre-
dentials of expert witnesses from other branches of science. Since there were
no schools for legal mineralogy, it was not obvious how mineralogists should
operate in the courtroom. For all these reasons, we should look beyond the
gushing report in the *Moniteur universel* for the details of the Legigand case.

Fortunately, other documents relating to the case have survived. It is possible to piece together Legigand's side of the story from the trial documents
and from letters exchanged between Legigand, royal officials, and the scientists concerned.[114] Legigand's story begins with the purchase of the stones
from a merchant in Rouen, who had acquired them from a ship's captain as
payment for merchandise the captain had sold in Brazil.[115] So claimed Legigand, through his lawyer, in a document printed in 1820 or 1821. The merchant had bought these stones without knowing what they were—to him they
were simply "pierres de Brésil." Legigand found that they were very transparent, hard, and covered in a fine film, and that they weighed 3.6 times more
than the same volume of water. In all these respects they resembled "the
stones that authors describe in their works under the name of 'diamans du
Brésil.'" Legigand also had the stones examined by a certain Mr. Vitalis, a
professor of chemistry at Rouen, whose only reservation was that the stones
did not burn in his furnace as diamonds should.[116] Legigand read widely—he
mentioned works by Bomare, Buffon, and the chemists Jean-Antoine Chaptal
and Antoine-François Fourcroy.[117] He learned that the combustion of diamond was a notoriously unreliable procedure, and that no colorless topaz had
yet been discovered in Brazil. In any case, Legigand later saw his diamonds
burned by a certain Mr. Pierlot, a chemist in Paris. These proofs helped to
persuade the Comte de Courcy-Montmorin—the unnamed count in the
newspaper reports—to enter into a business partnership with Legigand to
the tune of several thousand pounds. The count took his own precautions,
"examining the stones for several days, most probably having them examined
by artisans, and . . . consulting the relevant authors."[118]

114. The following is based on the Legigand dossier, AN F/12/2274, dossier 10.

115. "Mémoire concernant l'affaire des diamans du Brésil," in Legigand dossier.

116. Vitalis's profession is given in *Journal des débats politiques et littéraires*, Nov. 24, 1819,
p. 2.

117. Legigand did not give full citations, but he appears to have consulted these passages:
Bomare, *Dictionnaire raisonné*, vol. 6, 684–87; Buffon, *Histoire naturelle des minéraux*, vol. 4,
262–86; Antoine-François Fourcroy, *Encyclopédie méthodique: Chymie, pharmacie et métallurgie*, vol. 4 (Paris, 1805), 152–55, and vol. 5 (Paris, 1808), 618–19; and Jean-Antoine Chaptal, *Chimie
appliquée aux arts* (Paris, 1807), vol. 2, 342–56.

118. "Mémoire concernant l'affaire des diamans du Brésil," 11.

Legigand was unmoved by the evidence presented at the 1819 trial. He voiced his concerns at the trial itself, when he engaged Haüy in "scientific discussions" (*discussions scientifiques*).[119] He expanded on these concerns in a letter written after the trial.[120] Before the trial he had asked Haüy why he believed there were white topazes in Brazil. Haüy had referred vaguely to "an English author." At the trial he gave a different answer, saying that his source was a letter from a "London savant" that he received many years before. A single report, and an outdated one at that! And Haüy had "misled the public," Legigand maintained, by asserting the existence of colorless Brazilian topaz in an article in the *Commercial Dictionary of Paris*. As for the double refraction experiment, the multiple images that Haüy saw in Legigand's stone could be explained away by the fact that the stone had far more facets than the diamond that only showed one image. It was the facets, Legigand argued, not the substance of the stone itself, that multiplied the images. Legigand's stones would refract light in the same way as Haüy's diamond if they were cut in the same way—something that he had verified immediately after the trial. These scientific objections were reinforced by legal ones. For one thing, the court had never given Legigand an opportunity to respond with his own experiments. Surely the right of reply applied just as well to experiments as it did to verbal arguments? Also, the point of the trial was to determine whether Legigand had behaved fraudulently, not to determine the identity of his stones. So, strictly speaking, the court had not delivered a judgment on the latter question. In all this, Legigand presented himself as an honest merchant who was up against a cabal of arrogant elites, including Haüy. The crystallographer's judgment was clouded, Legigand wrote, not just by his "extreme ignorance about diamonds and white topaz" but also by "his sympathy for the interests of Mr. de Courcey, my opponent."

Dissatisfied with the first trial of his stones, Legigand spent many months seeking a second trial. In October 1821, he wrote to the Minister of the Interior to put his case.[121] His aim now was to involve the Consulting Committee of Arts and Manufactures, a body set up during the French Revolution to evaluate inventions for the purpose of distributing prizes and subsidies.[122] The minister was sufficiently impressed to write a spirited letter to the consulting committee, repeating Legigand's arguments about the identity of the

---

119. *Journal des débats politiques et littéraires*, Nov. 24, 1819, p. 3.

120. This paragraph and the next are based on Legigand to Minister of the Interior, Dec. 6, 1821, in Legigand dossier.

121. Legigand to Interior Minister, Oct. 9, 1821, in Legigand dossier.

122. Gillispie, *Science and Polity in France: The Revolutionary and Napoleonic Years*, 199–209.

stones and pointing out that tens of millions of pounds worth of diamonds would be a boon for the French economy.[123] With so much at stake, it was worth taking a second look. But the committee—which included the chemists Louis-Jacques Thénard and Joseph-Louis Gay-Lussac—was unmoved.[124] Its members reminded the minister of Haüy's original demonstration, adding that Legigand had since visited Thénard at his laboratory and asked him to test the stones. Thénard had tried and failed to burn the stones. "The proof was decisive," the Committee reported. Once again, however, Legigand had a reply.[125] Thénard had *not* done the experiment! Legigand knew because he was there. Thénard's laboratory assistant had done the experiment, and he had not done it properly—he had heated the stone too briefly and with too weak a flame. Legigand knew how to do it because he had recently seen it done by three jewelers in Paris, who had succeeded in burning two of his stones, as chemists had done on earlier occasions in Rouen, Evreux and Paris. It seemed that Thénard's experiment was no more a fair trial than Haüy's had been.

In subsequent letters, Legigand and his lawyer explained what *would* count as a fair trial.[126] It would be done in Legigand's presence. It would not be done by a legal tribunal, which was competent to judge legal questions but not to judge mineralogical ones. It would not be done by Haüy or Thénard, who would never back down "in the interest of the honor of their supporters and of their so-called knowledge." It would be done "in the presence of the authorities," a point that Legigand insisted upon after meeting two mineralogists who were friends of Haüy and who told him he could "make all the representations he liked, yet we will always agree with Haüy." The same mineralogists claimed that Legigand's sources were outdated, that Buffon, Chaptal and so on were dead, and that a work dies with its author. But this was sheer partiality, Legigand argued, since these mineralogists would certainly not treat Haüy's works in this way. "Haüy is on the verge of death, but I ask you, Monseigneur, will they bury his books the day they bury him?" Finally, a fair trial would be done by "scientists convoked in a formal manner and united by a higher authority." Only a full, formal, impartial inquiry under the auspices of the consulting committee would settle the question. Alas for Legigand, no such inquiry took place. He was tried again in 1823, convicted

---

123. Interior Minister to Consulting Committee, Oct. 20, 1821, in Legigand dossier.

124. Consulting Committee to Interior Minister, Nov. 20, 1821, in Legigand dossier.

125. Legigand to Interior Minister, Dec. 6, 1821. Cf. Legigand to Interior Minister, Dec. 12, 1821. Both in Legigand dossier.

126. Legigand to Chauveau-Lagarde (Legigand's lawyer), Nov. 24, 1821; Chauveau-Lagarde, "Mémoire au roi, concernant les pierres fines ou diamans du Brésil," n.d., sent by Chauveau-Lagarde to Monsieur le Duc, Sept. 21, 1822. Both in Legigand dossier.

of fraud and sentenced to a fine and six months in prison.[127] Fifteen years later he was before the Parisian courts once more, calling himself a "traveler trading in Brazilian diamonds."[128]

It is easy to dismiss Legigand as a liar and a fraud, as the courts eventually did. But this does not explain his repeated requests for a scientific inquiry into the identity of the stones, requests that he made *after* he had been cleared of fraud in 1819. He had genuine doubts about the decisiveness of Haüy's tests. These doubts reflected the difficulty of applying the science of precious stones in a legal context. The difficulties were partly due to the tests themselves, which were not as clear-cut as the author of the article in the *Moniteur universel* made out. The test that the journalist considered most decisive—a comparison of the primitive forms of topaz and diamond—was not mentioned in any of the other documents relating to the case. Legigand was right that experiments on the combustion of diamonds were delicate procedures that often failed to give the expected result. He was right that observations of the double refraction of gems were often confounded by the facets of the gems.[129] And he was right that many naturalists of the period denied that there were any white topazes in Brazil—Haüy himself made no mention of this variety of topaz in his 1801 treatise.

Haüy did mention this variety in his 1817 treatise, but this raised the awkward question of why the court should follow the 1817 treatise rather than the 1801 one, and more generally why the works of great eighteenth-century naturalists were no longer trustworthy.[130] The answer was obvious to anyone involved in Haüy's mineralogical program, the whole point of which was to build progressively on the revolution brought about by his 1801 treatise. To anyone outside that program, however, it was not obvious why a nineteenth-century observation was better than an eighteenth-century one. Nor was it obvious that the experiments done by Haüy at the museum—not to mention the experiments he did afterward in private meetings, or the letter he cited from a London naturalist—were admissible in a court of law.[131] To naturalists at the museum, these acts met the highest standards of evidence in the science of mineralogy. To Legigand and his lawyer, they fell well short of a fair trial. The application of the science of gems in the legal realm was as contentious as its application in the realm of jewelers and connoisseurs.

127. *Journal de Paris*, June 16, 1825, pp. 2–3; *Journal des débats politiques et littéraires*, June 16, 1825, p. 4; *Moniteur universel*, July 23, 1825, p. 3.

128. *Journal de la ville de Saint-Quentin et de l'arrondissement*, Dec. 8, 1838.

129. Brard, *Minéralogie appliquée aux arts*, vol. 3, 151–54.

130. Haüy, *Traité de minéralogie*, 509–10; idem, *Pierres précieuses*, 236.

131. "Private meetings" are mentioned in *Moniteur universel*, Dec. 13, 1819, p. 3.

Value-Free Evaluation

The end of gems and the beginning of gemology were two sides of the same coin. There was a need for an applied science of gems precisely because mineralogy was no longer about gems. This was partly because mineralogy no longer recognized gems as a natural category, but also because of a more general divergence between the learned study of minerals and the traditional audience for that study. Mineralogy was now a "hieroglyphic science," as Pujoulx put it, written in an obscure language that only savants could understand.[132] Mineralogical treatises were now very different from craft manuals; research collections were distinct from industrial and ornamental ones; tests based on crystal form and chemical composition were of little use to the practicing jeweler. No sooner had these gaps appeared, however, than bridges were thrown across them. A new generation of lapidaries translated the mineralogy into a language that others could understand. Curators found ways to combine research and ornamentation without conflating the two. The careers of Pujoulx, Caire-Morant, and Brard lay somewhere between the professor of mineralogy on the one hand, and the cutter, jeweler and autodidact on the other. There was plenty of disagreement about how this bridging should be done, with debates on everything from the nomenclature of corundum to the utility of electroscopes. But there was a broad agreement that gaps had opened up and that bridges ought to be built.

This was a new development in the study of gems, though the novelty of it should not be overstated. There had always been gaps, whether institutional or intellectual. In the seventeenth century, for example, Anselmus Boethius de Boodt was a university-trained physician who claimed that jewelers had a muddled nomenclature for their wares. By contrast, Robert de Berquen was a shop-trained goldsmith who believed that his daily experience of gems gave him knowledge about gems that a physician like Boodt could never have. These gaps were real, but Berquen took himself to be doing essentially the same thing in writing his *Marvels of the Indies* as Boodt had done in writing his *History of Gems and Stones*. They were both writing about the natures and qualities of gems. Berquen did not see himself as an interpreter of Boodt but as a rival, as someone who had a different but better classification of gems. By contrast, neither Brard, Pujoulx nor Caire-Morant had any intention of improving on Haüy's 1801 treatise when they wrote their own lapidaries. Instead, they aimed to interpret the treatise for whichever audience they were addressing, whether jewelers, cutters, or the reading public. It was no longer

132. Pujoulx, *Minéralogie*, 5–6.

possible to give a complete account of the natures *and* qualities of gems in the same book. Haüy required two books, *Treatise on Mineralogy* and *Treatise on Precious Stones*, to do what Boodt had done in one book.

It is tempting to trace these changes to the French Revolution, with its well-documented consequences for the scientific world.[133] It is surely no co-incidence that most of the main developments in the French applied science of gems happened early in the nineteenth century, from the publication of Haüy's *Treatise on Mineralogy* in 1801 to the publication of Brard's *Popular Mineralogy* two decades later. Many of these developments were in some sense a legacy of the revolution, such as the existence of well-resourced and reform-minded scientific institutions along the lines of the House of Mines, the École Polytechnique, and the museum. The career of someone like Haüy, with his single-minded focus on the science of mineralogy, is hard to imag-ine without these institutions. The revolution led to the reign of Napoleon Bonaparte, and along with it a science-friendly culture that helps to explain Pujoulx's success as a science writer and journalist. The very idea of "science applied to arts" has been traced to the revolutionary commitment to both industry and education.[134]

Yet the applied science of gems was the culmination of trends that had begun before 1789, from the emergence of crystallography in the 1770s to the creation of the School of Mines in 1783 to the establishment of Caire-Morant's rock crystal manufactory in 1778. Caire-Morant's *Science of Precious Stones Applied to the Arts* may have been published in 1826, but the first draft was completed before the revolution began.[135] Looking forward rather than back-ward, we find that several significant events in this chapter happened after the fall of Napoleon in 1815. The return to France of the Comte de Bournon, the émigré crystallographer, was a direct result of the restoration of the Bourbon monarchy. It is a good approximation to say that the applied science of gems was a product of the revolution and the Napoleonic years, but it is an ap-proximation only.

In any case, my aim in this chapter has not been to link science to this or that political event. It has been to explain how value-free evaluation is pos-sible. The explanation is not that value-free science is a myth. There is an

133. Major surveys are Dhombres and Dhombres, *Naissance d'un nouveau pouvoir*; Gillispie, *Science and Polity in France at the End of the Old Regime*; Robert Fox, *The Savant and the State: Science and Cultural Politics in Nineteenth-Century France* (Baltimore, MD: Johns Hopkins Uni-versity Press, 2012), 9–28.

134. Robert Bud, "'Applied Science': A Phrase in Search of a Meaning," *Isis* 103, no. 3 (2012): 542–43.

135. Caire-Morant, *Pierres précieuses*, viii.

important sense in which mineralogy did become value-free after 1800. There is also an important sense in which it continued to make judgments about the value of gems. The key is to see that mineralogy did not make these judgments on its own. It did so with the help of a whole host of bridging practices, from the construction of new types of hydrostatic balance to the writing of guides to museums. These practices were sometimes done by savants such as Haüy. But usually they were done by a wider cast of characters that included jewelers, goldsmiths, cutters, collectors, connoisseurs, journalists, entrepreneurs, lawyers and government officials. The science of mineralogy may have become more exclusive, but the use of mineralogy to evaluate gems was as inclusive as it had always been. If there is a myth to debunk, it is not that science is value-free, but that value-free science makes its own evaluations.

Conclusion

There was a science of gems in early modern Europe. The idea that gems are a purely human category, a product of commercial ambitions and aesthetic preferences, was alien to the many naturalists who studied them before about 1800. And gems were not a marginal category. Well into the eighteenth century, stones and metals were the most important classes of mineral in European mineralogical treatises, while precious stones were the most important class of stone. As a result, they were central to many of the new sciences that emerged between 1500 and 1800. Renaissance natural history, the codification of the arts, experimental philosophy, experimental physics, Enlightenment chemistry and mineralogy, and the birth pangs of gemology—gems shed light on all these developments. They also show that materialists need to take evaluation and transmaterialism into account. Evaluation drove early modern science just as much as production did. Likewise, the interaction between material worlds mattered as much as the interaction between the material world and the mental world.

So far, I have illustrated these points by taking each of the new sciences separately. It is time to take them together. What were the main stages of the long-term development of the science of gems? How do transmaterialism and material evaluation help make sense of these changes? And what are the prospects for extending these ideas beyond early modern Europe? I take these questions in turn.

A Brief History of Garnet

The science of gems was first and foremost a science of classification. And gem classification was not invented in the eighteenth century. Nor was it confined to learned treatises on mineralogy. The practice of placing gems into groups, and expressing those groups in writing, was common to a wide range of people: ancient encyclopedists, medieval theologians, Renaissance travelers, seventeenth-century goldsmiths, and the authors of nineteenth-century museum guides, as well as natural historians, natural philosophers, chemists, and mixed mathematicians. Pliny the Elder, Robert de Berquen, and Jean-Baptiste Pujoulx are as much a part of the history of gem classification as well-known naturalists such as Georg Agricola and René-Just Haüy. This makes for a complex history, but one that can usefully be summarized as the rise and fall of the transparency-color-locality (TCL) scheme. The TCL scheme was invented in the ancient world, revised in the Middle Ages, and consolidated and challenged in the early modern period, before collapsing at the end of the eighteenth century—only to be reinvented a few years later. Let me review this history in more detail, drawing on the preceding chapters and using garnets as a test case.

The origins of the TCL scheme go back to at least the first century AD, when it was sketched out by Pliny the Elder in his *Natural History*. There we find a division of gems (*gemmae*) into broad classes (*genera*) based on their color. Each gem was then divided into varieties according to their place of origin. For example, the first of the broad classes was made up of "fiery red gemstones"; the first item in this class was *carbunculus*; and carbunculus was divided into stones that came from India, Carthage, Ethiopia, and Alabanda.[1] Pliny also used other properties to group his gems, especially transparency, but these other properties were overshadowed by color and locality.

Pliny's scheme persisted into the Middle Ages, but with notable revisions. Albert the Great's treatment of *granatus* illustrates both trends.[2] Like Pliny, Albert played up color and locality in his description of this stone. It was slightly darker than the carbunculus, he wrote; there was a violet variety as well as a red variety; and these stones were found in Ethiopia and near Tyre. But Albert added other observations that would have annoyed Pliny, such as that granatus can "gladden the heart and dispel sorrow." And the philosopher gave no broad classification of gems, simply listing them in alphabetical order—granatus comes between *gerachidem* and *hiena* in his scheme. It would be a mistake to conclude that Albert did not classify gems. He distinguished carefully between different species and between varieties within species, often in ways that anticipated later mineralogists. For example, he clearly distinguished between granatus and other red stones, such as carbunculus.

1. Pliny, *Natural History*, vol. 10, bk. 37, 238–42.
2. Albert, *Book of Minerals*, 96.

The result is that his granatus is recognizable as an ancestor of today's garnet. The same cannot be said for Pliny's carbunculus, which included red corundum, red spinel, and red marble, as well as garnets.

The TCL scheme was consolidated between the middle of the sixteenth century and the middle of the eighteenth century. In this period the scheme acquired the trappings of systematic mineralogy, from the lists of distinguishing characteristics in Agricola's *On the Nature of Fossils* (1546) to the synoptic tables of mineral species in Bomare's *Mineralogy* (1762). Transparency became more significant, adding an extra layer to classification schemes from Boodt's *History of Gems and Stones* (1609) onward. The distinction between oriental and occidental stones emerged out of the ancient practice of naming stones after their place of origin. Boodt's treatment of granatus was characteristic of the period.[3] He grouped this stone with other red stones, which were part of the large class of transparent stones. There are two kinds of garnet, Boodt argued, the oriental and the occidental. And these come in three colors: dark red, yellow, and purple. This scheme had not changed significantly when Bomare published the second edition of his *Mineralogy* in 1774. Like Boodt, Bomare divided garnet into stones of three different colors and into oriental and occidental varieties. He placed garnet among the "precious stones and crystals," a category that contained many of the same species as Boodt's "transparent gems." This was not a revolution. It was a slow elaboration of a scheme originally sketched out by Pliny.

The early modern challenges to this scheme came from many quarters: the gem trade, gem cutting, botany, mining, chemistry, and experimental philosophy, among others. These sources drew attention to new criteria for classifying gems and raised questions about the old criteria. Traders noted that gems of different colors could occur in the same mine, which suggested to some that color was not as decisive a criterion as once thought. Cutters raised a similar problem when they observed that gems of different colors can be equally hard. The experimental philosophy suggested a whole new approach to gems, one based on the form of their crystals, the nature of their chemical components, their optical and electrical behavior, and measurements of their density. These properties had not been entirely ignored in earlier periods—one thinks of Pliny's description of the geometry of diamond, and Albert the Great's assertion that all gemstones attract light bodies when rubbed. But such hints were developed much further in the decades around 1700, especially in Boyle's *Origine and Virtues of Gems* (1672) and in work inspired by it, such as Henckel's *Origin of Stones* (1734) and Dufay's exhaustive

3. Boodt, *Gemmarum*, 152–54.

study of gemstones in the same decade. Yet these new approaches made little difference to the natural history of minerals for much of the eighteenth century. Again, Bomare's *Mineralogy* is a useful indicator. The only reference to chemical composition, in Bomare's description of garnet, was in a footnote. He did give a list of seven types of garnet crystals; but this, too, was incidental to his distinction between *granatus orientalis* and *granatus occidentalis*. The newer sciences of gems had not yet made contact with the older ones.

When they did make contact, in the last quarter of the eighteenth century, the results were dramatic. In 1772, Lavoisier declared that diamond was a combustible substance, not a vitrifiable stone. Five years later, Bergman showed how to separate gems into their components, concluding that garnet was not a precious stone but a schorl. Meanwhile, in 1783, Romé de l'Isle gave a new classification of gems based on crystal form, hardness, and double refraction. On these criteria, oriental ruby, oriental sapphire, and oriental topaz were no longer separate species but differently colored varieties of the same species. Romé de l'Isle borrowed his data on gem density from Brisson, who went on to publish his precise and comprehensive table of densities in 1787. Gem species were now being defined by both their "internal" properties (with the help of chemistry) and their "external" properties (with the help of physics and crystallography). These two approaches sometimes clashed, but they increasingly gave concordant results. An early example was the identification of zircon as a distinct species, first by the crystallographer Romé de l'Isle in 1783 and then by the chemist Klaproth in 1789. More concordances followed from the work of Klaproth and, in the 1790s, from his French follower Vauquelin. The publication of Haüy's *Treatise on Mineralogy* in 1801 was a sign of the times. There, garnet was no longer defined in terms of its color, transparency, or locality.[4] Each of those properties was relegated to the category of "accident" or "annotation." Instead, chemistry and crystallography took up the bulk of Haüy's description of the stone. And Haüy's garnet was not classified as a "precious stone," or even as a "stone." It was one of the fifty-five species in the category "earthy substances," meaning substances that contain an earth but no acid. Tellingly, this category included talc and asbestos but not diamond or turquoise. The category of gemstones had collapsed under the weight of physics, chemistry, and crystallography.

Outside mineralogical treatises, however, the category was quickly reinvented. Brard's *Precious Stones* (1813) was typical of a new generation of books on gems that appeared in France in the first three decades of the nineteenth century. Brard did not claim to make any new contributions to the science of

4. Haüy, *Traité de minéralogie*, vol. 2, 540–59.

mineralogy. He did not wish to compete with Haüy's 1801 textbook. Nor did he wish to impose Haüy's taxonomy on cutters, jewelers, and collectors. Instead, he hoped to build a bridge between the two, so that the insights of the new mineralogy could be brought to bear on the arts. Brard's book was soon followed by similar works by Pujoulx, Caire-Morant, Drée, and by Haüy himself. Each of these authors was looking for a "concordance" between science and the arts, to use Haüy's term. Hence Brard's decision to retain the category "precious stone" and to divide it by hardness, a property that was recognized by both cutters and mineralogists as a mark of gem identity. In Brard's scheme, garnet was one of the hard precious stones. It was distinguished from the soft precious stones, such as tourmaline and lapis lazuli, by its ability to scratch quartz. It was subdivided both by its colors and by its crystal form, which carried equal weight in this scheme. This was in a book explicitly dedicated to *pierres précieuses*, as the title shows. Brard and his contemporaries helped to reinvent the literary genre of the lapidary. At the same time, they reinvented the category of gems.

From Materialism to Transmaterialism

How do transmaterialism and material evaluation help to make sense of these developments? Take transmaterialism first. This is a name for the habit of paying attention to the relationship between different parts of material life. This can mean various things for historians, as the history of gem science illustrates. One is that different material substances have different histories. Gems do not have the same history as plants, for example. Gems did not undergo a species explosion in sixteenth-century Europe, even though plants did. With regards to gems, the real novelty was not the number of new species known to European naturalists but the number of new localities for the species already known. And gems differed among themselves: the key to determining the composition of diamond was intense heat, whereas the decomposition of ruby and emerald relied on the careful use of acids.

We have also seen that the differences between materials were not merely chronological. The history of gem classification is not simply the history of plant classification with a delay of a few decades. Crystallography may have been modeled on botany, especially the botany of Carl Linnaeus: "Crystals are the flowers of the mineral kingdom," wrote Haüy. But Haüy's crystallography was more experimental and more mathematical than Linnaean botany had been.[5] In the history of chemistry, gems lent themselves to one approach to

5. Haüy, *Tableau comparatif*, xix ("flowers"). For details, see Bycroft, "Neo-Positivist Theory of Scientific Change," 134–36.

composition, metallic ores to another approach, porcelain to yet another. This gives new significance to the interaction between different materials. When materials merged, so did the ideas and methods associated with them. We have seen that the separation of gems into their components late in the eighteenth century was the result of multiple mergers of this kind—metals with stones, porcelain with diamonds, drugs with glass. On a smaller scale, the migration of methods across materials helps to make sense of the careers of individuals. Dufay took a method of inquiry that academicians had already applied to plants, mineral waters, metals, and earths, and extended this method to gems. Moreover, Dufay's method was itself concerned with the differences between materials—between metal and glass, glass and amber, diamond and marble, and so on. These examples concern material substances, but the same approach can be applied to material hardware, to material needs, and to anything else studied under the heading of materialism and materiality.

Transmaterialism may sound abstract, but it does real historical work. In particular, it puts a new spin on familiar themes of the Scientific Revolution. The eighteenth century takes on new significance as a period when old methods were adapted to new matters. Hence the two-stage history of both quantification and classification. For gems, this meant taking classification by color, locality and transparency and extending it to crystal form, chemical composition, and physical properties such as density. The practices of gem classification changed in the process, just as the practices of quantification changed when they migrated from planets to electricity. Similarly, natural history takes on a larger role when we think about the material culture of natural history in relation to that of experimental philosophy and experimental physics. Early modern physics was a form of stamp collecting, in the sense that it relied on large collections of carefully chosen objects, whether a heterogeneous collection such as Boyle's or a more ordered collection such as Dufay's. These collections were studied with tools normally associated with experimental physics, such as air pumps and electroscopes. The same collections were used for quantification as well as classification. Different trades, too, had different material worlds. The tools of gem cutters drew attention to the hardness of gems; the prices of gems, so important to merchants, drew attention to their geographical distribution. Tools and prices came together in Boodt's effort to classify gems by their hardness and locality. Félibien, Boyle, Dufay, Buffon, and Lavoisier all benefited from hand-hand coordination, the interaction between different mechanical arts. Finally, these interactions had a geographical dimension because material expertise was unevenly distributed across Europe and across the globe. Boodt drew on the material specialties of at least three regions: corundum mining in Ceylon, hard-stone carving in Florence, and the mining of metallic ores in central and eastern Europe. These specialties converged, by various routes, on the imperial court in Prague. Once we see that matters matter—note the plural!—we are in a better position to see that the eighteenth century, natural history, geography, and merchants and artisans also mattered.

From Production to Evaluation

Material evaluation is the second major theme of this history. The science of gems existed, in large part, to determine the goodness of gems. It also existed for other reasons, of course, including discovering, extracting, enhancing, and fabricating gems. But evaluation was as important as production, judging as important as making. Evaluation mattered at each stage in the history of the TCL scheme. Early classifications of gems were explicit attempts to rank gems in order of their goodness. In his *Natural History*, Pliny the Elder promised to describe "gemstones that are acknowledged as such, beginning with the finest."[6] He saw gems as luxury objects, and his choice of characters (color, transparency, and brilliance) reflected this conception. Much had changed by the thirteenth century, when Albert the Great and Marbode of Rennes wrote their influential lapidaries. What did not change was the link between classification and evaluation. The value of gems was now understood in terms of their virtue (*virtus*), a term that bundled together properties that would later be separated out into medical, moral, commercial, and physical components.

Evaluation remained central to Pliny's scheme when it was revived and consolidated in the early modern period. Boodt, no less than Pliny, wore his judgments on his sleeve. He aimed to describe gems in the order of their value, starting with the "rarest and most expensive" and proceeding "by degrees to the meanest ones."[7] This caused him to divide gems into transparent, semitransparent, and opaque classes, and to describe them in that order. He divided gems into varieties in the same spirit, distinguishing between "oriental" and "occidental" varieties and making it clear that, in most cases, the oriental varieties were the more valuable ones. These two taxonomic principles—the transparent/opaque distinction and the oriental/occidental distinction—persisted for well over a century after the publication of Boodt's book. They were reinforced by the writings of seventeenth-century goldsmiths and cutters, such as Rosnel and Berquen, who made no secret of their desire to distinguish pricey specimens from cheaper ones. The same

6. Pliny, *Natural History*, vol. 10, bk. 37, 205.
7. Boodt, *Gemmarum*, "Ad lectorum."

principles were codified by eighteenth-century mineralogists, despite their rhetorical disdain for jewelers and goldsmiths. Argenville may have shunned the phrase "precious stone," on the grounds that it was too commercial for natural history, but his classification of gems differed little from that of Berquen and Rosnel. As late as 1774, Bomare was still giving each species of gem a "rank" based on its hardness. He was still making careful distinctions between "oriental garnet," and "occidental garnet," "oriental ruby" and "occidental ruby," and so on. The systematization of mineralogy did not make the discipline any less evaluative. It just made the evaluations more systematic.

Meanwhile, challenges to Pliny's scheme were tied to different forms of evaluation. Gem traders drew attention to the geographical distribution of gems, principally because they knew that different mines produced gems of different qualities. Cutters drew attention to hardness because they knew that the hardness of a gem was related to its brilliance when polished. Experimental philosophers, no less than goldsmiths and merchants, were dedicated to the evaluation of useful objects. Boyle's study of gems was part of a wider pattern of research in which he searched for new techniques to determine the purity of gold, the strength of mineral waters, the salubrity of the air, and the saltiness of the sea. Dufay's work on gems was a continuation of the academy's long-running interest in the evaluation of plants and animals (for medical purposes) and minerals (for industrial purposes). The quantification of gem science in the eighteenth century did not eliminate the question of their quality. On the contrary: the search for more precise measurements of preciousness was the motive for Daubenton's work on gem color, Buffon's on gem optics, and Brisson's on gem density. Meanwhile, chemists sought to break gems into their components while claiming that the components of substances were the basis for their value. This claim was dubious with regards to gems, but it made good sense for mineral waters, metallic ores, and porcelain. Techniques used to analyze these substances were adapted to gems in the second half of the eighteenth century.

In the long run, these techniques drove a wedge between the taxonomies of mineralogists on the one hand, and those of jewelers and cutters on the other. In an important sense, mineralogy ceased to be concerned with the evaluation of gems around 1800. The fields favored by mineralogists—chemistry, crystallography, experimental physics—could lead to classifications that made little commercial or aesthetic sense. Expensive stones were lumped together with cheap ones; the most precious stones were no longer grouped together but scattered through several different classes. In other ways, however, evaluation continued to drive the science of gems. There were some surprising affinities between the old taxonomy and the new one, such as Haüy's observation that *télésie* (gem-grade corundum to us) could just as well be called "pierre gemme orientale." Hardness was a property that bridged the old and the new; it was therefore an obvious choice by people (like Brard) who sought a middle ground between them. Mineralogists continued to rely on collectors, connoisseurs, cutters, and merchants. They tried to explain the optical phenomena that made certain gems (especially diamond, opal, and iris) more valuable than others. Above all, they tried to repurpose the new mineralogy as a means of evaluation. They could not evaluate gems directly, but they could do so indirectly, by translating mineralogists' categories into jewelers' ones. They could then use the mineralogists' tests to make the distinctions that mattered to jewelers—between "garnet" and "oriental ruby," for example, or between "spinel" and "red tourmaline." Applied mineralogy continued to be evaluative, even when mineralogy itself no longer was.

As this survey suggests, gem appraisal was a very wide-ranging practice. It was both qualitative and quantitative. It was linked to chemistry and physics as well as to medicine and natural history. It was done by humble diamond cutters such as Berquen as well as by aristocrats such as Dufay. It was tied to many of the engines of intellectual change in medieval and early modern Europe: the Christianization of natural history in the thirteenth century; the integration of Europe into global trading networks in the sixteenth century; the expansion of royal courts in the seventeenth; and the growth of state-sponsored science in the eighteenth. Gem appraisal was an aspect of diplomacy and law as well as commerce, as shown by Réaumur's work on Persian turquoise (diplomacy) and Haüy's on Brazilian topaz (law). These two examples also capture the spatial dimension of gem species, the fact that the natures of gems were defined in terms of their native lands. Material evaluation had an experimental dimension as well, one that may be summarized by the types of hardware used to do it: balances for weighing gems, furnaces for assaying ores, alembics for distilling mineral waters, cabinets for surveying many objects at a glance. Gem appraisal was bound up with metrology, assaying, quality control, medical testing, connoisseurship, and materials science. The history of gem science is an argument for treating material evaluation as a distinct and general practice, on a par with material production. The difference is that the historical study of material production has been around for centuries, whereas the historical study of material evaluation has only just begun.

Gems Beyond the Scientific Revolution

This book has been organized around the idea of the Scientific Revolution, a useful shorthand for the collapse of Aristotelean natural philosophy in

Europe from roughly 1500 to 1800. But the book has been episodic rather than encyclopedic. Many episodes have been left out. There was plenty of gem science outside France, for example, from the diamond expertise of the Dutch East India Company to the early experiments on gem combustion at the courts of the Grand Duke of Tuscany and the Holy Roman Emperor.[8] The sixteenth century has only been considered briefly here, with some major traditions sidelined, among them natural magic and transmutational alchemy, both of which had a place for gems. Nor have I done justice to the persistence of ancient and medieval ideas about gems into the modern period. Even in the high Enlightenment, the experimental study of gems went hand in hand with the interpretation of ancient texts on the topic.[9] Even today, in retail crystal shops, modern chemistry and geology rub shoulders with ancient ideas about the healing power of crystals. Regarding semiprecious stones, an entire book could be written about the formation of the category "quartz," the gradual realization that substances as diverse as onyx, flint, and amethyst all have the same nature. Gender, empire, and natural kinds could each be the basis of an alternative history of gem science, as noted in the introduction. The full story of gems in the Scientific Revolution is yet to be told.

By the same token, there is much more to the history of gem science than the Scientific Revolution. Gems are part of modern science as well as medieval science, even if the category of gems is no longer an official member of the club of mineralogical entities. Equally obviously, gems have been part of the study of natural world outside Europe in all periods. If there was such a thing as gem science in early modern Europe, as I have been arguing, the question arises of how it was related to gem science outside Europe, whether in the modern or the premodern period. I shall end with some thoughts on this vast question, drawing on recent secondary literature as well as on the expanded form of materialism defended in the preceding chapters.

The hydrostatic balance is a good place to start, partly because it is often presented as the starting point of science itself. It was a product of the original eureka moment, which was also a moment of material evaluation: Archimedes's discovery that a gold crown could be assayed by weighing it in water.[10] We have seen that the hydrostatic balance became a tool for identifying

8. Dutch EIC in Jan de Laet, *De gemmis et lapidibus* (Leyden, 1647), cf. Sabel, "Rare Earth," passim. On gem combustion, see chap. 6, note 28.

9. Note the use of ancient authorities, such as Pliny, in Bomare, *Minéralogie*, vol. 1, 228–65; and the experimental philology in Caire-Morant, *Pierres précieuses*, 185–92.

10. For an example of the symbolic power of this story, see the frontispiece to Wootton, *Invention of Science.*

gemstones in seventeenth-century Europe, a tool for classifying them by the end of the eighteenth century, and the preferred instrument of the applied science of gems early in the nineteenth century. It is worth adding that Galileo Galilei created a hydrostatic balance for identifying gemstones several decades before Boyle did so, as Annibale Mottana has recently shown.[11] The question is, how was this European tradition of using hydrostatics to identify gems related to the Arabic tradition of doing the same thing, a tradition that goes back to the work of al-Bīrūnī and al-Khāzini in the eleventh and twelfth centuries?[12] One possibility is that there was a line of transmission, however indirect, between al-Bīrūnī and al-Khāzini on the one hand, and Galileo and Boyle on the other. Another possibility is that the Arabic texts came into contact with European mineralogy when those texts were studied by orientalists of the nineteenth century. In fact, this is more than a possibility. The French scholar Jean-Jacques Clément-Mullet, in a study of al-Bīrūnī's text published in 1858, compared the density data in this text with the data in mineralogical textbooks, thereby using the hydrostatic balance as a measure of Arabic civilization.[13]

Arash Khazeni's work on the history of turquoise suggests another way of connecting the dots in the Eurasian history of gem appraisal. Khazeni focuses on the *jivahirnama*, a tradition of Persian-language texts on the natural history of gems that emerged from Islamic courts between the fifteenth and nineteenth centuries. These texts were especially rich in information about turquoise, most of which came from the town of Nishapur, now in Iran. Khazeni maps out the trade in turquoise across Eurasia, paying particular attention to the way the meaning of these stones changed in early modern Europe. The stone "was demystified and lots much of its value," in Khazeni's words. Even its color was reduced to physics and chemistry.[14] The one thing I would add to this otherwise comprehensive account is that Arabic and European accounts of turquoise had an important point in common, which was a concern for the stone's place of origin. Khazeni's own account of the *jivahirnama* shows that the authors were sensitive to the variable qualities of turquoise that came from different mines around Nishapur. In the words of the fifteenth-century author Ibn Mansur: "The one expert in knowing that jewel [turquoise] knows as soon as he sees it what mine it originates from."[15]

11. Mottana, "Galileo as Gemmologist."
12. See introduction, notes 22, 23, and 24.
13. Clément-Mullet, *Recherches sur l'histoire naturelle*, 5–7, 10–11, 23–30.
14. Khazeni, *Sky Blue Stone*, 15, and chap. 4.
15. Ibid., 35.

As we have seen, there was the same link between quality and geography in European natural history, a link that persisted deep into the eighteenth century and that underlay the use of physical and chemical criteria to identify gems. Species were spatial, in Paris as in Nishapur. The spatial understanding of gems was "common ground," in Claire Sabel's evocative phrase.[16] It was a basis for communication between cultures that otherwise disagreed on the meaning of gems.

Two other kinds of gem, diamond and jade, help to bring the story up to the present. Much has been written about the history of diamond trade and production in the modern world, from the discovery of diamonds in South Africa in the 1860s to the expansion of the industry into Russia, Canada, and Australia since the middle of the twentieth century. The most recent histories of this topic are also histories of empire, gender, capitalism, coerced labor, and environmental decay.[17] The challenge is to bring evaluation into this picture without flattening it out. It is all too easy to reduce modern diamond appraisal to the "four Cs" beloved of jewelry companies: cut, color, carat, and clarity. The reality of diamond appraisal is more interesting than the standard formula suggests. A recent visit to a diamond workshop in the Birmingham Jewellery Quarter revealed a motley set of techniques: a wall-chart showing the international standards for diamond colors and sizes; the eyes, hands, and tools of the goldsmiths who put this chart into practice; the loupe used by a woman in charge of the quality control of finished jewels; a space-age machine, worth tens of thousands of pounds, for distinguishing real diamonds from synthetic ones; and the trained eyes of the workers who interpret the images generated by this machine.[18] These processes become public when a large diamond is discovered, such as the 1,758-carat Sewelô diamond, found at a mine in Botswana in 2019. A report of this discovery may be read as a set of evaluative techniques, from the act of sliding a diamond down a glove to the use of medical scanners to see inside the stone.[19] The history of these practices is complex. It includes the gemological associations that sprung up in Europe and North America in the 1930s, the corporations that fought for the right to appraise the stones, the laborers who graded them at the mine,

16. Sabel, "Rare Earth," 11–21, passim.

17. See Vanneste, *Blood, Sweat and Earth.*

18. Personal observation.

19. Ed Caesar, "The Woman Shaking Up the Diamond Industry," *New Yorker*, Jan. 27, 2020, https://www.newyorker.com/magazine/2020/02/03/the-woman-shaking-up-the-diamond -industry.

the technicians who built ever more sophisticated instruments to analyze them, and so on.[20] The struggle for control over the world's diamonds was also the struggle for control over the means of evaluating them.

Of course, different gem species have different histories, in the modern period as well as the early modern period. Jade has a different geography and a different set of material properties than diamond. Historically important deposits of the stone include those in China, Central America, and Aotearoa New Zealand. Jade is softer than diamond but much tougher, an ideal material for making carved hand tools. Its coloring is more obvious and more variable than that of diamond. These factors shaped the classifications of jade developed by people who worked it historically, such as the indigenous Māori people of New Zealand. Māori had an elaborate scheme for naming and identifying different kinds of jade before the colonial era, a scheme based on the color of the stone, its places of origin, and its suitability for various sorts of tool and ornament.[21] This expertise is now being used in tandem with nuclear physics to distinguish between jade found in New Zealand deposits and jade found elsewhere, a question of considerable cultural and commercial importance to Ngāi Tahu, the Māori iwi (tribe) that now owns the deposits in question.[22] Meanwhile, there have been admirable efforts to combine Māori and European approaches to jade in public presentations of the stone and in scientific research about it.[23] It seems likely that these efforts are not entirely new—that Māori expertise has long fed into European ideas about the stone,

20. On gemological institutes, see introduction, note 20. There are glimpses of the wider history of modern diamond appraisal in Steven Press, *Blood and Diamonds: Germany's Imperial Ambitions in Africa* (Cambridge, MA: Harvard University Press, 2021).

21. G. L. Pearce, *The Story of New Zealand Jade, Commonly Known as Greenstone* (Auckland: Collins, 1971), chaps. 3, 4, and 5; Barry Brailsford, *Greenstone Trails: The Māori and Pounamu* (Hamilton, New Zealand: Stoneprint Press, 1996); Russell Beck, Maika Mason, and Andris Apse, *Pounamu: The Jade of New Zealand* (North Shore: Penguin, 2010), 17–20. The history of jade nomenclature elsewhere, especially in China, is the subject of Liu Shang-i, Richard W. Hughes, Zhou Zhengyu and Kaylan Khourie, *Broken Bangle: The Blunder-Besmirched History of Jade Nomenclature* (Lotus Gemology: Bangkok, 2024).

22. Christopher Adams, Hamish J. Campbell, and Russell Beck, "Characterisation and Origin of New Zealand Nephrite Jade Using Its Strontium Isotopic Signature," *Lithos* 97, no. 3 (2007): 307–22. The practicalities of authenticating jade are explored in John Reid and Matthew Rout, *Tribal Economies—Ngāi Tahu: An Examination of the Historic and Current Tītī and Pounamu Economic Frameworks* (Christchurch: Ngāi Tahu Research Centre, 2019).

23. For example, Beck et al., *Pounamu*; and the Tough Pounamu project led by Nick Mortimer and Simon Cox, a collaboration between GNS Science, the University of Otago, and Ngāi Tahu.

in New Zealand and beyond.[24] It seems equally likely that the expertise in question was evaluative as well as productive.

These are suggestions for further research rather than hard findings. But the examples do capture some of the richness of gem appraisal in the modern world. It is certainly a more varied activity than "four Cs" formula would suggest. It is both more scientific than that, and less scientific. It relies on the latest research in geology, chemistry, and nuclear physics; but it also relies on pleasure, spontaneity, manual skill, visual judgment, and local knowledge of gem deposits. The two points go together. Once we see that science is a means of evaluation, we can also see that it is one means among many.

---

24. There is some evidence for this in Rodney Grapes, "Pounamu: Nomenclature, First European Impressions, Acquisition, Chemical Analyses and Mineralogical Characterisation," *Geoscience Society of New Zealand: Journal of Historical Studies* 77 (2023): 1–31.

Acknowledgments

One of the pleasures of completing a long project—and this book was not a short project—is the chance to thank the people who made it possible.

I am forever grateful to Philip Catton and Clemency Montelle for introducing me to the history and philosophy of science at the University of Canterbury, New Zealand. This was an ideal home for undergraduates who strayed across the faculties, as were the HPS departments at the University of Toronto and the University of Cambridge. In Toronto, Mark Solovey and Marga Vicedo showed me the ropes of the academic life and made that life seem attainable as well as attractive.

In Cambridge, Hasok Chang and Simon Schaffer were ideal mentors. If there is any clarity or concision in this book, it owes a great deal to Hasok and to his seemingly infinite patience when commenting on garbled earlier versions of the project. If there is any genuine materialism in the book, it has a lot to do with Simon's wit and erudition. Pierre Crépel served as a kind of unofficial adviser to the project, introducing me to the rigors of archival research and to the community of French historians of science in Paris, Lyon, and beyond.

I was equally fortunate in my mentors at the Max Planck Institute for the History of Science in Berlin and the Global History and Culture Centre at the University of Warwick. In Berlin, Sven Dupré saw the potential of gems as a topic for historical research at the intersection of art and science. Warwick proved to be an ideal environment to draw out the global and economic dimensions of the topic. Maxine Berg, Giorgio Riello, and Anne Gerritsen were mentors before they became equally valuable, morale-boosting colleagues.

None of this would have been possible without the generosity of institutions both private and public. Scholarships from Tower Insurance, the University of Toronto, and the Gates Foundation helped to pay the bills during different stages of my studies. The book you are reading is the product of a project that was funded for one year by the Max Planck Institute and for three further years by the Leverhulme Foundation and the University of Warwick. The project would then have died a premature death, if not for an appointment in the History Department at the University of Warwick that I have held since 2017. I have been extremely lucky, not only to have a job that allows time for research, but also to have colleagues and students who make the job a pleasure rather than a chore.

The book took a long time to write, but it would have taken longer without the close attention of other scholars. Two anonymous referees for the University of Chicago Press scrutinized the text of multiple drafts over the course of several years. They saved me from numerous embarrassments and the book from numerous digressions. They know who they are, even if I do not. Separately, Claire Sabel read the whole typescript and made many useful suggestions and corrections, drawing on her own research on the history of Eurasian gem knowledge. Claire kindly shared a copy of her exciting doctoral thesis in the latter stages of the writing of the present book. James Evans, too, read a full draft, generously sharing his gemological expertise. Natacha Postel-Vinay wisely insisted that the book should have one or two theses rather than eight or nine. James Poskett and Alexander Wragge-Morley made incisive comments on early drafts of the introduction, comments that led to substantial changes in the overall argument. Samir Boumediene, Guido van Meersbergen, Michael Hunter, Mark Philp, Larry Principe, and Giorgio Riello each read one or more chapters and gave timely and erudite feedback. Christine Schwab and Laura Tsitlidze meticulously edited the typescript for the press. No doubt some errors remain, for which I alone am responsible.

A book is more than the sum of its words. Karen Darling, my editor at the University of Chicago Press, has shepherded the project for over a decade. She intervened at crucial moments to keep the project afloat, as did Maxine Berg and Larry Principe. Sean Dyde averted a disaster in Berlin. Irène Passeron and the team at ENCCRE taught me how to read the *Encyclopédie*. Florence Greffe guided me through the archives of the Paris Academy of Sciences. François Farges, professor at the National Museum of Natural History in Paris, has been extremely generous with his time and expertise. The same goes for Robin Hansen at the Natural History Museum in London, who initiated me into the dark arts of mineral photography.

Nicholas Yiannarakis showed me how to use a gem-cutting machine, with interesting consequences. Stefan Nicolescu opened my eyes to the world of present-day gemology by inviting me to the wonderful annual mineral symposium at the Yale Peabody Museum. Brendan Laurs and Richard Drucker, of

Gem-A, introduced me to the equally colorful community of British gemology. Adrian Wilson has been a stimulating interlocutor in the last few years, always ready to share ideas about the eighteenth-century problem. Anna Ferrari, Tony Gill, Sarah Hammond, Robin Hansen, Herbert Horovitz, Andreas Massanek, Dan Pemberton, Jeff Skovil, and John Ward were a great help during the scramble to secure image permissions.

Numerous archivists, librarians, and curators have contributed to the book, often anonymously. Special thanks go to the enlightened administrators of libraries and archives who have made high-resolution digital copies of their holdings available for free (or for a nominal fee) online. Numerous audiences, too many to list here, heard versions of the argument in this book at workshops, conferences, and seminars. Special thanks go to the participants in a series of workshops entitled "Gems in Transit" in 2015 and 2016, and to the co-organizers of those workshops, Sven Dupré and Marta Ajmar. The list of scholars whose emails and conversations have fed into the book includes David Armstrong, Stefan Bauer, Spike Bucklow, Catherine Constable, Sherril Dixon, Sean Dyde, Rémi Franckowiak, Susannah Gibson, Josh Nall, Francis Neary, Simon Nightingale, Madeleine Pinault Sorensen, Marcia Pointon, Chitra Ramalingam, Jenny Rampling, Minwoo Seo, Emma Spary, Mary Terrall, Xiaona Wang, Caitlin Wylie, Sergei Zotov, and no doubt others I have overlooked.

Two less conventional mentions are in order. Perhaps more than any other generation, mine was shaped by blogs. Richard Chappell, Will Thomas, and Justin Smith-Ruiu have done more to form my mind than most other scholars, simply by publishing their ideas at *Philosophy, et cetera*, *Ether Wave Propaganda*, and *The Hinternet*. Blogs are old news, of course. Now we have artificial intelligence, driven by large language models. It seems appropriate to record that, to my knowledge, LLMs played no part in the research and writing of this book.

Academic families are one thing, actual families another. Olivier and Catherine Postel-Vinay made my many visits to Paris easy, cheap, and enjoyable. David and Thomas did not make anything cheaper, but they made everything more meaningful. Natacha helped me to see the point of this book, and of much else besides. My parents, Christine and Trevor, are the authors of the author. I dedicate the book to them.

Appendix 1: *Diamonds Used in the Argument of Boyle's* Gems

A table of passages in Robert Boyle's *Essay About the Origine and Virtues of Gems* (1672) in which he referred to diamonds he had seen. This is an extract of a larger table covering all species of gems in all Boyle's published and unpublished writings. Text in quotation marks is from Boyle's text, whereas text without quotation marks is a paraphrase of his text. Page numbers refer to Hunter and Davis, *The Works of Robert Boyle*, vol. 7, 3–72. The column "Place in argument" indicates how the diamond supported the two main theses in *Gems*, i.e., that gems have a fluid origin, and that their medical virtues are due to mineral impurities. The only diamond in the table that did *not* support one of these theses was the one on page 28.

Description	Owner and location	What Boyle did	What Boyle observed	Place in argument	Page
"a pretty large one that was rough . . ."	Boyle	"in my Collection of Minerals"	Observed only	Surface has triangular planes meeting at points, each plane containing smaller triangular planes	Gems have regular shapes, so have fluid origin	14
"several other [i.e., different from pretty large rough one] rough Diamonds [with] Angular and determinate shapes"	Unknown	Observed only	Have angular and determinate shapes	Gems have regular shapes, so have fluid origin	14
"a rough Diamond"	Unknown	Observed only	Flakes in one plane not parallel to those in another plane, so flakes must make angles in body of stone	Gems have internal texture resembling that of crystals formed in fluid	18
"a pretty big Diamond unrought"	Unknown	Observed under microscope	Surfaces show edges of parallel flakes that run through body of stone	Gems have internal texture resembling that of crystals formed in fluid	18
"rough . . . either Blewish or Greenish . . . And I particularly contemplated one Stone [among these], which . . . was so Green, that I should have taken it for an Emerald"	Not Boyle	Boyle "has seen [these stones, which] came directly out of the Indies, and were soon after bought by Traders in Diamonds for such"	Observed only color	Color [see Description]	Gem color is variable, so due to mineral impurities	22
"some, that were Yellowish"	Unknown	Observed only	Color [see Description]	Gem color is variable, so due to mineral impurities	22
"others [i.e., other diamonds] that were more Yellow"	Unknown	Observed only	Color [see Description]	Gem color is variable, so due to mineral impurities	22

Quotation	Owner	Provenance	Test	Color [see Description]	Inference	Page
"so perfectly Yellow, that I at first took it for a fair Topaz, though it were a Diamond valued at near three pound weight of Gold"	Unknown		Observed only		Gem color is variable, so due to mineral impurities	22
"a Diamond that seem'd to me to be about the bigness of two ordinary pease or less"	Not Boyle	showed to Boyle by "an ancient Cutter of Diamonds of great Practice and Experience"	Weighed in air [by cutter, not Boyle]	The owner observed that diamonds of this size sometimes differ by a carat in weight	Gems have inclusions, so formed in fluid medium	26
"an extraordinary Diamond [that has] odd Clouds"	Unknown		Weighed in air and water	No numerical data given	Gems have inclusions, so formed in fluid medium	26
"an uncut Diamond"	Unknown		Weighed in air and water	Diamond lighter than some other (unspecified) stones	Corrects an author's claim that diamonds are hardest and heaviest of all stones	28
"Diamonds, newly brought from the Indies, and some of them very fair ones"	Unknown	Boyle "observed" it	Observed only external form	Great variety in the area of their surfaces, in their shapes, and in their number of corners	Gems of same species are different, so gems not formed by seminal principle	35
"a Diamond [that] upon a little friction attracts, vigorously enough to be wonderd at by the Spectators"	Boyle	"I keep by me"	Rubbed gently	Attracted light bodies	Minerals in gems can act outside gem, so can explain medical virtues of gem	47
"a Diamond, whose Electrical faculty may be excited not only by rubbing, but, without it, by a languid degree of adventitious heat"	Boyle	"I have"	Heated gently	Attracted light bodies	Minerals in gems can act outside gem, so can explain medical virtues of gem	48
"a Diamond, which by Water, made a little more than Luke-warm, I could bring to shine in the dark"	Boyle	"I have in my keeping"	Placed in lukewarm water	Shone in dark	Minerals in gems can act outside gem, so can explain medical virtues of gem	49

Appendix 2: Gem Specimens from the Regent's Survey, 1714–1719

Specimens sent to Paris between 1714 and 1719 as part of the regent's survey (*enquête du regent*). Based on correspondence between René Réaumur and royal officials (*intendants*) in each province (*généralité*) of France. The crystals (*cristaux*) and transparent stones (*pierres transparentes*) were probably transparent varieties of quartz. Where data are unavailable, cells have been left blank. Marble and marcasites are omitted from the table. Based on correspondence transcribed in Demeulenaere-Douyère and Sturdy, *Enquête du Régent*.

Type	Samples	Province	Sent to Paris
agate		Languedoc	
agate	two samples	Poitiers	1716
agate	samples of all colors	Poitiers	3/7/1717
agate		Aix-en-Provence	2/17/1716
amber		Perpignan or Roussillon	8/12/1717
amber		Aix-en-Provence	
Armenian stone		Languedoc	
colored stones		Riom or Auvergne	10/7/1718
congelations	multiple samples	Aix-en-Provence	< 5/15/1715
congelations	multiple samples	Aix-en-Provence	4/8/1716
coral	two samples	Aix-en-Provence	3/28/1716
coral	multiple samples	Aix-en-Provence	
crystal		Navarre and Béarn	4/15/1716
crystal	four sets of samples	Bretagne	5/21/1717
crystal	one sample	Languedoc	8/17/1717
crystal	multiple samples	Aix-en-Provence	4/3/1716
crystal	multiple samples	Grenoble or Dauphiné	11/7/1717

(*continued*)

(continued)

Type	Samples	Province	Sent to Paris
crystal	three samples	Navarre and Béarn	4/15/1716
crystal	two samples	Languedoc	8/17/1717
crystal	two stones with crystals	Grenoble or Dauphiné	< 11/7/1717
crystal—amethyst		Aix-en-Provence	2/17/1716
crystal—amethyst		Perpignan or Roussillon	1716
crystal—amethyst	two samples	Bourges	10/25/1716
crystal—amethyst		Perpignan or Roussillon	1716
crystal—amethyst		Riom or Auvergne	10/7/1718
crystal—colored	two samples	Aix-en-Provence	1/2/1716
crystal—green		Aix-en-Provence	2/17/1716
crystal—rock crystal	one sample	Lyon	11/24/1716
crystal—rock crystal		Bretagne	
crystal—transparent	three samples	Bretagne	end of 1717
crystal—white		Aix-en-Provence	2/17/1716
crystallization		Aix-en-Provence	3/28/1715
crystallization		Aix-en-Provence	3/28/1715
crystallization	one of each kind	Alsace	2/1/1717
figured stone		Languedoc	< June 1717
figured stone	multiple samples	Languedoc	8/17/1717
flint	one sample	Poitiers	1716
flint	three samples	La Rochelle / Pays d'Aunis	3/26/1716
flint	multiple samples	Lyon	1718
jet		Languedoc	11/4/1715
jet	several kinds listed	Aix-en-Provence	1/2/1716
jet		Aix-en-Provence	2/17/1716
jet		Aix-en-Provence	< 4/8/1716
lapis lazuli	a few small pieces	Aix-en-Provence	2/19/1714
pearl		La Rochelle / Pays d'Aunis	3/26/1716
pearl	fifteen samples	Bretagne	4/10/1718
pearl	two samples	Bretagne	
polishable stones		Grenoble or Dauphiné	11/7/1717
transparent stones	multiple samples	Languedoc	< 6/22/1717
transparent stones	two packets	Grenoble or Dauphiné	9/13/1716
transparent stones		Perpignan or Roussillon	1716
turquoise		Languedoc	
turquoise	one sample	Languedoc	6/25/1717
turquoise		Montauban	< 12/4/1715
turquoise	multiple samples	Navarre and Béarn	1717
turquoise	two samples	Navarre and Béarn	6/13/1717
turquoise	multiple samples	Navarre and Béarn	9/10/1717
turquoise	multiple samples	Navarre and Béarn	9/12/1717
turquoise	24 livres (about 12 kg)	Navarre and Béarn	11/29/1717
turquoise		Navarre and Béarn	4/5/1719

Appendix 3: Gems in Dufay's Experiments

Species of gem that Dufay used as experimental samples in his research as an academician. The columns on the left list all the species of fine stone given in Argenville, *Oryctology* (1755), as well as some other species of interest not listed there. Asterisked species are those that Dufay used in one or more investigations. An X represents a reference to a gem named as an experimental sample in Dufay's articles. The years in the top row of the table correspond to the following sources:

Bernard le Bovier de Fontenelle, "Sur une pierre de Berne qui est une espèce de phosphore," *HAS* 1724 (1726): 58–71 (Dufay read this paper at meetings in 1725)

Dufay, "Mémoire sur la teinture et la dissolution de plusieurs espèces de pierres," *MAS* 1728 (1730): 50–67

Dufay, "Mémoire sur un grand nombre de phosphores nouveaux," *MAS* 1730 (1732): 524–35

Dufay, "Second mémoire sur la teinture des pierres," *MAS* 1732 (1735): 169–81

Dufay, "M2. Quels sont les corps qui sont susceptibles d'électricité," *MAS* 1733 (1735): 73–84

Dufay, "M6. Quel rapport il y a entre l'électricité et la faculté de rendre de la lumière," *MAS* 1734 (1736): 503–26

Dufay, "Recherches sur la lumière des diamants et de plusieurs autres matières," *MAS* 1735 (1738): 347–72

Dufay, "Sur le cristal d'Islande" [unpublished paper read to the academy in 1738]

		Year of paper							
		1725	1728	1730[a]	1732	1733[a]	1734[a]	1735	1738[b]
TRANSPARENT AND SEMITRANSPARENT FINE STONES	aventurine								X
	chrysolite								X
	chrysoprase								X
	girasol								X
	iris								X
	sardony*				X				X
	sardonyx								X
	oriental vermillion								X
	aquamarine[c]*							X	X
	beryl								X
	chalcedony*		X		X				X
	garnet*	X				X			X
	heliotrope								X
	peridot*					X			X
	agate*		X		X	X		X	X
	amethyst*		X			X			X
	carnelian*		X		X	X			X
	dendrites*		X						X
	diamond*					X	X	X	X
	emerald*					X		X	X
	hyacinth								X
	opal*	X				X			X
	cat's eye*					X			X
	rock crystal*		X	X		X	X	X	X
	ruby*	X				X		X	X
	sapphire*	X	X			X		X	X
	topaz*			X		X		X	X
OPAQUE FINE STONES	lapis lazuli*		X					X	
	malachite*	X	X						
	alabaster*		X						
	Amazonian*		X						
	granite*		X			X			
	jade*		X						
	jasper*	X	X	X	X	X		X	
	marble*		X		X	X		X	
	porphyry*		X	X		X			
	turquoise*		X						
SELECTED OTHER STONES	armagnac*		X						
	Berne phosphor*	X				X		X	
	Bologna phosphor*	X	X					X	
	common amethyst*	X						X	
	common emerald*	X	X					X	
	common topaz*							X	
	Iceland spar*	X	X	X					
	jacinth*	X				X			

[a] In these papers Dufay stated that he tried all or most "precious stones," without naming all the individual stones.

[b] In this research Dufay tried all "transparent stones" but did not name any.

[c] Dufay mentioned aquamarine in a manuscript linked to his 1735 paper, but not in the paper itself.

Appendix 4: Comparative Table
of Enlightenment Gem Taxonomies

Covers major categories containing gems in three key books. In each category, the order of the stones reflects the extent to which their classification varies across the three authors. Asterisked stones are listed by all three authors in one of the categories given here. A dash indicates a stone that is not so listed by an author. Up and down arrows indicate movements across categories, e.g., the first down arrow in the "Bomare" column indicates that girasol can be found in one of Bomare's lower categories, namely "Agates, or semitransparent pebbles." Data from Rosnel, *Indian Mercury* (1667), Argenville, *Lithology* (1742), and Bomare, *Mineralogy* (1762).

Rosnel 1667	Argenville 1742	Bomare 1762
True precious stones	Transparent crystalline stones	Precious stones and crystals
diamond*	diamond*	diamond*
ruby*	ruby*	ruby*
sapphire*	sapphire*	sapphire*
topaz*	topaz*	topaz*
emerald*	emerald*	emerald*
amethyst*	amethyst*	amethyst*
aquamarine*	aquamarine*	aquamarine*
hyacinth*	hyacinth*	hyacinth*
chrysolite*	chrysolite*	chrysolite*
garnet*	garnet*	garnet*
girasol*	girasol*	↓
opal*	↓	↓
↓	rock crystal*	rock crystal*
iris	iris	—
peridot	peridot	—

(*continued*)

Rosnel 1667	Argenville 1742	Bomare 1762
vermeil	vermeil	—
turquoise	↓	—
amandine	—	—
smaragdoprase	—	—
carbuncle	—	—
—	beryl	beryl
—	chrysoprase	—
—	—	tourmaline
Agates	**Semitransparent crystalline stones**	**Agates, or semitransparent pebbles**
agate [various]*	agate [various]*	agate [two kinds]*
sardonyx*	sardonyx*	sardonyx*
chalcedony*	chalcedony*	chalcedony*
jade*	jade*	jade*
carnelian*	carnelian*	carnelian*
jasper*	jasper*	↓
↑	opal*	opal*
↑	↑	girasol*
heliotrope	heliotrope	—
nephritic stone	↓	—
lapis lazuli	↓	—
malachite	↓	—
onyx	—	onyx
serpentine	—	—
Armenian stone	—	—
aventurine	—	—
—	dendrite	—
—	cat's eye	cat's eye
—	oculus mundi	oculus mundi
—	—	cacholong
[no comparable category]	**Opaque stones that take a polish**	**Composite stones**
↑	↑	jasper*
↑	turquoise	—
↑	malachite	—
↑	lapis lazuli	—
↑	nephritic stone	—
—	granite	granite
—	porphyry	porphyry
—	alabaster	—
—	marble	—
—	—	simple sandy rocks
—	—	composite rock, or pebble
Lesser precious stones	**[no comparable category]**	**[no comparable category]**
rock crystal*	↑	↑
coral	—	—
amber	—	—
pearl	—	—

Appendix 5: Refraction Data from Buffon and Rochon

The meanings of the headings are as follows. Classification = a simplified version of Buffon's mineral classification scheme. Species = species name given by Buffon. Number = number of refracted rays. Index = whether index of refraction is given in text. Strength = size of the angle that separates the ordinary and extraordinary rays. Sense = whether the number of refractions depends on the orientation of the crystal with respect to the observer. Checkmarks = data present in Buffon's work. Data that Buffon attributes to Rochon are in bold. The source of the rest of the data is unknown, but it may have been from Rochon as well. Data from Buffon, *Natural History of Minerals*, vols. 3 (1785) and 4 (1786).

Classification	Species	Number	Index	Strength	Sense
Vitreous—quartz	rock crystal	2	✓	**low**	no
Vitreous—quartz	Cornish and Alençon diamonds	2			
Vitreous—quartz	amethyst	2			
Vitreous—quartz	Bohemian topaz	2			
Vitreous—quartz	chrysolite	2			
Vitreous—quartz	aquamarine	2			
Vitreous—quartz	crystal-topaz	2			
Vitreous—schorl	crystallized stalactites of schorl	2			
Vitreous—schorl	emerald	2			
Vitreous—schorl	peridot	2	✓	**medium**	**no**
Vitreous—schorl	Brazilian ruby	2			
Vitreous—schorl	Saxon topaz	2			
Vitreous—schorl	garnet	2			
Vitreous—schorl	hyacinth	2			

(continued)

(*continued*)

Classification	Species	Number	Index	Strength	Sense
Calcareous	Iceland spar	**2+**	✓	**high**	yes
Calcareous	other calcareous spars, inc. gypsum	**2+**			
Calcareous	fluorspar	2			
Plants and animals	barite	1			
Plants and animals	diamond	1	✓		
Plants and animals	ruby	**1**	✓		
Plants and animals	topaz	**1**	✓		
Plants and animals	oriental sapphire	**1**	✓		
Plants and animals	ballas ruby	1			
Plants and animals	spinel ruby	1			
Plants and animals	vermillion	1			
Plants and animals	girasol	1	✓		

Bibliography

Articles from early modern dictionaries and encyclopedias are cited in full in the footnotes. Only the dictionaries and encyclopedias themselves are cited here. Where a second or later edition of a work has been used, the date of the first edition in the original language has been given in square brackets (in the footnotes) or in an expanded form (in the list below). For modern editions of ancient works, the approximate date of the earliest known manuscript is given.

Archives

Archives de la Manufacture nationale de Sèvres (Sèvres)
    Y/39—porcelain manuscripts linked to René Réaumur
Archives de la Ville de Paris (Paris)
    Dossiers de faillite
Archives de l'Académie Royale des sciences (Paris)
    Biographical dossiers of René Réaumur, Charles Dufay, and René-Just Haüy
    69J—René Réaumur papers
    Procès-verbaux
Archives Nationales (Paris)
    Minutier central des notaires de Paris
    Sous-série AJ/15—Muséum national d'histoire naturelle
    Sous-série F/7—Police
    Sous-série F/12—Commerce et industrie
    Sous-série O/1—Maison du roi sous l'ancien régime
    Sous-série T/1490—Archives de la communauté d'orfèvres de Paris
Beinecke Rare Book and Manuscript Library (New Haven, CT)
    Jean Chardin Correspondence and Documents
Bibliothèque Municipale de Caen (Caen)
    MS 171—Jean Hellot notebooks
Bibliothèque nationale de France (Paris)
    Fonds Clairambault, Ms 499
    Département des Cartes et Plans (see chap. 2 for citations of individual maps)

British Library (London)
 Hans Sloane manuscripts
Royal Society of London (London)
 Boyle papers

Printed Primary Sources

Académie française. *Dictionnaire de l'Académie française*. Paris, 1694.

Acosta, José de. *Natural and Moral History of the Indies*. Edited by Jane E. Mangan. Translated by Frances M. López-Morillas. Durham, NC: Duke University Press, 2002. Originally published in London, 1590.

Agricola, Georg. *De natura fossilium*. 2nd ed. Basel, 1558. Originally published in 1546.

Agricola, Georg. *De natura fossilium (Textbook of Mineralogy)*. Translated by Mark Chance Bandy and Jean A. Bandy. Boulder, CO: Geological Society of America, 1955. Originally published in 1546.

Albert the Great. *Albertus Magnus' Book of Minerals*. Translated by Dorothy Wyckoff. Oxford: Clarendon Press, 1967. Originally published in 1569.

Albert the Great. *Liber mineralium*. Cologne, 1569.

Aldrovandi, Ulisse. *Musaeum metallicum*. Bologna, 1648.

Anon. *Bergwerk- und Probierbüchlein: A Translation from the German of the Bergbüchlein, a Sixteenth-Century Book on Mining Geology, and of the Probierbüchlein, a Sixteenth-Century Work on Assaying*. Edited by Cyril Stanley Smith. Translated by Anneliese Grünhaldt Sisco. New York: American Institute of Mining and Metallurgical Engineers, 1949. Originally published ca. 1500 (*Bergbüchlein*); 1578 (*Probierbüchlein*).

Anon. "Bersuchet, welche mit einigen Edelgesteinen, sowol im feuer, als auch vermittelst eines Tschirnhausischen Brennglases angestellet worden." *Hamburgisches Magazin* 18 (1757): 164–80.

Anon. *Catalogue des diamans, pierres de couleurs, pierres précieuses et boîtes, composant la superbe collection du Citoyen Daugny*. Paris, 1798.

Anon. "Esperienze fatte con lo specchio ustorio di Firenze sopra le gemme, e le pietre dure." *Giornale de' letterati d'Italia* 8 (1711): 221–309.

Anon. *Table générale raisonnée des matières contenues dans les trente premiers volumes des Annales de chimie*. Paris, 1801.

Arcet, Jean de. *Mémoire sur l'action d'un feu egal, violent et continué pendant plusieurs jours sur un grand nombre de terres, de pierres [et] de chaux métalliques*. Paris, 1766.

Arcet, Jean de. *Second mémoire sur l'action d'un feu égal, violent, et continué pendant plusieurs jours*. Paris, 1771.

Arfe y Villafane, Juan de. *Quilatador de la plata, oro, y piedras*. Valladolid, 1572.

Argenville, Antoine-Joseph Dezallier d'. *Conchyliologie nouvelle et portative*. Paris, 1767.

Argenville, Antoine-Joseph Dezallier d'. *La conchyliologie, ou histoire naturelle des coquilles de mer, d'eau douce, terrestres et fossiles*. Paris, 1780.

Argenville, Antoine-Joseph Dezallier d'. *L'histoire naturelle éclaircie dans deux de ses parties principales: La lithologie et la conchyliologie*. Paris, 1742.

Argenville, Antoine-Joseph Dezallier d'. *L'histoire naturelle éclaircie dans une de ses parties principales, l'oryctologie*. Paris, 1755.

Barbosa, Duarte. *The Book of Duarte Barbosa, an Account of the Countries Bordering on the*

Indian Ocean and Their Inhabitants. Edited by Mansel Longworth Dames. Farnham: Hakluyt Society, 1918. Originally published in 1563.

Barbosa, Duarte. *A Description of the Coasts of East Africa and Malabar.* Edited and translated by Henry E. J. Stanley. Farnham: Hakluyt Society, 2010. Originally published in 1563.

Beccaria, Giambattista. "Observations sur la double réfraction du cristal de roche." *Journal de physique* 2 (1772): 504–10.

Bergman, Torbern. *Manuel du minéralogiste, ou Sciagraphie du règne mineral.* Translated and edited by Jean-André Mongez. Paris, 1784. Originally published in 1782.

Bergman, Torbern. *Manuel du minéralogiste, ou Sciagraphie du règne mineral.* Translated by Jean-André Mongez. Edited by Jean-Claude Delamétherie. 2 vols. Paris, 1792. Originally published in 1782.

Bergman, Torbern. *Opuscules chimiques et physiques.* Translated by Louis-Bernard Guyton de Morveau and Claudine Picardet. Dijon, 1780–1785. Originally published in 1779–1780.

Bergman, Torbern. "Recherches chimiques sur la terre des Pierres précieuses ou gemmes." *Journal de physique* 14 (October 1779): 257–80.

Bergman, Torbern. *Sciagraphia regni mineralis.* Leipzig, 1782.

Bergman, Torbern. "La terre des gemmes." In Bergman, *Opuscules chimiques et physiques,* vol. 2, 78–124.

Bergman, Torbern. "Variae crystallarum formae, e spatho ortae." *Nova Acta Regiae Societatis Scientiarum Upsaliensis* 1 (1773): 150–55.

Bernier, François. *Abrégé de la philosophie de Gassendi.* Edited by Sylvia Murr and Geneviève Stefani. Paris, 1992. Originally published in 1674–1675.

Bernier, François. *Un libertin dans l'Inde moghole: Les voyages de François Bernier, 1656–1669.* Edited by Frédéric Tinguely, Adrien Paschoud, and Charles-Antoine Chamay. Paris: Chandeigne, 2008. Originally published in 1699.

Berquen, Robert de. *La liste des messieurs les gardes et anciens gardes de l'orphevrerie de Paris.* Paris, 1655.

Berquen, Robert de. *Le livre d'allois en or et en argent, ou brève instruction pour répondre par devant mes seigneurs de la Cour des monnoyes en l'interrogatoire qui sera faitte et sur les alloiemans qui seront donné aux prétendans maistres en l'art d'orfèvrerie de Paris.* Paris, 1671.

Berquen, Robert de. *Merveilles des Indes Orientales et Occidentales.* Paris, 1661.

Bīrūnī, Abū Rayḥān al-. *Al-Beruni's Book on Mineralogy: The Book Most Comprehensive in Knowledge on Precious Stones.* Edited and translated by Hakim Mohammad Said. Islamabad: Pakistan Hijara Council, 1989. Originally written ca. 1050 AD.

Blegny, Nicolas de. *Les adresses de la ville de Paris, avec le Tresor des almanachs.* Paris, 1692.

Boizard, Jean. *Traité des monoyes, de leurs circonstances & dépendances.* Paris, 1692.

Bomare, Jacques-Christophe Valmont de. *Catalogue du cabinet d'histoire naturelle de M. Bomare de Valmont comprenant les minéraux, végetaux, animaux, & quelques productions, tant de la nature que de l'art.* Paris, 1758.

Bomare, Jacques-Christophe Valmont de. *Dictionnaire raisonné universel d'histoire naturelle.* 2nd ed. Paris, 1775. Originally published in 1764.

Bomare, Jacques-Christophe Valmont de. *Minéralogie, ou Nouvelle exposition du règne minéral.* Paris, 1762.

Bomare, Jacques-Christophe Valmont de. *Minéralogie, ou Nouvelle exposition du règne minéral.* 2nd ed. Paris, 1774.

Boodt, Anselmus Boethius de. *Gemmarum et lapidum historia.* Hanover, 1609.

Boodt, Anselmus Boethius de. *Gemmarum et lapidum historia*. Edited by Adrian Toll. Leiden, 1636. Originally published in 1609.

Boodt, Anselmus Boethius de. *Le parfait joaillier ou Histoire des pierreries*. Edited by Adrian Toll. Translated by Jean Bachou. Lyon, 1644. Originally published in 1609.

Bournon, Louis-Jacques, Comte de. *Catalogue de la collection minéralogique du Comte de Bournon*. London, 1813.

Bournon, Louis-Jacques, Comte de. *Collection minéralogique particulière du roi*. Paris, 1817.

Bournon, Louis-Jacques, Comte de. *Traité de minéralogie*. London: 1808.

Boyle, Robert. "An Account of the Honourable Robert Boyle's Way of Examining Waters as to Freshness and Saltness." *PT* 17, no. 197 (1693): 627–41.

Boyle, Robert. *Certain Physiological Essays* (1661). In *Works*, vol. 2, 3–203. (*CPE*)

Boyle, Robert. "Chap. II. Containing Various Observations About Diamonds." In *Experimenta et Observationes Physicae*, 385–91. ("Diamonds")

Boyle, Robert. "Chap. V. Containing Experiments and Observations Solitary; in Two Pentades." In *Experimenta et Observationes Physicae*, 411–18. ("Observations Solitary")

Boyle, Robert. "A Conjecture at the Causes of the Reall Virtues of Gems & Medicinall Stones" (from unpublished MS). In *Works*, vol. 7, xiv–xv. ("Gems and Medicinal Stones")

Boyle, Robert. *A Continuation of New Experiments Physico-Mechanical, Touching the Spring and Weight of the Air, and Their Effects* (1669). In *Works*, vol. 6, 27–187. (*Spring, 1st Continuation*)

Boyle, Robert. *A Continuation of New Experiments Physico-Mechanical* (1682). In *Works*, vol. 9, 121–263. (*Spring, 2nd Continuation*)

Boyle, Robert. *A Discourse of the Things Above Reason* (1681). In *Works*, vol. 9, 361–93. (*Things Above Reason*)

Boyle, Robert. *An Essay About the Origine and Virtues of Gems* (1672). In *Works*, vol. 7, 3–72. (*Gems*)

Boyle, Robert. *An Essay of the Great Effects of Even Languid and Unheeded Motion* (1685). In *Works*, vol. 10, 251–349. (*Languid Motion*)

Boyle, Robert. *Essays of Effluviums* (1673). In *Works*, vol. 7, 227–336. (*Effluviums*)

Boyle, Robert. *Experimenta et Observationes Physicae* (1691). In *Works*, vol. 11, 367–439. (*Exp. Obs. Physicae*)

Boyle, Robert. *Experiments and Considerations About the Porosity of Bodies* (1684). In *Works*, vol. 10, 103–54. (*Porosity*)

Boyle, Robert. *Experiments and Considerations Touching Colours* (1664). In *Works*, vol. 4, 3–203. (*Colours*)

Boyle, Robert. *Experiments and Notes About the Mechanical Origine or Production of Electricity* (1675). In *Works*, vol. 8, 510–23. (*Electricity*)

Boyle, Robert. *Experiments and Notes About the Producibleness of Chemical Principles* (1680). In *Works*, vol. 9, 19–120. (*Producibleness*)

Boyle, Robert. *Experiments, Notes, &c. About the Mechanical Origine of Qualities* (1675). In *Works*, vol. 8, 315–523. (*Mechanical Origin of Qualities*)

Boyle, Robert. "Material Relating to the Usefulness of Natural Philosophy" (from unpublished MS). In *Works*, vol. 13, 289–361. ("Usefulness, Unpublished")

Boyle, Robert. *Medicina Hydrostatica* (1690). In *Works*, vol. 11, 199–281. (*Medicina Hydrostatica*)

Boyle, Robert. *Memoirs for the Natural History of the Human Blood* (1684). In *Works*, vol. 10, 3–101. (*Human Blood*)

Boyle, Robert. "New Experiments About the Superficial Figures of Fluids, Especially of Liquors Contingent to Other Liquors." In *Works*, vol. 8, 568–84. ("Figures of Fluids")

Boyle, Robert. "New Experiments About the Weaken'd Spring, And Some Un-observ'd Effects of the Air." *PT* 10, no. 120 (1675): 467–76. ("Weakened Spring")

Boyle, Robert. "New Experiments Concerning the Relation Between Light and Air," *PT* 2, nos. 31–32 (January 6 and February 10, 1668): 581–600, 605–12. In *Works*, vol. 6.

Boyle, Robert. *New Experiments Physico-Mechanical, Touching the Spring of the Air and Its Effects* (1660). In *Works*, vol. 1, 141–306. (*Spring*)

Boyle, Robert. "Notes upon the 27th Section" (1660s). In *Works*, vol. 14, 77–87. ("27th Section")

Boyle, Robert. *Observations About Mr. Clayton's Diamond* (1664). In *Works*, vol. 4, 185–203. (*Clayton's Diamond*)

Boyle, Robert. *Occasional Reflections upon Several Subjects* (1665). In *Works*, vol. 5, 1–187. (*Occasional Reflections*)

Boyle, Robert. *Of Absolute Rest in Bodies* (1669). In *Works*, vol. 6, 190–213. (*Absolute Rest*)

Boyle, Robert. "Of the Atmospheres of Consistent Bodies." In *A Continuation of New Experiments Physico-Mechanical, Touching the Spring and Weight of the Air, and Their Effects* (1669). In *Works*, vol. 6, 165–79. ("Atmospheres")

Boyle, Robert. "Of the Atomicall Philosophy" (from unpublished MS). In *Works*, vol. 13, 225–35. ("Atomical Philosophy")

Boyle, Robert. *Of the Reconcileableness of Specifick Medicines to the Corpuscular Philosophy* (1685). In *Works*, vol. 10, 351–435. (*Specific Medicines*)

Boyle, Robert. *The Origin of Forms and Qualities* (1666). In *Works*, vol. 5, 281–491. (*Forms and Qualities*)

Boyle, Robert. "Papers on Petrifaction and Mineralogy" (from unpublished MS). In *Works*, vol. 13, 363–423. ("Petrifaction")

Boyle, Robert. "Paralipomena" (from unpublished MS). In Michael Hunter, Harriet Knight, and Charles Littleton, "Robert Boyle's Paralipomena: An Analysis and Reconstruction," in *The Boyle Papers: Understanding the Manuscripts of Robert Boyle*, edited by Michael Hunter, 177–218. Aldershot: Ashgate, 2007. ("Paralipomena")

Boyle, Robert. "Queries About Gems (c. 1660s–1670s)." In Michael Hunter, "Robert Boyle's 'Heads' and 'Inquiries.'" *Robert Boyle Occasional Papers* 1, 28–30. London: University of London, 2005. ("Queries About Gems")

Boyle, Robert. *The Sceptical Chymist* (1661). In *Works*, vol. 2, 205–379. (*Sceptical Chymist*)

Boyle, Robert. "Shewing the Occasion of Making This New Essay-Instrument." *PT* 10, no. 115 (1675): 329–48.

Boyle, Robert. *Short Memoirs for the Natural Experimental History of Mineral Waters* (1685). In *Works*, vol. 10, 205–49. (*Mineral Waters*)

Boyle, Robert. *Some Considerations Touching the Usefulness of Experimental Natural Philosophy* (1663). In *Works*, vol. 3, 189–561. (*Usefulness I*)

Boyle, Robert. *Tracts Written by the Honourable Robert Boyle (Cosmical Qualities)* (1670). In *Works*, vol. 6, 259–364.

Boyle, Robert. *Tracts Written by the Honourable Robert Boyle (Flame and Air)* (1672). In *Works*, vol. 7, 73–227.

Boyle, Robert. *Tracts Written by the Honourable Robert Boyle (Saltness of the Sea)* (1673). In *Works*, vol. 7, 337–451.

Boyle, Robert. *The Usefulness of Natural Philosophy, II*, sec. 2 (1671). In *Works*, vol. 6, 389–541. (*Usefulness II*)

Boyle, Robert. *The Workdiaries of Robert Boyle*. Edited by Michael Hunter. Centre for Editing Lives and Letters. http://www.livesandletters.ac.uk/wd.

Boyle, Robert. *The Works of Robert Boyle.* Edited by Michael Hunter and Edward Davis. London: Pickering and Chatto, 1999.

Brard, Cyprien-Prosper. *Description historique d'une collection de minéralogie appliquée aux arts.* Paris, 1833.

Brard, Cyprien-Prosper. *Manuel du minéralogiste et du géologue voyageur.* Paris, 1808.

Brard, Cyprien-Prosper. *Minéralogie appliquée aux arts.* Paris, 1821.

Brard, Cyprien-Prosper. *Minéralogie populaire.* Paris, 1826.

Brard, Cyprien-Prosper. *Traité des pierres précieuses, des porphyres, granits, marbres, albâtres, et autres roches, propres à recevoir le poli et à orner les monumens publics et les édifices particuliers.* Paris, 1808.

Brice, Germain. *Description nouvelle de la ville de Paris.* Paris, 1698.

Brisson, Mathurin-Jacques. *Pesanteur spécifique des corps: Ouvrage utile à l'histoire naturelle, à la physique, aux arts et au commerce.* Paris, 1787.

Buffon, Georges-Louis Leclerc, Comte de. *Histoire naturelle des minéraux.* Paris, 1783-1788.

Buffon, Georges-Louis Leclerc, Comte de. *Histoire naturelle générale et particulière, Supplément: Des élémens,* vol. 1. Paris, 1774.

Buffon, Georges-Louis Leclerc, Comte de. "Premier discours." In Buffon, *Histoire naturelle, générale et particulière,* vol. 1. Paris, 1749.

Buffon, Georges-Louis Leclerc, Comte de. "Septième mémoire: Observations sur les couleurs accidentelles, & sur les ombres colorées." In Buffon, *Histoire naturelle, générale et particulière, Supplément,* vol. 1. Paris, 1774.

Buffon, Georges-Louis Leclerc, Comte de. "Sixième mémoire: Expériences sur la lumière, et sur la chaleur qu'elle peut produire." In Buffon, *Histoire naturelle, générale et particulière, Supplément,* vol. 1. Paris, 1774.

Caire-Morant, Antoine. *La science des pierres précieuses, appliquées aux arts.* Paris, 1826.

Cardan, Jerome. *The De Subtilitate of Girolamo Cardano.* Translated by John Forrester. Tempe, AZ: Arizona Center for Medieval and Renaissance Studies, 2013. Originally published in 1560.

Cellini, Benvenuto. *I trattati dell'oreficeria e della scultura.* Edited by Carlo Milanesi. Florence, 1857. Originally published in 1569.

Cellini, Benvenuto. *The Treatises of Benvenuto Cellini on Goldsmithing and Sculpture.* Translated by Charles R. Ashbee. New York: Dover Publications, 1967. Originally published in 1569.

Cesalpino, Andrea. *De metallicis libri tres.* Rome, 1596.

Chambers, Ephraim. *Cyclopaedia, or An Universal Dictionary of Arts and Sciences.* London, 1728.

Chaptal, Jean-Antoine. *Chimie appliquée aux arts.* Paris, 1807.

Chardin, Jean. *Voyages en Perse.* Edited by Louis Langlès. Paris, 1811. Originally published in 1711.

Clayton, John. "Observations of the Comparative, Intensive or Specific Gravities of Various Bodies." *PT* 17, no. 199 (1693): 694-95.

Clément-Mullet, Jean-Jacques. *Recherches sur l'histoire naturelle et la physique chez les Arabes.* Paris, 1858.

Coutre, Jacques de. *The Memoirs and Memorials of Jacques de Coutre: Security, Trade and Society in 16th- and 17th-Century Southeast Asia.* Edited by Peter Borschberg. Translated by Roopanjali Roy. Singapore: NUS Press, 2014. Originally written in the 1620s.

Cronstedt, Axel Fredrik. *Essai d'une nouvelle minéralogie.* Translated by Dreux fils. Paris, 1771. Originally published in 1758.

Daubenton, Louis-Jean-Marie. "De la connoissance des pierres précieuses." *MAS* 1750 (1754): 28-38.

Delambre, Jean-Baptiste Joseph. "Notice sur la vie et les ouvrages de M. Rochon." *MAS* 1817 (1819): lxii–lxxxii.

Deleuze, Joseph. *Histoire et description du Muséum Royal d'histoire naturelle.* Paris, 1823.

Démeste, Jean. *Lettres au Dr Bernard sur la chymie, la docimasie, la cristallographie, la lithologie, la minéralogie et la physique en general.* Paris, 1779.

Diderot, Denis, and Jean le Rond d'Alembert, eds. *Encyclopédie, ou dictionnaire raisonné des sciences, des arts et des métiers.* Paris, 1751–1772.

Drée, Etienne-Gilbert, Marquis de. *Catalogue des huit collections qui composent le musée minéralogique de Et. de Drée.* Paris, 1811.

Drée, Etienne-Gilbert, Marquis de. *Description des objets composant les 4 collections du Marquis de Drée.* Paris, 1816.

Dufay, Charles-François de Cisternay. "A Letter Concerning Electricity." *PT* (1734–1735): 258–66.

Dufay, Charles-François de Cisternay. "Mémoire sur la teinture et la dissolution de plusieurs espèces de pierres." *MAS* 1728 (1730): 50–67.

Dufay, Charles-François de Cisternay. "Mémoire sur un grand nombre de phosphores nouveaux." *MAS* 1730 (1732): 524–35.

Dufay, Charles-François de Cisternay. "M1. L'histoire de l'électricité." *MAS* 1733 (1735): 23–35.

Dufay, Charles-François de Cisternay. "M2. Quels sont les corps qui sont susceptibles d'électricité." *MAS* 1733 (1735): 73–84.

Dufay, Charles-François de Cisternay. "M3. Des corps qui sont les plus vivement attirés par les matières électriques, et de ceux qui sont les plus propres à transmettre l'électricité." *MAS* 1733 (1735): 233–54.

Dufay, Charles-François de Cisternay. "M4. L'attraction et la répulsion des corps électriques." *MAS* 1733 (1735): 457–76.

Dufay, Charles-François de Cisternay. "M5. Des nouvelles découvertes sur cette matière, faites depuis peu par M. Gray." *MAS* 1734 (1736): 341–61.

Dufay, Charles-François de Cisternay. "M6. Quel rapport il y a entre l'électricité et la faculté de rendre de la lumière." *MAS* 1734 (1736): 503–26.

Dufay, Charles-François de Cisternay. "M7. Quelques additions aux mémoires précédants," *MAS* 1737 (1740): 86–100.

Dufay, Charles-François de Cisternay. "M8." *MAS* 1737 (1740): 307–25.

Dufay, Charles-François de Cisternay. "Recherches sur la lumière des diamants et de plusieurs autres matières." *MAS* 1735 (1738): 347–72.

Dufay, Charles-François de Cisternay. "Second mémoire sur la teinture des pierres." *MAS* 1732 (1735): 169–81.

Dufay, Charles-François de Cisternay. "Sur quelques expériences de catoptrique." *MAS* 1726 (1728): 165–75.

Dutens, Louis. *Des pierres précieuses et des pierres fines.* 2nd ed. Florence, 1783. Originally published in 1776.

Ellicott, John. "A Letter from Mr. John Ellicott, F.R.S. to the President, Concerning the Specific Gravity of Diamonds." *PT* 43, nos. 472–77 (1744): 468–72.

Félibien, André. *Des principes de l'architecture, de la sculpture, de la peinture, et des autres arts qui en dependent.* Paris, 1676.

Fontenelle, Bernard le Bovier de. "Éloge de M. Dufay." *HAS* 1739 (1741): 73–83.

Fontenelle, Bernard le Bovier de. "Sur l'électricité." *HAS* 1733 (1735): 4–13.

Fontenelle, Bernard le Bovier de. "Sur l'origine des pierres." *HAS* 1716 (1718): 8–16.

Fontenelle, Bernard le Bovier de. "Sur une pierre de Berne qui est une espèce de phosphore." *HAS* 1724 (1726): 58–71.

Fourcroy, Antoine-François. *Encyclopédie méthodique: Chymie, pharmacie et métallurgie.* Vol. 4. Paris, 1805. Vol. 5. Paris, 1808.

Gellert, Christlieb Ehregott. *Chimie métallurgique.* Translated by Paul-Henri Thiry, Baron d'Holbach. Paris, 1758. Originally published in 1751.

Gersaint, Edmé-François. *Catalogue raisonné de coquilles et autres curiosités naturelles.* Paris, 1736.

Gersaint, Edmé-François. *Catalogue raisonné des différens effets curieux & rares contenus dans le cabinet de feu M. le Chevalier de la Roque.* Paris, 1745.

Guyton de Morveau, Louis-Bernard. "Mémoire sur l'Hyacinthe de France, congénère à celle de Ceylan, et sur la nouvelle terre simple qui entre dans sa composition." *Annales de chimie* 22 (1797): 72–95.

Hakluyt, Richard. *The Principal Navigations Voyages Traffiques and Discoveries of the English Nation.* Cambridge, UK: Cambridge University Press, 1903–1905. Originally published in 1589–1600.

Haüy, René-Just. "Analyse de la chrysolite de M. Romé de l'Isle." *Journal des mines* 6, no. 33 (1797): 688–91.

Haüy, René-Just. "Description de la Cymophane avec quelques réflexions sur les couleurs de gemmes." *Journal des mines* 4, no. 21 (1796): 5–16.

Haüy, René-Just. "Émeraude et béril." *Journal des mines* 6, no. 33 (1797): 686–88.

Haüy, René-Just. *Essai d'une théorie sur la structure des cristaux.* Paris, 1784.

Haüy, René-Just. *Exposition raisonnée de la théorie de l'électricité et du magnétisme.* Paris, 1787.

Haüy, René-Just. "Extrait d'un mémoire sur la structure des cristaux de grenat." *Journal de physique* 19 (1782): 366–70.

Haüy, René-Just. "Mémoire sur la double réfraction du spath d'Islande." *MAS* 1788 (1791): 34–61.

Haüy, René-Just. "Mémoire sur la structure du cristal de roche." *MAS* 1786 (1788): 78–93.

Haüy, René-Just. "Mémoire sur l'électricité des minéraux." *Annales de chimie et de physique* 8 (1818): 383–401.

Haüy, René-Just. "Mémoire sur les tourmalines de Sibérie." *Annales du Muséum national d'histoire naturelle* 3 (1804): 233–44.

Haüy, René-Just. "Observations sur des cristaux trouvés parmi des pierres de Ceylan et qui paroissent appartenir à l'espèce de Corindon vulgairement spath adamantin." *Memoires de la Société d'histoire naturelle de Paris* (1799): 55–58.

Haüy, René-Just. "Observations sur l'électricité des minéraux." *Annales du Muséum national d'histoire naturelle* 15 (1810): 1–8.

Haüy, René-Just. "Suite de l'extrait du traité de minéralogie du Citoyen Haüy." *Journal des mines* 6, no. 33 (1797): 655–92.

Haüy, René-Just. "Sur la cristallisation de l'émeraude." *Journal des mines* 2, no. 19 (1795): 72–74.

Haüy, René-Just. "Sur la double réfraction de plusieurs substances minérales." *Annales de chimie* 17 (1793): 140–55.

Haüy, René-Just. "Sur la double réfraction du cristal de roche." *Journal d'histoire naturelle* 1 (1792): 406–8.

Haüy, René-Just. "Sur la double réfraction du spath calcaire transparent." *Journal d'histoire naturelle* 1 (1792): 63–80.

Haüy, René-Just. "Sur l'arragonite." *Annales du Muséum national d'histoire naturelle* 11 (1808): 237–42.

Haüy, René-Just. "Sur le diamant." *Journal d'histoire naturelle* 1 (1792): 377–84.

Haüy, René-Just. "Sur l'électricité produite dans les minéraux à l'aide de la pression." *Mémoires du Muséum national d'histoire naturelle* 3 (1817): 223–28.

Haüy, René-Just. "Sur les couleurs de l'agathe opaline nommé communément opale." *Journal d'histoire naturelle* 2 (1792): 9–18.

Haüy, René-Just. "Sur les hydrophanes." *Journal d'histoire naturelle* 1 (1792): 294–99.

Haüy, René-Just. "Sur les pierres appelées jusqu'ici Hyacinthe et Jargon de Ceylon, leurs différences, leurs caractères physiques et géométriques." *Journal des mines* 5, no. 26 (1796–1797): 83–96.

Haüy, René-Just. *Tableau comparatif des résultats de la cristallographie et de l'analyse chimique, relativement à la classification des minéraux.* Paris, 1809.

Haüy, René-Just. *Traité de minéralogie.* Paris, 1801.

Haüy, René-Just. *Traité de minéralogie.* 2nd ed. Paris, 1822. Originally published in 1801.

Haüy, René-Just. *Traité des caractères physiques des pierres précieuses.* Paris, 1817.

Haüy, René-Just. *Traité élémentaire de physique.* Paris, 1803.

Henckel, Johann Friedrich. "Dissertation sur une véritable topase [1737]." In Henckel, *Pyritologie, ou Histoire naturelle de la pyrite,* 500–503.

Henckel, Johann Friedrich. *Idée générale de l'origine des pierres* [1734]. Translated by Paul-Henri Thiry, Baron d'Holbach. In *Pyritologie, ou Histoire naturelle de la pyrite,* 395–455.

Henckel, Johann Friedrich. *Introduction à la minéralogie.* Translated by Paul-Henri Thiry, Baron d'Holbach. Paris, 1756. Originally published in 1747.

Henckel, Johann Friedrich. *Pyritologie, ou Histoire naturelle de la pyrite.* Translated by Paul-Henri Thiry, Baron d'Holbach. Paris, 1760. Originally published in 1725.

Homberg, Wilhelm. "Manière de copier sur verre coloré les pierres gravées." *MAS* 1712 (1714): 187–94.

Hugard, J.-A. *Galerie de minéralogie et de géologie.* Paris, 1855.

Jeffries, David. *A Treatise on Diamonds and Pearls.* 2nd ed. London, 1751. Originally published in 1750.

Klaproth, Martin Heinrich. *Beiträge zur chemischen Kenntnis der Mineralkörper.* Posen and Berlin, 1795–1815.

Klaproth, Martin Heinrich. "Chemical Examination of the Circon, or Jargon of Ceylon." In Klaproth, *Analytical Essays Towards Promoting the Chemical Knowledge of Mineral Substances,* vol. 1, translated by Gruber, 175–94. London, 1801–1804. Originally published in 1795–1815.

Laet, Jan de. *De gemmis et lapidibus.* Leyden, 1647.

Lavoisier, Antoine-Laurent. "Mémoire sur l'effet que produit sur les pierres précieuses un degré de feu très-violent." *MAS* 1782 (1785), 476–85.

Lavoisier, Antoine-Laurent. "Premier mémoire sur la destruction du diamant." *MAS* 1772 2e partie (1776): 564–91.

Lavoisier, Antoine-Laurent. "Second mémoire sur la destruction du diamant." *MAS* 1772 2e partie (1776): 591–616.

le Maire, Charles. *Paris ancien et nouveau.* Paris, 1685.

Lehmann, Johann Gottlob. *L'art des mines.* Translated by Paul-Henri Thiry, Baron d'Holbach. Paris, 1759. Originally published in 1750–1758.

Leonardi, Camillo. *The Mirror of Stones.* London, 1750. Originally published in 1502.

Leonardi, Camillo. *Speculum lapidum.* Venice, 1502.

Leroy, Pierre. *Statuts et privilèges du corps des marchands Orfèvres-Joyailliers de la ville de Paris.* Paris, 1734.

Linden, Diederich Wessel. *Lettres sur la minéralogie et la métallurgie pratiques*. Paris, 1752.

Linschoten, Jan Huyghen van. *Voyage of John Huyghen van Linschoten to the East Indies*. Edited by Arthur C. Burnell. Translated by William Phillip. Cambridge, UK: Cambridge University Press, 1885. Originally published in 1596.

Lister, Martin. *A Journey to Paris in the Year 1698*. London, 1699.

Lucas, Jean-André-Henri. *Tableau méthodique des espèces minérales*. Paris, 1806–1813.

Macquer, Pierre-Joseph. *Dictionnaire de chimie*. 2nd ed. Paris, 1778. Originally published in 1766.

Marbode of Rennes. *Marbode of Rennes' (1035–1123) De lapidibus*. Translated by C. W. King. Wiesbaden: Steiner, 1977. Originally written ca. 1096.

Mawe, John. *A Treatise on Precious Stones*. London, 1813.

Nicols, Thomas. *Lapidary; or, The History of Pretious Stones, with Cautions for the Undeceiving of All Those That Deal with Pretious Stones*. Cambridge, UK, 1652.

Nollet, Jean-Antoine. *Leçons de physique expérimentale*. Paris, 1753.

Nollet, Jean-Antoine. "Observations sur quelques nouveaux phénomènes d'électricité." *MAS* 1746 (1751): 1–23.

Orta, Garcia da. *Aromatum, et simplicium aliquot medicamentorum apud Indos nascentium*. Translated by Carolus Clusius. Antwerp, 1567. Originally published in 1563.

Orta, Garcia da. *Colloquies on the Simples and Drugs of India*. Translated by Clements Markham. London: Henry Sotheran, 1913. Originally published in 1563.

Orta, Garcia da. *Colóquios dos simples e drogas he cousas medicinais da Índia*. Goa, 1563.

Pliny the Elder. *Natural History*. Vol. 10. Translated by David E. Eichholz. Cambridge, MA: Harvard University Press, 1989. Originally written ca. AD 80.

Pott, Johann Heinrich. *Lithogéognosie*. Translated by Didier d'Arclais de Montamy. Paris, 1753. Originally published in 1746.

Pujoulx, Jean-Baptiste. *Minéralogie à l'usage des gens du monde*. Paris, 1813.

Pujoulx, Jean-Baptiste. *Paris à la fin du XVIIIe siècle* (1801). In *Tableaux de Paris*, edited by Sophie Lefay. Paris: Société Française d'étude du dix-huitième siècle, 2016.

Pujoulx, Jean-Baptiste. *Promenades au Jardin des plantes, à la ménagerie et dans les galeries du Muséum d'histoire naturelle*. Paris, 1803.

Réaumur, René-Antoine Ferchault de. "Art de faire une nouvelle espèce de porcelaine, par les moyens extrêmement simples & faciles." *MAS* 1739 (1741): 370–88.

Réaumur, René-Antoine Ferchault de. "De la nature de la terre en générale, et du caractère des différentes espèces de terre." *MAS* 1730 (1732): 243–83.

Réaumur, René-Antoine Ferchault de. "Description d'une mine de fer du pays de Foix; avec quelques réflexions sur la manière dont elle a été formée." *MAS* 1718 (1720): 139–42.

Réaumur, René-Antoine Ferchault de. *Fabrique des ancres, lue à l'Académie en juillet 1723*. Paris, 1764.

Réaumur, René-Antoine Ferchault de. "Idée générale des différentes manières de faire la porcelaine." *MAS* 1727 (1729): 185–203.

Réaumur, René-Antoine Ferchault de. "Mémoire sur la matière qui colore les perles fausses." *MAS* 1716 (1718): 229–44.

Réaumur, René-Antoine Ferchault de. *Réaumur's Memoirs on Steel and Iron*. Translated by Anneliese G. Sisco. Edited by Cyril S. Smith. Chicago: University of Chicago Press, 1956. Originally published in 1722.

[Réaumur, René-Antoine Ferchault de?]. "Réflexions sur l'utilité dont l'Académie des sciences pourroit être au royaume, si le royaume luy donnoit les secours dont elle a besoin." In Ernest

Maindron, *L'Académie des sciences: Histoire de l'Académie, fondation de l'Institut national*, 103–10. Paris, 1888. Originally written ca. 1720.

Réaumur, René-Antoine Ferchault de. "Second mémoire sur la porcelaine." *MAS* 1729 (1731): 325–46.

Réaumur, René-Antoine Ferchault de. "Sur la nature et la formation des cailloux." *MAS* 1721 (1723): 255–76.

Réaumur, René-Antoine Ferchault de. "Sur la rondeur qui semblent affecter certaines espèces de pierres, & entr'autres sur celle qu'affectent les cailloux." *MAS* 1723 (1725): 273–84.

Réaumur, René-Antoine Ferchault de. "Sur les mines de turquoises du Royaume; sur la nature de la matière qu'on y trouve, & sur la manière dont on lui donne la couleur." *MAS* 1715 (1717): 174–93.

Rochon, Alexis-Marie de. *Recueil de mémoires sur la mécanique et la physique*. Paris, 1783.

Rochon, Alexis-Marie de. *Voyages à Madagascar, à Maroc, et aux Indes Orientales*. Paris, 1801.

Romé de l'Isle, Jean-Baptiste Louis. *Catalogue raisonné des minéraux, pierres fines et cristallisées, pétrifications, coquilles, madrépores, et autres curiosités de la nature et de l'art: Qui composent le cabinet de M. Galois*. Paris, 1780.

Romé de l'Isle, Jean-Baptiste Louis. *Cristallographie, ou Description des formes propres à tous les corps du règne minéral*. Paris, 1783.

Romé de l'Isle, Jean-Baptiste Louis. *Des caractères extérieurs des minéraux*. Paris, 1784.

Romé de l'Isle, Jean-Baptiste Louis. *Essai de cristallographie*. Paris, 1772.

Romé de l'Isle, Jean-Baptiste Louis, and Fabien Gautier d'Agoty. *Histoire naturelle ou Exposition générale de toutes ses parties gravées et imprimées en couleurs naturelles*. Paris, 1781.

Romé de l'Isle, Jean-Baptiste Louis, and Abbé Dugaut. *Catalogue systématique et raisonné des curiosités de la nature et de l'art, qui composent le cabinet de Mr. Davila*. Paris, 1767.

Rosnel, Pierre de. *Mercure Indien, ou Le trésor des Indes*. Paris, 1667.

Rosnel, Pierre de. *Traité sommaire de l'institution du corps et communauté des marchands orfèvres*. Paris, 1672.

Sage, Balthasar-Georges. *Analyse chimique et concordance des trois règnes*. Paris, 1786.

Sage, Balthasar-Georges. *Description des objets d'art de la collection de B. G. Sage de l'Institut de France*. Paris, 1807.

Sage, Balthasar-Georges. *Description méthodique du Cabinet de l'École royale des mines*. Paris, 1784.

Sage, Balthasar-Georges. *Supplément à la Description méthodique du Cabinet de l'École royale des mines*. Paris, 1787.

Savary des Brûlons, Jacques. *Dictionnaire universel de commerce*. Paris, 1726.

Savary des Brûlons, Jacques. *Le parfait négociant, ou instruction générale pour ce qui regarde le commerce*. Paris, 1675.

Smith, Adam. *An Inquiry into the Nature and Causes of the Wealth of Nations*. Indianapolis, IN: Liberty Fund and Oxford University Press, 1981. Originally published in 1776.

Swebach Desfontaines, François-Louis. *Manuel cristallographe, ou Abrégé de la cristallographie de M. Romé de l'Isle*. Paris, 1792.

Tavernier, Jean-Baptiste. *Les six voyages de Jean-Baptiste Tavernier, qu'il a fait en Turquie, en Perse et aux Indes*. Paris, 1676.

Tavernier, Jean-Baptiste. *Travels in India*. Edited by Valentine Ball. London: Macmillan, 1889. Originally published in 1676.

Theophrastus. *Theophrastus on Stones: Introduction, Greek Text, English Translation, and Commentary*. Edited by Earle Radcliffe Caley and John F. C. Richards. Columbus, OH: Ohio State University Press, 1956. Originally written ca. 300 BC.

Thévenot, Christian. *Jean de Thévenot.* Saint-Denis: Edilivre, 2012.

Thévenot, Christian. *Melchisédech Thévenot: Bibliothécaire du roi, 1620–1692.* Saint-Denis: Edilivre, 2012.

Thévenot, Jean. *Les voyages aux Indes orientales.* Edited by Françoise de Valence. Paris: H. Champion, 2008. Originally published in 1684.

Varthema, Ludovico de. *The Travels of Ludovico di Varthema.* Edited by George Percy Badger. Translated by John Winter Jones. Cambridge, UK: Cambridge University Press, 1863. Originally published in 1510.

Vauquelin, Louis-Nicolas. "Analyse de la tourmaline de Ceylan." *Journal des mines* 9, no. 54 (1799): 477–79.

Vauquelin, Louis-Nicolas. "Analyse de l'aigue-marine ou beril, et découverte d'une terre nouvelle dans cette pierre." *Journal des mines* 8, no. 43 (1798): 553–64.

Vauquelin, Louis-Nicolas. "Analyse de l'émeraude du Pérou." *Journal des mines* 8, no. 37 (1797): 93–97.

Vauquelin, Louis-Nicolas. "Analyse des grenats noirs du pic d'Erès-Lids." *Journal des mines* 8, no. 44 (1798): 571–73.

Vauquelin, Louis-Nicolas. "Analyse des topazes de Saxe, de Sibérie et du Brésil." *Annales du Muséum national d'histoire naturelle* 6 (1805): 21–25.

Vauquelin, Louis-Nicolas. "Analyse du rubis spinelle." *Journal des mines* 8, no. 37 (1797): 81–92.

Vauquelin, Louis-Nicolas. "Analyses comparées des hyacinthes de Ceylan et d'Expailly." *Journal des mines* 5, no. 26 (1796): 97–118.

Vauquelin, Louis-Nicolas. "De la chrysolite des jouailliers." *Journal des mines* 7, no. 37 (1797): 19–26.

Vauquelin, Louis-Nicolas. "De la topase blanche de Saxe." *Journal des mines* 4, no. 24 (1796): 1–4.

Vauquelin, Louis-Nicolas. "Des grenats rouges du même pic." *Journal des mines* 8, no. 44 (1798): 574–75.

Vauquelin, Louis-Nicolas. "Du Béril de Saxe, dans lequel M. Tromsdorf a annoncé l'existence d'une terre nouvelle qu'il a nommée Agustine." *Journal des mines* 15, no. 86 (1803): 81–87.

Vauquelin, Louis-Nicolas. "Du péridot du commerce." *Journal des mines* 4, no. 37 (1796): 37–44.

Vauquelin, Louis-Nicolas. "Expériences sur les grenats blancs ou leucite des volcans." *Journal des mines* 5, no. 27 (1796): 201–8.

Vauquelin, Louis-Nicolas. "Expériences sur les topases." *Journal des mines* 16, no. 96 (1804): 469–74.

Vauquelin, Louis-Nicolas. "Réflexions sur l'analyse des pierres, et résultats de plusieurs de ces analyses." *Annales de chimie* 30 (1799): 66–106.

Wallerius, Jean Gotschalk. *Minéralogie, ou Description générale des substances du règne minéral.* Translated by Paul-Henri Thiry, Baron d'Holbach. Paris, 1753. Originally published in 1747.

Secondary Sources

Adams, Christopher, Hamish J. Campbell, and Russell Beck. "Characterisation and Origin of New Zealand Nephrite Jade Using Its Strontium Isotopic Signature." *Lithos* 97, no. 3 (2007): 307–22.

Adams, Frank. *The Birth and Development of the Geological Sciences.* Baltimore, MD: Williams & Wilkins, 1938.

Anguissola, Anna, and Andreas Grüner, eds. *The Nature of Art: Pliny the Elder on Materials.* Turnhout: Brepols, 2020.

Anstey, Peter. *The Philosophy of Robert Boyle.* New York: Taylor & Francis, 2000.

Anstey, Peter, and Alberto Vanzo. *Experimental Philosophy and the Origins of Empiricism.* Cambridge, UK: Cambridge University Press, 2023.

Ashworth, William. "'Between the Trader and the Public': British Alcohol Standards and the Proof of Good Governance." *Technology and Culture* 42, no. 1 (2001): 27–50.

Ashworth, William. *Customs and Excise: Trade, Production, and Consumption in England, 1640–1845.* Oxford: Oxford University Press, 2003.

Ashworth, William. "Natural History and the Emblematic World View." In *Reappraisals of the Scientific Revolution,* edited by David C. Lindberg and Robert S. Westman, 303–32. Cambridge, UK: Cambridge University Press, 1990.

Ashworth, William. "Quality and the Roots of Manufacturing 'Expertise' in Eighteenth-Century Britain." *Osiris* 25, no. 1 (2010): 231–54.

Baghdiantz-McCabe, Ina. *Orientalism in Early Modern France: Eurasian Trade, Exoticism and the Ancien Regime.* Oxford: Berg, 2008.

Baker, Tawrin. "Color and Contingency in Robert Boyle's Works." In *Early Modern Colour Worlds,* edited by Tawrin Baker, Sven Dupré, Sachiko Kusukawa, and Karin Leonhard, 248–73. Leiden: Brill, 2015.

Ball, Sydney H. *A Roman Book on Precious Stones.* Los Angeles, CA: Gemological Institute of America, 1950.

Ball, Valentine. "Introduction." In Jean-Baptiste Tavernier, *Travels in India,* vol. 1, xi–xxxviii. London: Macmillan, 1889.

Banchetti-Robino, Marina Paola. *The Chemical Philosophy of Robert Boyle: Mechanicism, Chymical Atoms, and Emergence.* Oxford: Oxford University Press, 2020.

Bandy, Mark Chance, and Jean A. Bandy. "Foreword." In Georg Agricola, *De natura fossilium (Textbook of Mineralogy),* translated by Mark Chance Bandy and Jean A. Bandy, v–x. Boulder, CO: Geological Society of America, 1955.

Bapst, Germain. *Histoire des joyaux de la couronne de France.* Paris, 1889.

Barrera-Osorio, Antonio. *Experiencing Nature: The Spanish American Empire and the Early Scientific Revolution.* Austin, TX: University of Texas Press, 2006.

Beasley, Faith E. *Versailles Meets the Taj Mahal: François Bernier, Marguerite de la Sablière, and Enlightening Conversations in Seventeenth-Century France.* Toronto: Toronto University Press, 2018.

Beck, Pieter T. L. "Strong Foundations: Petrus van Musschenbroek's Experimental Research on the Strength of Materials." *Historical Studies in the Natural Sciences* 53, no. 2 (2023): 109–46.

Beck, Russell, Maika Mason, and Andris Apse. *Pounamu: The Jade of New Zealand.* North Shore: Penguin, 2010.

Becquerel, Jean. "Chaire de physique appliquée à l'histoire naturelle." *Archives du Muséum national d'histoire naturelle* (1935): 82–104.

Belich, James. *The World the Plague Made: The Black Death and the Rise of Europe.* Princeton, NJ: Princeton University Press, 2022.

Beretta, Marco. "Collected, Analysed, Displayed: Lavoisier and Minerals." In *From Private to Public: Natural Collections and Museums,* edited by Marco Beretta, 113–40. Sagamore Beach, MA: Science History Publications, 2005.

Berkeley, Edmund, and Dorothy Smith Berkeley. *John Clayton: Pioneer of American Botany.* Chapel Hill, NC: University of North Carolina Press, 1963.

Bertomeu Sánchez, José Ramón, and Agustí Nieto-Galan, eds. *Chemistry, Medicine, and Crime: Mateu J. B. Orfila (1787–1853) and His Times.* Sagamore Beach, MA: Science History Publications, 2006.

Bertucci, Paola. *Artisanal Enlightenment: Science and the Mechanical Arts in Old Regime France.* New Haven, CT: Yale University Press, 2017.

Biggs, Norman. "Mathematics at the Mint: A Seventeenth-Century Saga." *British Numismatic Journal* 87 (2017): 151–61.

Bimbenet-Privat, Michèle. *Les orfèvres et l'orfèvrerie de Paris au XVIIe siècle.* Paris: Commission des travaux historiques de la ville de Paris, 2002.

Birembaut, Arthur. "L'enseignement de la minéralogie et des techniques minières." In *Enseignement et diffusion des sciences en France au dix-huitième siècle*, 2nd ed., edited by René Taton and Yves Laissus, 365–418. Paris: Hermann, 1986.

Bleichmar, Daniela. "Learning to Look: Visual Expertise Across Art and Science in Eighteenth-Century France." *Eighteenth-Century Studies* 46, no. 1 (2012): 85–111.

Blondel, Christine. "Haüy et l'électricité: De la démonstration-spectacle à la diffusion d'une science newtonienne." *Revue d'histoire des sciences* 50, no. 3 (1997): 265–82.

Boklund, Uno. "Wallerius, John Gottschalk." In *DSB*, vol. 14, 144–45.

Bol, Marjolijn. "The Emerald and the Eye: On Sight and Light in the Artisan's Workshop and the Scholar's Study." In *Perspective as Practice: Renaissance Cultures of Optics*, edited by Sven Dupré, 77–101. Turnhout: Brepols, 2019.

Bol, Marjolijn. "Gems in the Water of Paradise: The Iconography and Reception of Heavenly Stones in the Ghent Altarpiece." In *Van Eyck Studies: Papers Presented at the Eighteenth Symposium for the Study of Understanding and Technology in Painting*, edited by Christina Currie, Bart Fransen, Valentine Henderiks, Cyriel Stroo and Dominique Vanwijnsberghe, 34–58. Leuven: Peeters Publishers, 2017.

Bol, Marjolijn. "Polito et Claro: The Art and Knowledge of Polishing, 1100–1500." In *Gems in the Early Modern World: Materials, Knowledge and Global Trade, 1450–1800*, edited by Michael Bycroft and Sven Dupré, 223–57. London: Palgrave Macmillan, 2019.

Boldrini, Federica. "All That Glitters Is Not Gold: False Jewellery and Its Juridical Regulation in Italy Between the Late Middle Ages and the Early Modern Period." In *Faking It! The Performance of Forgery in Late Medieval and Early Modern Culture*, edited by Philip Lavender and Matilda Amundsen Bergström, 52–75. Leiden: Brill, 2023.

Bonnassieux, Pierre, and Eugène Lelong. *Conseil de commerce et Bureau du commerce 1700–1791: Inventaire analytique des procès-verbaux.* Paris: Imprimerie nationale, 1900.

Boumediene, Samir. *La colonisation du savoir: Une histoire des plantes médicinales du Nouveau monde, 1492–1750.* Vaulx-en-Velin: Les Éditions des Mondes à faire, 2019.

Boumediene, Samir, and Valentina Pugliano. "La route des succédanés: Les remèdes exotiques, l'innovation médicale et le marché des substituts au XVIe siècle." *Revue d'histoire moderne et contemporaine* 3, no. 66.3 (2019): 24–54.

Bourdier, Franck. "Origines et transformations du cabinet du Jardin Royal des Plantes." *Revue générale des sciences pures et appliquées* 18 (1962): 36–50.

Bourguet, Marie Noëlle, Christian Licoppe, and H. Otto Sibum, eds. *Instruments, Travel and Science: Itineraries of Precision from the Seventeenth to the Twentieth Century.* London: Routledge, 2002.

Brailsford, Barry. *Greenstone Trails: The Māori and Pounamu.* Hamilton, New Zealand: Stoneprint Press, 1996.

Brock, William H. *The Fontana History of Chemistry.* London: Fontana, 1992.

Brunet, Pierre. *Les physiciens hollandais et la méthode expérimentale en France au XVIIIe siècle.* Paris: Albert Blanchard, 1926.

Buchwald, Jed. "Experimental Investigations of Double Refraction from Huygens to Malus." *Archive for the History of Exact Sciences* 21, no. 4 (1980): 311–73.

Buchwald, Jed, and Robert Fox. *Oxford Handbook of the History of Physics.* Oxford: Oxford University Press, 2013.

Bud, Robert. "'Applied Science': A Phrase in Search of a Meaning." *Isis* 103, no. 3 (2012): 537–45.

Buettner, Brigitte. *The Mineral and the Visual: Precious Stones in Medieval Secular Culture.* University Park, PA: Penn State University Press, 2022.

Burke, John G. "Bournon, Comte de, Louis-Jacques." In *DSB*, vol. 2, 355.

Burke, John G. *Origins of the Science of Crystals.* Berkeley: University of California Press, 1966.

Burke, John G. "Valmont de Bomare, Jacques-Christophe." In *DSB*, vol. 13, 565–66.

Butters, Suzanne. *The Triumph of Vulcan: Sculptors' Tools, Porphyry, and the Prince in Ducal Florence.* Florence: Olschki, 1996.

Butters, Suzanne. "'Una pietra eppure non una pietra': Pietre dure e le botteghe medicee nella Firenze del Cinquecento." In *Arti fiorentine: La grande storia dell'Artigianato*, vol. 3: *Il cinquecento*, 133–85. Florence: Giunti Gruppo, 2000.

Bycroft, Michael. "Anselmus Boethius de Boodt and the Emergence of the Oriental/Occidental Distinction in European Mineralogy." In *Gems in the Early Modern World*, edited by Michael Bycroft and Sven Dupré, 149–72. London: Palgrave Macmillan, 2019.

Bycroft, Michael. "Dossier critique de l'article PIERRES PRÉCIEUSES" (*Encyclopédie*, t. XII, p. 593b–595a). *Édition numérique collaborative et critique de l'Encyclopédie.* https://enccre .academie-sciences.fr/encyclopedie/article/v12-1449-47. Accessed June 26, 2020.

Bycroft, Michael. "Experiments on Collections at the Royal Society of London and the Paris Academy of Sciences, 1660–1740." In *The Institutionalization of Science in Early Modern Europe*, edited by Mordechai Feingold and Giulia Giannini, 236–65. Leiden: Brill, 2019.

Bycroft, Michael. "La fin des pierres précieuses?" In *Avoir une âme pour les pierres: Arts, sciences et minéralité, du tournant des lumières au crépuscule du romanticisme*, edited by Pierre Glaudes, Anouchka Vasak and Baldine Saint Girons, 189–202. Rennes: Presses universitaires de Rennes, 2024.

Bycroft, Michael. "The Hand of the Connoisseur: Gems and Hardness in Enlightenment Mineralogy." *History of Science* 60, no. 4 (2022): 517–21.

Bycroft, Michael. "Iatrochemistry and the Evaluation of Mineral Waters in France, 1600–1750." *Bulletin of the History of Medicine* 91, no. 2 (2017): 303–30.

Bycroft, Michael. "Introduction: Science Beyond the Enlightenment." *Journal of Early Modern Studies* 12, no. 1 (2023): 9–31.

Bycroft, Michael. "A Neo-Positivist Theory of Scientific Change." *British Journal for the History of Science* 9 (2024): 129–48.

Bycroft, Michael. "Physics and Natural History in the Eighteenth Century: The Case of Charles Dufay." PhD diss. University of Cambridge, 2013.

Bycroft, Michael. "Regulation and Intellectual Change at the Paris Goldsmiths' Guild, 1660–1740." *Journal of Early Modern History* 22, no. 6 (2018): 500–527.

Bycroft, Michael. "Robert Boyle's Restless Gems." In *Ingenuity in the Making: Matter and Technique in Early Modern Europe*, edited by Richard J. Oosterhoff, José Ramón Marcaida, and Alexander Marr, 36–49. Pittsburgh, PA: University of Pittsburgh Press, 2021.

Bycroft, Michael. "Style and Substance in Rococo Science." *Journal of Interdisciplinary History* 48, no. 3 (2018): 359–84.

Bycroft, Michael. "What Difference Does a Translation Make? The *Traité des vernis* (1723) in the Career of Charles Dufay." In *Translation and the Circulation of Knowledge in Early Modern Science*, edited by Sietske Fransen and Niall Hodson, 66–90. Leiden: Brill, 2017.

Bycroft, Michael. "Wonders in the Academy: The Value of Strange Facts in the Experimental Research of Charles Dufay." *Historical Studies in the Natural Sciences* 43, no. 3 (2013): 334–70.

Bycroft, Michael, and Sven Dupré, eds. *Gems in the Early Modern World: Materials, Knowledge and Global Trade, 1450–1800*. London: Palgrave Macmillan, 2019.

Bycroft, Michael, and Alexander Wragge-Morley, "Science and Connoisseurship in the European Enlightenment." *History of Science* 60, no. 4 (2022): 439–57.

Caesar, Ed. "The Woman Shaking Up the Diamond Industry." *New Yorker*, January 27, 2020. https://www.newyorker.com/magazine/2020/02/03/the-woman-shaking-up-the-diamond-industry.

Cage, E. Claire. *The Science of Proof: Forensic Medicine in Modern France*. Cambridge, UK: Cambridge University Press, 2022.

Cantor, Geoffrey. "The Eighteenth Century Problem." *History of Science* 20 (1982): 44–63.

Castelluccio, Stéphane. *Le garde-meuble de la Couronne et ses intendants du XVIe au XVIIIe siècle*. Paris: Comité des travaux historiques et scientifiques, 2004.

Castelluccio, Stéphane. *Les meubles de pierres dures de Louis XIV et l'atelier des Gobelins*. Paris: Faton, 2007.

Castelluccio, Stéphane. *Le prince et le marchand: Le commerce de luxe chez les marchands merciers parisiens pendant le règne de Louis XIV*. Paris: SPM, 2014.

Chabrand, Jean-Armand. *Antoine Cayre-Morand: Fondateur de la Manufacture de Cristal de Roche de Briançon*. Grenoble, 1874.

Chang, Hasok. "The Chemical Revolution Revisited." *Studies in History and Philosophy of Science* 49 (2015): 91–98.

Chang, Hasok. *Is Water H2O? Evidence, Realism and Pluralism*. Springer: Dordrecht, 2012.

Chuang, Grace. "The Role of the Savant and the Académie Royale des Sciences in Porcelain Research and Development in France, 1715 to 1772." Master's thesis, Bard Graduate Center, New York, 2010.

Clericuzio, Antonio. *Elements, Principles and Corpuscles: A Study of Atomism and Chemistry in the Seventeenth Century*. Dordrecht: Springer, 2000.

Cohen, Floris H. *How Modern Science Came into the World: Four Civilizations, One 17th-Century Breakthrough*. Amsterdam University Press, 2010.

Cohen, Floris H. "Postscript 2012 to *The Scientific Revolution: A Historiographical Inquiry*." July 2012. https://hfloriscohen.wordpress.com/wp-content/uploads/2020/06/postscript-chinese-srhi.pdf.

Cohen, Floris H. *The Scientific Revolution: A Historiographical Inquiry*. Chicago: University of Chicago Press, 1994.

Cohen, I. Bernard. *Franklin and Newton: An Inquiry into Speculative Newtonian Experimental Science and Franklin's Work in Electricity as an Example Thereof*. Philadelphia: American Philosophical Society, 1956.

Coley, Noel. "'Cures Without Care': 'Chemical Physicians' and Mineral Waters in 17th Century English Medicine." *Medical History* 23 (1979): 191–214.

Combet, Michel, and Anne-Sylvie Moretti. *La Dordogne de Cyprien Brard*. Périgueux: Archives départmentales de la Dordogne, 1995.

Content, Derek J. *Ruby, Sapphire & Spinel: An Archaeological, Textural, and Cultural Study.* Turnhout: Brepols, 2016.

Cook, Harold. *Matters of Exchange: Commerce, Medicine, and Science in the Dutch Golden Age.* New Haven, CT: Yale University Press, 2007.

Cook, Harold, Amy R. W. Meyers, and Pamela Smith, "Introduction: Making and Knowing." In *Ways of Making and Knowing: The Material Culture of Empirical Knowledge*, edited by Pamela Smith, Amy R. W. Meyers, and Harold J. Cook, 1–16. Ann Arbor: University of Michigan Press, 2014.

Cooper, Alix. *Inventing the Indigenous: Local Knowledge and Natural History in Early Modern Europe.* Cambridge, UK: Cambridge University Press, 2007.

Copenhaver, Brian. *Magic in Western Culture: From Antiquity to the Enlightenment.* Cambridge, UK: Cambridge University Press, 2015.

Coulardot, Lisa. "Les lumières au banc d'essai: Science, économie et environment autour de Jean Hellot (1685–1766)." PhD diss. European University Institute, 2024.

Crespo, Hugo. "The Plundering of the Ceylonese Royal Treasure, 1551–1553: Its Character, Cost, and Dispersal." In *Gems in the Early Modern World: Materials, Knowledge and Global Trade, 1450–1800*, edited by Michael Bycroft and Sven Dupré, 35–64. London: Palgrave Macmillan, 2019.

Crosland, Maurice. *Historical Studies in the Language of Chemistry.* London: Heineman, 1962.

Crowley, Patrick R. "Factitious Gems and the Matter of Facts in Pliny's Natural History." In *The Nature of Art: Pliny the Elder on Materials*, edited by Anna Anguissola and Andreas Grüner, 246–59. Turnhout: Brepols, 2020.

Cruysse, Dirk van der. *Chardin le Persan.* Paris: Fayard, 1998.

Cunningham, Andrew, and Perry Williams, "De-centring the 'Big Picture': 'The Origins of Modern Science' and the Modern Origins of Science." *British Journal for the History of Science* 26, no. 4 (1993): 407–32.

Curry, Helen, Nicholas Jardine, James Secord, and Emma Spary, eds. *Worlds of Natural History.* Cambridge, UK: Cambridge University Press, 2018.

Dames, Mansel Longworth. "Introduction." In Duarte Barbosa, *The Book of Duarte Barbosa, an Account of the Countries Bordering on the Indian Ocean and Their Inhabitants*, xxxiii–lxxii. Farnham: Hakluyt Society, 1918.

Damme, Stéphane van. "Un ancien régime des sciences et des savoirs." In *Histoire des sciences et des savoirs*, vol. 1: *De la Renaissance aux Lumières*, edited by Stéphane van Damme. Paris: Seuil, 2015.

Daston, Lorraine. "The Cold Light of Facts and the Facts of Cold Light: Luminescence and the Transformation of the Scientific Fact, 1600–1750." *Early Modern France* 3 (1997): 1–27.

Daston, Lorraine. "The Factual Sensibility." *Isis* 79, no. 3 (1988): 452–67.

Daston, Lorraine. "The Naturalistic Fallacy Is Modern." *Isis* 105, no. 3 (2014): 579–87.

Daston, Lorraine. "Philosophie de la nature et philosophie naturelle (1500–1750)." In *Histoire des sciences et des savoirs*, vol. 1: *De la Renaissance aux Lumières*, edited by Stéphane van Damme, 177–203. Paris: Seuil, 2015.

Daston, Lorraine. "Type Specimens and Scientific Memory." *Critical Inquiry* 31, no. 1 (2004): 153–82.

Daston, Lorraine, and Katharine Park, eds. *Cambridge History of Science*, vol. 3: *Early Modern Science.* Cambridge, UK: Cambridge University Press, 2006.

Daston, Lorraine, and Katharine Park. *Wonders and the Order of Nature, 1150–1750.* New York: Zone Books, 1998.

Delbourgo, James. *A Most Amazing Scene of Wonders: Electricity and Enlightenment in Early America*. Cambridge, MA: Harvard University Press, 2006.

Delépine, Marcel. "Ses oeuvres chimiques [de Vauquelin]." *Revue d'histoire de la pharmacie* 51, no. 177 (1963): 78–88.

Demeulenaere-Douyère, Christiane, and Eric Brian, eds. *Règlement, usages et science dans la France de l'absolutisme*. Paris: Éditions Tec & Doc, 2002.

Demeulenaere-Douyère, Christiane, and David Sturdy. *L'Enquête du Régent, 1716–1718: Sciences, techniques, et politique dans la France pré-industrielle*. Turnhout: Brepols, 2008.

Dew, Nicholas. *Orientalism in Louis XIV's France*. Oxford: Oxford University Press, 2009.

Dew, Nicholas. "Reading Travels in the Culture of Curiosity: Thévenot's Collection of Voyages." *Journal of Early Modern History* 10, no. 1 (2006): 39–59.

Dhombres, Nicole, and Jean Dhombres. *Naissance d'un nouveau pouvoir: Sciences et savants en France 1793–1824*. Paris: Payot, 1989.

Dietz, Bettina, and Thomas Nutz. "Collections Curieuses: The Aesthetics of Curiosity and Elite Lifestyle in Eighteenth-Century Paris." *Eighteenth-Century Life* 29, no. 3 (2005): 44–75.

DiMeo, Michelle. *Lady Ranelagh: The Incomparable Life of Robert Boyle's Sister*. Chicago: University of Chicago Press, 2021.

Distelberger, Rudolf. "Thoughts on Rudolfine Art in the 'Court Workshops' in Prague." In *Rudolf II and Prague: The Court and the City*, edited by Fučíková, Eliška, James M. Bradburne, Beket Bukovinska, Jaroslava Hausenblasová, Lumomír Konečný, Ivan Muchka, and Michal Šroněk, 188–208. London: Thames & Hudson, 1997.

Donkin, Robert A. *Beyond Price: Pearls and Pearl-Fishing, Origins to the Age of Discoveries*. Philadelphia: American Philosophical Society, 1998.

Dupré, Sven. "The Art of Glassmaking and the Nature of Stones: The Role of Imitation in Anselm de Boodt's Classification of Stones." In *Steinformen: Materialität, Qualität, Imitation*, edited by Isabella Augart, Maurice Saß, and Iris Wenderholm, 207–20. Berlin: De Gruyter, 2019.

Dwyer, Déirdre M. "Expert Evidence in English Law Courts, 1559–1800." *Journal of Legal History* 28, no. 1 (2007): 93–118.

Easterby-Smith, Sarah. *Cultivating Commerce: Cultures of Botany in Britain and France, 1760–1815*. Cambridge, UK: Cambridge University Press, 2017.

Eddy, Matthew. *The Language of Mineralogy: John Walker, Chemistry and the Edinburgh Medical School, 1750–1800*. London: Routledge, 2016.

Eddy, Matthew, and Ursula Klein, eds. *A Cultural History of Chemistry in the Eighteenth Century*. London: Bloomsbury, 2022.

Elisa, Michaeli Maria. "Agate: Fortunes and Misfortunes." In *The Nature of Art: Pliny the Elder on Materials*, edited by Anna Anguissola and Andreas Grüner, 288–97. Turnhout: Brepols, 2020.

Essig, Mark. "Poison Murder and Expert Testimony: Doubting the Physician in Late Nineteenth-Century America." *Yale Journal of Law and the Humanities* 14, no. 1 (2002): 177–210.

Evans, James. "The First Identification of Spinel." *Gemmology Bulletin* (2020): 2–7.

Evans, James. "A History of Gemmology, 1912–1972." *Gemmology Bulletin* (2024): 1–14.

Evans, James. "Rediscovering Manufactured Ruby, Part 1." *Gemmology Bulletin* (2020): 1–10.

Evans, Joan. *Magical Jewels of the Middle Ages and the Renaissance, Particularly in England*. Oxford: Clarendon Press, 1922.

Evans, Joan, and Mary Sergeantson. *English Mediaeval Lapidaries*. London: Oxford University Press, 1933.

Evans, Joan, and Paul Studer. *Anglo-Norman Lapidaries.* Paris: Champion, 1924.

Evans, Robert J. W. *Rudolf II and His World: A Study in Intellectual History, 1576–1612.* Oxford: Clarendon Press, 1973.

Farges, François. "Les grands diamants de la couronne de François I à Louis XVI." *Versalia* 16 (2014): 55–78.

Farges, François, and Johan Kjellman. "Bicentenaire du décès de René-Just Haüy: Les dernières découvertes au Muséum national d'histoire naturelle." *Le règne minéral* 165 (2022): 7–42.

Farges, François, and Olivier Segura. *Pierres précieuses: Guide visual.* Paris: Dunod, 2023.

Farges, François, Scott Sucher, Herbert Horovitz, and Jean-Marc Fourcault. "The French Blue and the Hope: New Data from the Discovery of a Historical Lead Cast." *Gems and Gemology* 45, no. 1 (2009): 4–19.

Fauque, Danielle. "Alexis-Marie Rochon (1741–1817): Savant astronome et opticien." *Revue d'histoire des sciences* 38, no. 1 (1985): 3–36.

Felten, Sebastian. *Money in the Dutch Republic: Everyday Practice and Circuits of Exchange.* Cambridge, UK: Cambridge University Press, 2022.

Ferrier, Ronald W. *A Journey to Persia: Jean Chardin's Portrait of a Seventeenth-Century Empire.* London: I. B. Tauris, 1996.

Feyel, Gilles. *La presse en France des origines à 1944: Histoire politique et matérielle.* Paris: Ellipses, 2007.

Findlen, Paula, ed. *Early Modern Things: Objects and Their Histories, 1500–1800.* New York: Routledge, 2013.

Findlen, Paula. "Early Modern Things: Objects in Motion, 1500–1800." In Findlen, *Early Modern Things,* 3–28.

Findlen, Paula, ed. *Empires of Knowledge: Scientific Networks in the Early Modern World.* London: Routledge, 2018.

Findlen, Paula. "Natural History." In *Cambridge History of Science,* vol. 3: *Early Modern Science,* edited by Lorraine Daston and Katharine Park, 438–68. Cambridge, UK: Cambridge University Press, 2006.

Findlen, Paula. *Possessing Nature: Museums, Collecting, and Scientific Culture in Early Modern Italy.* Berkeley: University of California Press, 1994.

Findlen, Paula. "Sites of Anatomy, Botany, and Natural History." In *Cambridge History of Science,* vol. 3: *Early Modern Science,* edited by Lorraine Daston and Katharine Park, 272–89. Cambridge, UK: Cambridge University Press, 2006.

Finnegan, Diarmid A. "The Spatial Turn: Geographical Approaches in the History of Science." *Journal of the History of Biology* 41 (2008): 369–88.

Fors, Hjalmar. "Elements in the Melting Pot: Merging Chemistry, Assaying and Natural History, c. 1730–1760." *Osiris* 29 (2014): 230–44.

Fors, Hjalmar. *The Limits of Matter: Chemistry, Mining and Enlightenment.* Chicago: University of Chicago Press, 2015.

Foucault, Michel. *Les mots et les choses: Une archéologie des sciences humaines.* Paris: Gallimard, 1966.

Fox, Robert. *The Savant and the State: Science and Cultural Politics in Nineteenth-Century France.* Baltimore, MD: Johns Hopkins University Press, 2012.

Frängsmyr, Tore, John L. Heilbron, and Robin E. Rider, eds. *The Quantifying Spirit in the Eighteenth Century.* Berkeley: University of California Press, 1990.

Freedman, Paul. *Out of the East: Spices and the Medieval Imagination.* New Haven, CT: Yale University Press, 2008.

Garber, Daniel. "Physics and Foundations." In *Cambridge History of Science*, vol. 3: *Early Modern Science*, edited by Lorraine Daston and Katharine Park, 21–69. Cambridge, UK: Cambridge University Press, 2006.

Garber, Daniel. "Remarks on the Pre-History of the Mechanical Philosophy." In *The Mechanization of Natural Philosophy*, edited by Sophie Roux and Daniel Garber, 3–26. Dordrecht: Springer, 2013.

Gaukroger, Stephen. *The Collapse of Mechanism and the Rise of Sensibility: Science and the Shaping of Modernity, 1680–1760*. Oxford: Oxford University Press, 2011.

Genet-Varcin, Émilienne, and Jacques Roger. "Bibliographie de Buffon." In *Buffon: Oeuvres philosophiques*, edited by Jean Piveteau, Maurice Fréchet, and Charles Bruneau, 513–70. Paris: Presses universitaires de France, 1954.

Germer, Stefan. *Art-pouvoir-discours: La carrière intellectuelle d'André Félibien dans la France de Louis XIV*. Translated by Aude Virey-Wallon. Paris: Éditions FMSH-Fondation Maison des sciences de l'homme, 2016.

Gillispie, Charles C. "The Natural History of Industry." *Isis* 48, no. 4 (1957): 398–407.

Gillispie, Charles C. *Science and Polity in France at the End of the Old Regime*. Princeton, NJ: Princeton University Press, 1980.

Gillispie, Charles C. *Science and Polity in France: The Revolutionary and Napoleonic Years*. Princeton, NJ: Princeton University Press, 2004.

Ginzburg, Carlo. "Morelli, Freud and Sherlock Holmes: Clues and Scientific Method." Translated by Anna David. *History Workshop Journal* 9 (1980): 5–36.

Giovannetti-Singh, Gianamar. "Galenizing the New World: Joseph-François Lafitau's 'Galenization' of Canadian Ginseng, ca. 1716–1724." *Notes and Records of the Royal Society Journal of the History of Science* 75, no. 1 (2020): 59–72.

Giovannetti-Singh, Gianamar, and Rory Kent. "Crises and the History of Science: A Materialist Rehabilitation." *British Journal for the History of Science* 9 (2024): 39–37.

Glorieux, Guillaume. *A l'enseigne de Gersaint: Edme-François Gersaint, marchand d'art sur le Pont Notre-Dame, 1694–1750*. Seyssel: Éditions Champ Vallon, 2002.

Golan, Tal. *Laws of Men and Laws of Nature: The History of Scientific Expert Testimony in England and America*. Cambridge, MA: Harvard University Press, 2004.

Graeber, David. *Toward an Anthropological Theory of Value: The False Coin of Our Own Dreams*. New York: Palgrave, 2001.

Grapes, Rodney. "Pounamu: Nomenclature, First European Impressions, Acquisition, Chemical Analyses and Mineralogical Characterisation." *Geoscience Society of New Zealand: Journal of Historical Studies* 77 (December 2023): 1–31.

Grodzinski, Paul. "The History of Diamond Polishing." *Industrial Diamond Review* 1, special supplement (1953): 1–13.

Guasparri, Andrea. "Explicit Nomenclature and Classification in Pliny's Natural History XXXII." *Studies in the History and Philosophy of Science Part A* 44, no. 3 (2013): 347–53.

Guerlac, Henry. *Lavoisier, the Crucial Year: The Background and Origin of His First Experiments on Combustion in 1772*. Ithaca, NY: Cornell University Press, 1961.

Guerlac, Henry. "Sage, Balthazar-Georges." In *DSB*, vol. 12, 63–69.

Guerlac, Henry. "Some French Antecedents of the Chemical Revolution." *Chymia* 5 (1959): 73–112.

Guerrini, Anita. *The Courtiers' Anatomists: Animals and Humans in Louis XIV's Paris*. Chicago: University of Chicago Press, 2015.

Guerrini, Anita. "The Material Turn in the History of Life Science." *Literature Compass* 13, no. 7 (2016): 469–80.

Guillaume, Paul. "Autobiographie de Caire-Morand, fondateur de la manufacture de cristal de roche de Briançon, en 1778." *Bulletin de la Société d'études des Hautes-Alpes* (1883): 142–70.

Gysel, Carlos. "A. de Boodt, lapidaire et médecine de Rodolphe II." *Vesalius* 3, no. 1 (1997): 33–42.

Habib, Irfan. *An Atlas of the Mughal Empire: Political and Economic Maps with Detailed Notes, Bibliography and Index*. Oxford: Oxford University Press, 1982.

Hacking, Ian. "The Contingencies of Ambiguity." *Analysis* 67, no. 4 (2007): 269–77.

Hagner, Arthur F. "Introduction." In Robert Boyle, *An Essay About the Origine and Virtues of Gems*. New York: Hafner Publishing Company, 1972.

Hahn, Roger. *The Anatomy of a Scientific Institution: The Paris Academy of Sciences, 1666–1803*. Berkeley: University of California Press, 1971.

Hahn, Roger. "Review of C. Demeulenaere-Douyère and D. J. Sturdy, *L'Enquête du Régent*." *Isis* 101, no. 2 (2010): 427–38.

Hall, Marie Boas. *Promoting Experimental Learning: Experiment and the Royal Society 1660–1727*. Cambridge, UK: Cambridge University Press, 1991.

Halleux, Robert. "L'oeuvre minéralogique d'Anselme Boèce de Boodt 1550–1632." *Histoire et nature* 14 (1979): 63–81.

Harkness, Deborah. *The Jewel House: Elizabethan London and the Scientific Revolution*. New Haven, CT: Yale University Press, 2007.

Harrison, Peter. *The Bible, Protestantism, and the Rise of Natural Science*. Cambridge, UK: Cambridge University Press, 2001.

Harvey, David. "Value in Motion." *New Left Review* 126 (November/December 2020): 99–126.

Harvey, Edmund. *A History of Luminescence from the Earliest Times Until 1900*. Philadelphia: American Philosophical Society, 1957.

Heilbron, John L. "Dufay, Charles-François de Cisternai." In *DSB*, vol. 4, 214–17.

Heilbron, John L. *Electricity in the 17th and 18th Centuries: A Study of Early Modern Physics*. Berkeley: University of California Press, 1979.

Heilbron, John L. "History of Science." In *The Oxford Companion to the History of Modern Science*. Oxford: Oxford University Press, 2003. https://doi.org/10.1093/acref/9780195112290.001.0001.

Heilbron, John L. "The Measure of Enlightenment." In *The Quantifying Spirit in the Eighteenth Century*, edited by Tore Frängsmyr, John L. Heilbron, and Robin E. Rider, 207–44. Berkeley: University of California Press, 1990.

Heilbron, John L. *Weighing Imponderables and Other Quantitative Science Around 1800*. Berkeley: University of California Press, 1993.

Henry, John. "The Scientific Revolution: Five Books About It." *Isis* 107, no. 4 (2016): 809–17.

Hersey, Mark D. and Jeremy Vetter, "Shared Ground: Between Environmental History and the History of Science." *History of Science* 57, no. 4 (2019): 403–40.

Hessenbruch, Arne. "The Spread of Precision Measurement in Scandinavia 1660–1800." In *The Sciences in the European Periphery in the Enlightenment*, edited by Kostas Gavroglu, 179–224. New York: Springer, 1999.

Hiller, Johannes Erich. "Anselmus de Boodt als Wissenschafter und Naturphilosoph." *Archeion* 15, no. 3 (1933): 348–68.

Hillman, Joshua. "From Coallery to the Natural History of Strata: Mining and the Spatial Sciences of the Earth in Britain, 1600–1800." PhD diss. University of Leeds, 2024.

Hillman, Joshua. "Invisible Labour in the Woodwardian Collection." *Museum & Society* 22, nos. 2–3 (2024): 43–58.

Hofmeester, Karin. "Shifting Trajectories of Diamond Processing: From India to Europe and Back, from the Fifteenth Century to the Twentieth." *Journal of Global History* 8, no. 1 (2013): 25–49.

Holmes, Frederic L. *Eighteenth-Century Chemistry as an Investigative Enterprise.* Berkeley: Office for the History of Science and Technology, 1989.

Holmes, Urban T. "Mediaeval Gem Stones." *Speculum* 9, no. 2 (1934): 195–204.

Home, Roderick W. "Aepinus, the Tourmaline Crystal, and the Theory of Electricity and Magnetism." *Isis* 67 no. 1 (1976): 21–30.

Home, Roderick W. *The Effluvial Theory of Electricity.* New York: Arno Press, 1981.

Hooykaas, Reijer. "Haüy, René-Just." In *DSB*, vol. 6, 178–83.

Hooykaas, Reijer. *La naissance de la cristallographie en France au XVIIIe siècle.* Paris: University of Paris, 1953.

Hooykaas, Reijer. "Romé de l'Isle (or Delisle), Jean-Baptiste Louis." In *DSB*, vol. 11, 520–24.

Hubicki, Wlodzimierz. "Boodt, Anselmus Boetius de." In *DSB*, vol. 2, 292–93.

Hufbauer, Karl. *The Formation of the German Chemical Community (1720–1795).* Berkeley: University of California Press, 1982.

Hunter, Michael. *Boyle: Between God and Science.* London: Yale University Press, 2009.

Hunter, Michael. "Boyle on the Application of Science." In *The Bloomsbury Companion to Robert Boyle*, edited by Jan-Erik Jones, 285–305. London: Bloomsbury, 2021.

Hunter, Michael, ed., with contributions by Edward Davis, Harriet Knight, Charles Littleton, and Lawrence M. Principe. *The Boyle Papers: Understanding the Manuscripts of Robert Boyle.* Aldershot: Ashgate, 2007.

Hunter, Michael. *Boyle Studies: Aspects of the Life and Thought of Robert Boyle (1627–91).* London: Taylor & Francis, 2016.

Hunter, Michael. "The Cabinet Institutionalised: The Royal Society's 'Repository' and Its Background." In *The Origins of Museums: The Cabinet of Curiosities in Sixteenth and Seventeenth Century Europe*, edited by Oliver Impey and Arthur MacGregor, 159–68. Oxford: Clarendon Press, 1985.

Hunter, Michael. "Robert Boyle and the Early Royal Society: A Reciprocal Exchange in the Making of Baconian Science." *British Journal for the History of Science* 40, no. 1 (2007): 1–23.

Hunter, Michael, ed. *Robert Boyle Reconsidered.* Cambridge, UK: Cambridge University Press, 1994.

Hutter, Michael, and David Stark. "Pragmatist Perspectives on Valuation: An Introduction." In *Moments of Valuation: Exploring Sites of Dissonance*, edited by Ariane Berthoin Antal, Michael Hutter, and David Stark, 1–12. Oxford: Oxford University Press, 2015.

Impey, Oliver, and Arthur MacGregor, eds. *The Origins of Museums: The Cabinet of Curiosities in Sixteenth and Seventeenth Century Europe.* Oxford: Clarendon Press, 1985.

Irish, Stephen T. "The Corundum Stone and Crystallographic Chemistry." *Ambix* 64, no. 4 (2017): 301–25.

Jardine, Nicholas, James Secord, and Emma Spary, eds. *Cultures of Natural History.* Cambridge, UK: Cambridge University Press, 1996.

Jardine, Nicholas, and Emma Spary. "Worlds of History." In *Worlds of Natural History*, edited by Helen Curry, Nicholas Jardine, James Secord, and Emma Spary, 3–13. Cambridge, UK: Cambridge University Press, 2018.

Johnson, Kristin. "Natural History as Stamp Collecting: A Brief History." *Archives of Natural History* 34 (2007): 244–58.

Kaspar, Caroline. "L'oeuvre minéralogique et pétrographique des pharmaciens du Muséum." *Revue d'histoire de la pharmacie* 93, no. 347 (2005): 403–12.

Khanykov, Nikolaï Vladimirovich. "Analysis and Extracts of كتاب ميزان الحكمة: Book of the Balance of Wisdom, an Arabic Work on the Water-Balance, Written by 'Al-Khâzinî in the Twelfth Century." *Journal of the American Oriental Society* 6 (1858): 1–128.

Khazeni, Arash. *Sky Blue Stone: The Turquoise Trade in World History*. Berkeley: University of California Press, 2014.

Kilburn-Toppin, Jasmine. "'A Place of Great Trust to Be Supplied by Men of Skill and Integrity': Assayers and Knowledge Cultures in Late Sixteenth- and Seventeenth-Century London." *British Journal for the History of Science* 52, no. 2 (2019): 197–223.

King, James E. *Science and Rationalism in the Government of Louis XIV, 1661–1683*. New York: Octagon Books, 1972.

Klein, Ursula. "Artisanal-Scientific Experts in Eighteenth-Century France and Germany." *Annals of Science* 69, no. 3 (2012): 303–6.

Klein, Ursula. "Origin of the Concept Chemical Compound." *Science in Context* 7, no. 2 (2008): 163–204.

Klein, Ursula. "A Revolution That Never Happened." *Studies in History and Philosophy of Science* 49 (2015): 80–90.

Klein, Ursula. *Technoscience in History: Prussia, 1750–1850*. Cambridge, MA: MIT Press, 2020.

Klein, Ursula, and Wolfgang Lefèvre. *Materials in Eighteenth-Century Science: A Historical Ontology*. Cambridge, MA: MIT Press, 2007.

Klein, Ursula, and Emma Spary, eds. *Materials and Expertise in Early Modern Europe: Between Market and Laboratory*. Chicago: University of Chicago Press, 2010.

Knight, David. *Voyaging in Strange Seas: The Great Revolution in Science*. New Haven, CT: Yale University Press, 2015.

Kohler, Robert. "A Generalist's Vision." *Isis* 96, no. 2 (2005): 224–29.

Kouřimský, Jiří. *Encyclopédie des minéraux*. Paris: Gründ, 1985.

Kristjánsson, Leó. *Iceland Spar and Its Influence on the Development of Science and Technology in the Period 1780–1930*. Reykjavík: University of Iceland, 2010.

Küçük, Harun. "Early Modern Ottoman Science: A New Materialist Framework." *Journal of Early Modern History* 21 (2017): 407–19.

Kuhn, Thomas. "Mathematical Versus Experimental Traditions in the Development of Physical Science." *Journal of Interdisciplinary History* 7, no. 1 (1976): 1–31.

Kula, Witold. *Measures and Men*. Translated by R. Szreter. Princeton, NJ: Princeton University Press, 1986. Originally published in 1970.

Kümin, Beat. *The European World 1500–1800: An Introduction to Early Modern History*. 4th ed. London: Taylor & Francis, 2022. Originally published in 2009.

Kunz, George Frederick, and Charles Hugh Stevenson. *The Book of the Pearl: Its History, Art, Science and Industry*. Mineola, NY: Courier Dover Publications, 2002. Originally published in 1908.

Kusukawa, Sachiko. *Picturing the Book of Nature: Image, Text, and Argument in Sixteenth-Century Human Anatomy and Medical Botany*. Chicago: University of Chicago Press, 2011.

Laboulais, Isabelle. *La maison des mines: La genèse révolutionnaire d'un corps d'ingénieurs civils, 1794–1814*. Rennes: Presses universitaires de Rennes, 2012.

Lacroix, Alfred. "La vie et l'oeuvre de l'abbé René-Just Haüy." *Bulletin de la Société française de minéralogie* 67 (1944): 15–226.

Lafont, Anne. *1740, un abrégé du monde: Savoirs et collections autour de Dezallier d'Argenville.* Paris: Fage, 2012.

Laissus, Yves. "Les cabinets d'histoire naturelle." In *Enseignement et diffusion des sciences en France au dix-huitième siècle,* 2nd ed., edited by René Taton and Yves Laissus, 659–712. Paris: Hermann, 1986.

Laissus, Yves. "Le Jardin du Roi." In *Enseignement et diffusion des sciences en France au dix-huitième siècle,* 2nd ed., edited by René Taton and Yves Laissus, 287–341. Paris: Hermann, 1986.

Lamy, Edouard. *Les cabinets d'histoire naturelle en France au XVIIIe siècle et le Cabinet du Roi (1635–1793).* Paris: E. Lamy, 1930.

Lane, Kris. *Colour of Paradise: The Emerald in the Age of Gunpowder Empires.* New Haven, NJ: Yale University Press, 2002.

LaPorte, Joseph. *Natural Kinds and Conceptual Change.* Cambridge, UK: Cambridge University Press, 2004.

Laudan, Rachel. *From Mineralogy to Geology: The Foundations of a Science, 1650–1830.* Chicago: University of Chicago Press, 1987.

Lehman, Christine. "Pierre-Joseph Macquer: An Eighteenth-Century Artisanal-Scientific Expert." *Annals of Science* 69, no. 3 (2012): 307–33.

Lehman, Christine. "What Is the 'True' Nature of Diamond?" *Nuncius* 31 (2016): 361–407.

Lenman, Bruce. "England, the International Gem Trade and the Growth of Geographical Knowledge from Columbus to James I." In *Renaissance Culture in Context: Theory and Practice,* edited by Jean R. Brink and William F. Gentrup, 86–99. Brookfield, VT: Ashgate, 1993.

Lenzen, Godehard. *The History of Diamond Production and the Diamond Trade.* London: Barrie & Jenkins, 1970.

Leong, Elaine. *Recipes and Everyday Knowledge: Medicine, Science and the Household in Early Modern England.* Chicago: University of Chicago Press, 2018.

Leong, Elaine, and Alisha Rankin. "Testing Drugs and Trying Cures: Experiment and Medicine in Medieval and Early Modern Europe." *Bulletin of the History of Medicine* 91, no. 2 (2017): 157–82.

Lesch, John. "Systematics and the Quantifying Spirit." In *The Quantifying Spirit in the Eighteenth Century,* edited by Tore Frängsmyr, John L. Heilbron, and Robin E. Rider. Berkeley: University of California Press, 1990.

Lespinasse, René de. *Les métiers et corporations de la ville de Paris.* Vol. 2. Paris, 1892.

Licoppe, Christian. *La formation de la pratique scientifique: Le discours de l'expérience en France et en Angleterre, 1630–1820.* Paris: La Découverte, 1996.

Limoges, Camille. "Daubenton, Louis-Jean-Marie." In *DSB,* vol. 3, 111–14.

Lindberg, David C., and Robert S. Westman, eds. *Reappraisals of the Scientific Revolution.* Cambridge, UK: Cambridge University Press, 1990.

Lindqvist, Svante. "Labs in the Woods: The Quantification of Technology During the Late Enlightenment." In *The Quantifying Spirit in the Eighteenth Century,* edited by Tore Frängsmyr, John L. Heilbron, and Robin E. Rider, 291–314. Berkeley: University of California Press, 1990.

Lindroth, Sten. "Linnaeus, Carl." In *DSB,* vol. 8, 374–81.

Llana, James. "A Contribution of Natural History to the Chemical Revolution in France." *Ambix* 32, no. 2 (1985): 71–91.

Llana, James. "Natural History and the *Encyclopédie.*" *Journal of the History of Biology* 33, no. 1 (2000): 1–25.

Long, Pamela. *Artisan/Practitioners and the Rise of the New Science, 1400–1600.* Oregon State University Press, 2011.

Long, Pamela. "The Openness of Knowledge: An Ideal and Its Context in 16th-Century Writings on Mining and Metallurgy." *Technology and Culture* 32, no. 2 (1991): 318–55.

Long, Pamela. "Trading Zones in Early Modern Europe." *Isis* 106, no. 4 (2015): 840–47.

Loveland, Jeff. "Another Daubenton, Another Histoire Naturelle." *Journal of the History of Biology* 39, no. 3 (2006): 457–91.

Loveland, Jeff. "Louis-Jean-Marie Daubenton and the *Encyclopédie.*" *Studies on Voltaire and the Eighteenth Century* 12 (2003): 173–219.

Lungren, Anders. "The New Chemistry in Sweden: The Debate That Wasn't." *Osiris* 4 (1988): 146–68.

MacGregor, Arthur. "The Cabinet of Curiosities in Seventeenth-Century Britain." In *The Origins of Museums: The Cabinet of Curiosities in Sixteenth and Seventeenth Century Europe*, edited by Oliver Impey and Arthur MacGregor, 147–58. Oxford: Clarendon Press, 1985.

Maddison, Robert. *The Life of the Honourable Robert Boyle, F.R.S.* London: Taylor & Francis, 1969.

Margócsy, Dániel. *Commercial Visions: Science, Trade, and Visual Culture in the Dutch Golden Age.* Chicago: University of Chicago Press, 2014.

Mattana, Alessio, and Giacomo Savani. "Introduction: The Antique and the Natural: Exploring the Eighteenth-Century Textual Network." *Journal for Eighteenth-Century Studies* 43, no. 4 (2020): 423–32.

Mauskopf, Seymour. "Crystals and Compounds: Molecular Structure and Composition in Nineteenth-Century French Science." *Transactions of the American Philosophical Society* 66, no. 3 (1976): 1–82

Mazzucato, Mariana. *The Value of Everything: Making and Taking in the Global Economy.* London: Penguin, 2018.

Meli, Domenico Bertoloni. *Thinking with Objects: The Transformation of Mechanics in the Seventeenth Century.* Baltimore, MD: Johns Hopkins University Press, 2006.

Metzger, Hélène. *La genèse de la science des cristaux.* Paris: Albert Blanchard, 1918.

Metzger, Hélène. "Une théorie curieuse de la double réfraction chez Buffon." *Bulletin de la société minéralogique* 37 (1914): 162–76.

Michaud, Louis-Gabriel, ed. *Biographie universelle ancienne et moderne.* 2nd ed. Paris, 1843. Originally published in 1811.

Mignolo, Walter D. "Introduction." In José de Acosta, *Natural and Moral History of the Indies*, edited by Jane E. Mangan, translated by Frances M. López-Morillas, xvii–xxviii. Durham, NC: Duke University Press, 2002.

Minard, Philippe. *La fortune du colbertisme: Etat et industrie dans la France des lumières.* Paris: Fayard, 1998.

Morel, Bernard. *Les joyaux de la couronne de France.* Paris: A. Michel Fonds Mercator, 1988.

Morton, Alan G. *History of Botanical Science: An Account of the Development of Botany from Ancient Times to the Present Day.* London: Academic Press, 1981.

Mottana, Annibale. "Galileo as Gemmologist: The First Attempt in Europe at Scientifically Testing Gemstones." *Journal of Gemmology* 34, no. 1 (2014): 24–31.

Mottana, Annibale. "Italian Gemology During the Renaissance: A Step Towards Modern Mineralogy." In *The Origins of Geology in Italy*, edited by Gian Battista Vai and W. G. E. Caldwell, 1–21. Boulder, CO: Geological Society of America, 2006.

Müller-Wille, Staffan. "Eighteenth-Century Classifications of Non-Living Nature." In *Spaces of Classification*, edited by Ursula Klein, 115–30. Berlin: Max Planck Institute for the History of Science, 2003.

Napolitani, Maddalena. "'Born with the Taste for Science and the Arts': The Science and the Aesthetics of Balthazar-Georges Sage's Mineralogy Collections, 1783–1825." *Centaurus* 60, no. 4 (2018): 238–56.

Newcomb, Sally. *The World in a Crucible: Laboratory Practice and Geological Theory at the Beginning of Geology*. Boulder, CO: Geological Society of America, 2009.

Newman, William R. "Alchemy, Assaying and Experiment." In *Instruments and Experimentation in the History of Chemistry*, edited by Frederic L. Holmes and Trevor H. Levere, 35–54. Cambridge, MA: MIT Press, 2000.

Newman, William R. *Atoms and Alchemy: Chymistry and the Experimental Origins of the Scientific Revolution*. Chicago: University of Chicago Press, 2006.

Newman, William R. "Mercury and Sulphur Among the High Medieval Alchemists: From Rāzī and Avicenna to Albertus Magnus and Pseudo-Roger Bacon." *Ambix* 61, no. 4 (2014): 327–44.

Nickel, Helmut. "The Graphic Sources for the Moor with the Emerald Cluster." *Metropolitan Museum Journal* 15 (1980): 203–10.

Nieto-Galan, Agustí. "Between Craft Routines and Academic Rules: Natural Dyestuffs and the 'Art' of Dyeing in the Eighteenth Century." In *Materials and Expertise in Early Modern Europe: Between Market and Laboratory*, edited by Ursula Klein and Emma Spary, 321–54. Chicago: University of Chicago Press, 2010.

Ogden, Jack. *Diamonds: An Early History of the King of Gems*. New Haven, CT: Yale University Press, 2018.

Ogilvie, Brian. *The Science of Describing: Natural History in Renaissance Europe*. Chicago: University of Chicago Press, 2006.

Oldroyd, David. "The Doctrine of Property-Conferring Principles in Chemistry: Origins and Antecedents." *Organon* 12–13 (1976–1977): 441–62.

Oldroyd, David. "Mineralogy and the 'Chemical Revolution.'" *Centaurus* 19, no. 1 (1975): 54–71.

Oldroyd, David. "A Note on the Status of A. F. Cronstedt's Simple Earths and His Analytical Methods." *Isis* 65, no. 4 (1974): 506–12.

Oldroyd, David. *Sciences of the Earth: Studies in the History of Mineralogy and Geology*. Brookfield, VT: Ashgate, 1998.

Oldroyd, David. "Some Eighteenth-Century Methods for the Chemical Analysis of Minerals." *Journal of Chemical Education* 50, no. 5 (1973): 337–40.

Oldroyd, David. "Some Phlogistic Mineralogical Schemes, Illustrative of the Evolution of the Concept of 'Earth' in the 17th and 18th Centuries." *Annals of Science* 31, no. 4 (1974): 269–305.

Oosterhoff, Richard J., José Ramón Marcaida, and Alexander Marr, eds. *Ingenuity in the Making: Matter and Technique in Early Modern Europe*. Pittsburgh, PA: University of Pittsburgh Press, 2021.

Ostergard, Derek E., ed. *The Sèvres Porcelain Manufactory: Alexandre Brongniart and the Triumph of Art and Industry, 1800–1847*. New Haven, CN: Yale University Press, 1997.

Osterhammel, Jürgen. "Global History." In *Debating New Approaches to History*, edited by Marek Tamm and Peter Burke, 21–35. London: Bloomsbury, 2018.

Outram, Dorinda. "New Spaces for Natural History." In *Cultures of Natural History*, edited by Nicholas Jardine, James Secord, and Emma Spary, 249–65. Cambridge, UK: Cambridge University Press, 1996.

Pabst, Adolf. "Charles-Francois du Fay, a Pioneer in Crystal Optics." *American Mineralogist* 17 (1932): 569–72.

Parker, Harold T. "French Administrators and French Scientists During the Old Regime and the Early Years of the Revolution." In *Ideas in History: Essays Presented to Louis Gottschalk by His Former Students*, edited by Harold T. Parker and Richard Herr. Durham, NC: Duke University Press, 1965.

Partington, James R. *A History of Chemistry.* London: Macmillan, 1960–1964.

Pastoureau, Mireille. *Les atlas français XVIe–XVIIe siècles: Répertoire bibliographique et étude.* Paris: Bibliothèque nationale de France, 1984.

Pearce, G. L. *The Story of New Zealand Jade, Commonly Known as Greenstone.* Auckland: Collins, 1971.

Perkins, John, ed. "Sites of Chemistry in the Eighteenth Century." Special issue, *Ambix* 60, no. 2 (2013).

Piccolino, Marco. *The Taming of the Ray: Electric Fish Research in the Enlightenment, from John Walsh to Alessandro Volta.* Florence: Olschki, 2003.

Pickstone, John. "Thinking over Wine and Blood: Craft-Products, Foucault, and the Reconstruction of Enlightenment Knowledges." *Social Analysis: The International Journal of Social and Cultural Practice* 41, no. 1 (1997): 97–105.

Pickstone, John. "Ways of Knowing: Towards a Historical Sociology of Science, Technology and Medicine." *British Journal for the History of Science* 26, no. 4 (1993): 433–58.

Pointon, Marcia. *Rocks, Ice and Dirty Stones: Histories of Diamonds.* London: Reaktion Books, 2017.

Pollard, Mark. "Letters from China: A History of the Origins of the Chemical Analysis of Ceramics." *Ambix* 62, no. 1 (2015): 50–71.

Pomian, Krzysztof. *Collectors and Curiosities: Paris and Venice 1500–1800.* Cambridge, UK: Polity Press, 1990. Originally published in 1987.

Pomian, Krzysztof. "Les Wunderkammer entre trésor et collection particulière." In *La licorne et le bézoard: Une histoire des cabinets de curiosités*, edited by Krzysztof Pomian, 17–27. Montreuil: Gourcuff Gradenigo, 2013.

Porter, Roy, ed. *The Cambridge History of Science*, vol. 4: *Eighteenth Century Science.* Cambridge, UK: Cambridge University Press, 2003.

Porter, Theodore. "The Promotion of Mining and the Advancement of Science: The Chemical Revolution of Mineralogy." *Annals of Science* 38, no. 5 (1981): 543–70.

Poskett, James. *Horizons: The Global Origins of Modern Science.* London: Penguin, 2022.

Poskett, James, and Gianamar Giovannetti-Singh. "Global History of Science." In *Debating Contemporary Approaches to the History of Science*, edited by Lukas M. Verburgt, 22–25. London: Bloomsbury, 2024.

Poulot, Dominique. *Musée, nation, patrimoine, 1789–1815.* Paris: Gallimard, 1997.

Present, Pieter. "Petrus van Musschenbroek (1692–1761) and the Early Leiden Jar: A Discussion of the Neglected Manuscripts." *History of Science* 60, no. 1 (2022): 103–29.

Press, Steven. *Blood and Diamonds: Germany's Imperial Ambitions in Africa.* Cambridge, MA: Harvard University Press, 2021.

Prince, Sue Ann, ed. *Of Elephants and Roses: French Natural History 1790–1830.* Philadelphia: American Philosophical Society, 2013.

Principe, Lawrence M. *The Transmutations of Chymistry: Wilhelm Homberg and the Académie Royale des Sciences.* Chicago: University of Chicago Press, 2020.

Principe, Lawrence M., and William Newman. *Alchemy Tried in the Fire: Starkey, Boyle, and the Fate of Helmontian Chymistry*. Chicago: University of Chicago Press, 2002.

Pugliano, Valentina. "Natural History in the Apothecary's Shop." In *Worlds of Natural History*, edited by Helen Curry, Nicholas Jardine, James Secord, and Emma Spary, 44–60. Cambridge, UK: Cambridge University Press, 2018.

Pugliano, Valentina. "Pharmacy, Testing, and the Language of Truth in Renaissance Italy." *Bulletin of the History of Medicine* 91, no. 2 (2017): 232–73.

"Pujoulx, Jean-Baptiste." In *Annuaire nécrologique [for 1821]*, edited by Alphonse-Jacques Mahul, 265–68. Paris, 1822.

Pumfrey, Stephen. "Who Did the Work? Experimental Philosophers and Public Demonstrators in Augustan England." *British Journal for the History of Science* 28, no. 2 (1995): 131–56.

Rankin, Alisha. *Wonder Drugs, Experiment, and the Battle for Authority in Renaissance Science*. Chicago: University of Chicago Press, 2021.

Ratcliff, Marc J. "Experimentation, Communication and Patronage: A Perspective on René-Antoine Ferchault de Réaumur (1683–1757)." *Biology of the Cell* 97, no. 4 (2012): 231–33.

Read, Peter G. *Gemmology*. 3rd ed. London: Elsevier, 2005. Originally published in 1991.

Rees, William. *Industry Before the Industrial Revolution*. Cardiff: University of Wales Press, 1968.

Reiche, Ina, Colette Vignaud, Bernard Champagnon, Gérard Panczer, Christian Brouder, Guillaume Morin, Vicente Armando Solé, Laurent Charlet, and Michel Menu. "From Mastodon Ivory to Gemstone: The Origin of Turquoise Color in Odontolite." *American Mineralogist* 86, no. 11–12 (2001): 1519–24.

Reid, John, and Matthew Rout. *Tribal Economies—Ngāi Tahu: An Examination of the Historic and Current Tītī and Pounamu Economic Frameworks*. Christchurch: Ngāi Tahu Research Centre, 2019.

Riddle, John M. "Introduction." In *Marbode of Rennes' (1035–1123) De lapidibus*, translated by C. W. King, 1–27. Wiesbaden: Steiner, 1977.

Riddle, John M. "Preface." In *Marbode of Rennes' (1035–1123) De lapidibus*, translated by C. W. King, ix–xii. Wiesbaden: Steiner, 1977.

Riello, Giorgio. "'Things Seen and Unseen': The Material Culture of Early Modern Inventories and Their Representation of Domestic Interiors." In *Early Modern Things: Objects and Their Histories, 1500–1800*, edited by Paula Findlen, 124–50. New York: Routledge, 2013.

Riello, Giorgio. "'With Great Pomp and Magnificence': Royal Gifts and the Embassies Between Siam and France in the Late Seventeenth Century." In *Global Gifts: The Material Culture of Diplomacy in Early Modern Eurasia*, edited by Zoltán Biedermann, Anne Gerritsen, and Giorgio Riello, 235–65. Cambridge, UK: Cambridge University Press, 2018.

Roberts, Lissa, and Simon Schaffer. "Preface." In *The Mindful Hand: Inquiry and Invention from the Late Renaissance to Early Industrialisation*, edited by Lissa Roberts, Simon Schaffer, and Peter Dear, xiii–xxvii. Amsterdam: Koninklijke Nederlandse Akademie van Wetenschappen, 2007.

Roberts, Lissa, Simon Schaffer, and Peter Dear, eds. *The Mindful Hand: Inquiry and Invention from the Late Renaissance to Early Industrialisation*. Amsterdam: Koninklijke Nederlandse Akademie van Wetenschappen, 2007.

Roberts, Lissa, and Joppe van Driel. "The Case of Coal." In Roberts and Werrett, *Compound Histories*, 57–84.

Roberts, Lissa, and Simon Werrett, eds. *Compound Histories: Materials, Governance and Production, 1760–1840*. Leiden: Brill, 2017.

Roger, Jacques. *Buffon: A Life in Natural History*. Translated by Sarah Lucille. Ithaca, NY: Cornell University Press, 1997. Originally published in 1989.

Roller, Duane, and Duane H. D. Roller. "The Development of the Concept of Electric Charge." In *Harvard Case Studies in Experimental Science*, vol. 2, edited by James B. Conant. Cambridge, MA: Harvard University Press, 1957.

Roos, Anna Marie. "Taking Newton on Tour: The Scientific Travels of Martin Folkes, 1733–1735." *British Journal for the History of Science* 50, no. 4 (2017): 569–601.

Rossi, Paolo. "'Parrem uno, e pur saremo dua': The Genesis and Fate of Benvenuto Cellini's Trattati." In *Benvenuto Cellini: Sculptor, Goldsmith, Writer*, edited by Margaret Gallucci and Paolo Rossi, 171–200. Cambridge, UK: Cambridge University Press, 2004.

Rudwick, Martin. *Bursting the Limits of Time: The Reconstruction of Geohistory in the Age of Revolution*. Chicago: University of Chicago Press, 2005.

Rudwick, Martin. *The Meaning of Fossils: Episodes in the History of Palaeontology*. 2nd ed. Chicago: University of Chicago Press, 1985. Originally published in 1972.

Sabel, Claire Conklin. "'Glass Worke': Precious Minerals and the Archives of the Early Modern Early Sciences." In *New Earth Histories: Geo-Cosmologies and the Making of the Modern World*, edited by Alison Bashford, Emily M. Kern, and Adam Bobbette, 145–62. Chicago: University of Chicago Press, 2023.

Sabel, Claire Conklin. "The Impact of European Trade with Southeast Asia on the Mineralogical Studies of Robert Boyle." In *Gems in the Early Modern World: Materials, Knowledge and Global Trade, 1450–1800*, edited by Michael Bycroft and Sven Dupré, 87–116. London: Palgrave Macmillan, 2019.

Sabel, Claire Conklin. "Rare Earth: Gemstones, Geohistory, and Commercial Geography, c. 1600–1750." PhD diss. University of Pennsylvania, 2024. ProQuest (31485274), with author's permission.

Salomon, Charlotte. "The Pocket Laboratory: The Blowpipe in Eighteenth-Century Swedish Chemistry." *Ambix* 66, no. 1 (2019): 1–22.

Samuel, Edgar. "Gems from the Orient: The Activities of Sir John Chardin (1643–1713) as a Diamond Importer and East India Merchant." *Proceedings of the Huguenot Society* 27, no. 3 (2000): 351–68.

Sargent, Rose-Mary. *The Diffident Naturalist: Robert Boyle and the Philosophy of Experiment*. Chicago: University of Chicago Press, 1995.

Sarton, George. *Introduction to the History of Science*. Vol. 1. Malabar, FL: Robert E. Krieger, 1975. Originally published in 1927.

Scarisbrick, Diana, and Benjamin Zucker. *Elihu Yale: Merchant, Collector and Patron*. New York: Thames & Hudson, 2014.

Schaffer, Simon. "Ceremonies of Measurement: Rethinking the World History of Science." *Annales* 70 (2015): 335–60.

Schaffer, Simon. "Golden Means: Assay Instruments and the Geography of Precision in the Guinea Trade." In *Instruments, Travel and Science: Itineraries of Precision from the Seventeenth to the Twentieth Century*, edited by H. Otto Sibum, Marie Noëlle Bourguet, and Christian Licoppe, 20–50. London: Routledge, 2003.

Schaffer, Simon. "Measuring Virtue: Eudiometry, Enlightenment, and Pneumatic Medicine." In *The Medical Enlightenment of the Eighteenth Century*, edited by Andrew Cunningham and Roger French, 281–318. Cambridge, UK: Cambridge University Press, 1990.

Schaffer, Simon. "Natural Philosophy and Public Spectacle in the Eighteenth Century." *History of Science* 21, no. 1 (1983): 1–43.

Schiebinger, Londa L. *Plants and Empire: Colonial Bioprospecting in the Atlantic World.* Cambridge, MA: Harvard University Press, 2004.

Schmidt, Benjamin. *Inventing Exoticism: Geography, Globalism, and Europe's Early Modern World.* Philadelphia: University of Pennsylvania Press, 2015.

Schnapper, Antoine. *Curieux du Grand Siècle: Collections et collectionneurs dans la France du XVIIe siècle: Oeuvres d'art.* Paris: Flammarion, 1994.

Schufle, Joseph A. "Torbern Bergman, Earth Scientist." *Chymia* 12 (1967): 58–97.

Schuh, Curtis P. *Mineralogy and Crystallography: An Annotated Biobibliography of Books Published 1469 Through 1919.* Tucson, Arizona, 2007.

Schuh, Curtis P. *Mineralogy and Crystallography: On the History of These Sciences from Beginnings Through 1919.* Tucson, Arizona, 2007.

Secord, James A. "Against Revolutions." *BJHS Themes* 9 (2024): 17–37.

Secord, James A. "Introduction." *British Journal for the History of Science* 26, no. 4 (1993): 387–89.

Secord, James A. "Inventing the Scientific Revolution." *Isis* 114, no. 1 (2023): 50–76.

Selin, Helaine, ed. *Encyclopaedia of the History of Science, Technology, and Medicine in Non-Western Cultures,* 3rd ed. Dordrecht: Springer, 2016.

Shang-i, Liu, Richard W. Hughes, Zhou Zhengyu, and Kaylan Khourie. *Broken Bangle: The Blunder-Besmirched History of Jade Nomenclature.* Bangkok: Lotus Gemology, 2024.

Shank, J. B. "After the Scientific Revolution: Thinking Globally About the Histories of the Modern Sciences." *Journal of Early Modern History* 21, no. 5 (2017): 377–93.

Shapin, Steven. "Pump and Circumstance: Robert Boyle's Literary Technology." *Social Studies of Science* 14, no. 4 (1984): 481–520.

Shapin, Steven. "The Sciences of Subjectivity." *Social Studies of Science* 42, no. 2 (2012): 170–84.

Shapin, Steven. *A Social History of Truth: Civility and Science in Seventeenth-Century England.* Chicago: University of Chicago Press, 1994.

Shapin, Steven, and Simon Schaffer. *Leviathan and the Air Pump: Hobbes, Boyle, and the Experimental Life.* 2nd ed. Princeton, NJ: Princeton University Press, 2011. Originally published in 1985.

Shapiro, Alan E. "The Gradual Acceptance of Newton's Theory of Light and Color, 1672–1727." *Perspectives on Science* 4 (1996): 59–140.

Siegfried, Robert. "From Elements to Atoms: A History of Chemical Composition." *Transactions of the American Philosophical Society* 92, no. 4 (2002): 1–278.

Simon, Jonathan. *Chemistry, Pharmacy and Revolution in France, 1777–1945.* Aldershot: Ashgate, 2005.

Simon, Jonathan. "Mineralogy and Mineral Collections in 18th-Century France." *Endeavour* 26, no. 4 (2002): 132–36.

Simon, Jonathan. "Taste, Order and Aesthetics in Eighteenth-Century Mineral Collections." In *From Private to Public: Natural Collections and Museums,* edited by Marco Beretta, 97–112. Sagamore Beach, MA: Science History Publications, 2005.

Simon, Jonathan. "The Values of the Mineral Kingdom and the French Republic." In *Ordering the World in the Eighteenth Century,* edited by Donald Diana and Frank O'Gorman, 163–89. Basingstoke: Palgrave Macmillan, 2006.

Sinkankas, John. *Gemology: An Annotated Bibliography.* Metuchen, NJ: Scarecrow Press, 1993.

Sloan, Phillip R. "Natural History, 1670–1802." In *Companion to the History of Modern Science,* edited by Robert C. Olby, Geoffrey Cantor, Jonathan Hodge, and John R. R. Christie, 295–313. London: Routledge, 1990.

Smith, Cyril S. "Introduction." In René-Antoine Ferchault de Réaumur, *Réaumur's Memoirs on Steel and Iron*, vii–xxxiv. Chicago: University of Chicago Press, 1956.

Smith, Cyril Stanley, and R. J. Forbes. "Metallurgy and Assaying." In *A History of Technology*, vol. 3, edited by Charles J. Singer, 27–71. Oxford: Clarendon Press, 1954–1978.

Smith, Justin E. H., and James Delbourgo. "In Kind: Species of Exchange in Early Modern Science." *Annals of Science* 70, no. 3 (2013): 299–304.

Smith, Pamela. *The Body of the Artisan: Art and Experience in the Scientific Revolution*. Chicago: University of Chicago Press, 2004.

Smith, Pamela, ed. *Entangled Itineraries: Materials, Practices, and Knowledges Across Eurasia*. Pittsburgh, PA: University of Pittsburgh Press, 2019.

Smith, Pamela. *From Lived Experience to the Written Word: Reconstructing Practical Knowledge in the Early Modern World*. Chicago: University of Chicago Press, 2022.

Smith, Pamela. "Making as Knowing: Craft as Natural Philosophy." In *Ways of Making and Knowing: The Material Culture of Empirical Knowledge*, edited by Pamela Smith, Amy R. W. Meyers, and Harold J. Cook, 17–47. Ann Arbor: University of Michigan Press, 2014.

Smith, Pamela. "Nodes of Convergence, Material Complexes, and Entangled Itineraries." In Smith, *Entangled Itineraries*, 5–24.

Smith, Pamela, and Paula Findlen, eds. *Merchants and Marvels: Commerce and the Representation of Nature in Early Modern Europe*. Hoboken, NJ: Routledge, 2001.

Smith, Pamela, Amy R. W. Meyers, and Harold J. Cook, eds. *Ways of Making and Knowing: The Material Culture of Empirical Knowledge*. Ann Arbor: University of Michigan Press, 2014.

Smith, Pamela, and Benjamin Schmidt. "Knowledge and Its Making." In *Making Knowledge in Early Modern Europe: Practices, Objects, and Texts, 1400–1800*, edited by Pamela Smith and Benjamin Schmidt, 1–16. Chicago: University of Chicago Press, 2007.

Soll, Jacob. *The Information Master: Jean-Baptiste Colbert's Secret State Intelligence System*. Ann Arbor: University of Michigan Press, 2011.

Spary, Emma. *Eating the Enlightenment: Food and the Sciences in Paris*. Chicago: University of Chicago Press, 2012.

Spary, Emma. *Feeding France: New Sciences of Food, 1760–1815*. Cambridge, UK: Cambridge University Press, 2020.

Spary, Emma. "Forging Nature at the Republican Muséum." In *The Faces of Nature in Enlightenment Europe*, edited by Lorraine Daston and Gianna Pomata, 163–80. Berlin: BWV-Berliner Wissenschafts-Verlag, 2003.

Spary, Emma. "Scientific Symmetries." *History of Science* 42 (2004): 1–46.

Spary, Emma. *Utopia's Garden: French Natural History from Old Regime to Revolution*. Chicago: University of Chicago Press, 2000.

Spencer St. Clair, Charles. "The Classification of Minerals: Some Representative Mineral Schemes from Agricola to Werner." PhD diss. University of Oklahoma, 1966.

Spiesser, Michel. "Nicolas Louis Vauquelin—La découverte de deux nouveaux éléments: Le chrome (1797) et le glucinium (béryllium 1798)." *Bulletin de l'Union des physiciens* 10, no. 807 (1998): 1403–16.

Stanziani, Alessandro. *La qualité des produits en France: XVIII–XXe siècles*. Paris: Belin, 2003.

Steinle, Friedrich. "Entering New Fields: Exploratory Uses of Experimentation." *Philosophy of Science* 64 (1997): S65–S74.

Steinle, Friedrich. "Exploratives experimentieren: Charles Dufay und die Entdeckung der zwei Elektrizitäten." *Physik Journal* 3, no. 6 (2004): 47–52.

Steinle, Friedrich. "Wissen, Technik, Macht: Elektrizität Im 18. Jahrhundert." In *Macht des Wissens: Die Entstehung der modernen Wissensgesellschaft*, edited by Richard van Dülmen, Sina Rauschenbach, and Meinrad von Engelberg, 515–37. Cologne: Böhlau, 2004.

Stigler, Stephen. "Eight Centuries of Sampling Inspection: The Trial of the Pyx." *Journal of the American Statistical Association* 72, no. 359 (1977): 493–500.

Stolberg, Michael. *Uroscopy in Early Modern Europe*. Farnham: Ashgate, 2015.

Stroup, Alice. *A Company of Scientists: Botany, Patronage, and Community at the Seventeenth-Century Parisian Royal Academy of Sciences*. Berkeley: University of California Press, 1990.

Strunz, Hugo, and Ernest Nickel. *Strunz Mineralogical Tables: Chemical-Structural Mineral Classification System*, 9th ed. Stuttgart: Schweizerbart, 2001. Originally published in 1941.

Sturdy, David. *Science and Social Status: The Members of the Académie des Sciences, 1666–1750*. Woodbridge: Boydell Press, 1995.

Subrahmanyam, Sanjay. *Europe's India: Words, People, Empires, 1500–1800*. Cambridge, MA: Harvard University Press, 2017.

Sumner, James. *Brewing Science, Technology and Print, 1700–1880*. Pittsburgh, PA: University of Pittsburgh Press, 2016.

Szabadváry, Ferenc. *History of Analytical Chemistry*. International Series of Monographs in Analytical Chemistry. Oxford: Pergamon Press, 1966. Originally published in 1960.

Taton, René. "Brisson, Mathurin-Jacques." In *DSB*, vol. 2, 473–75.

Taton, René, and Yves Laissus, eds. *Enseignement et diffusion des sciences en France au dix-huitième siècle*, 2nd ed. Paris: Hermann, 1986. Originally published in 1964.

Taylor, Kenneth. "Lamétherie, Jean-Claude de." In *DSB*, vol. 7, 602–4.

Terrall, Mary. *Catching Nature in the Act: Réaumur and the Practice of Natural History in the Eighteenth Century*. Chicago: University of Chicago Press, 2014.

Terrall, Mary. "Handling Objects in Natural History Collections." In *The Material Cultures of Enlightenment Arts and Sciences*, edited by Adriana Craciun and Simon Schaffer, 15–33. London: Palgrave Macmillan, 2016.

Thorndike, Lynn. *A History of Magic and Experimental Science*. Vol. 6. New York: Macmillan, 1941.

Tīfāshī, Aḥmad ibn Yūsuf al. *Arab Roots of Gemology: Ahmad ibn Yusuf al Tifaschi's Best Thoughts on the Best of Stones*. Edited and translated by Samar Najm Abul Huda. London: Scarecrow Press, 1998.

Tillander, Herbert. "The Carat Weight." *Journal of Gemmology* 17, no. 8 (1981): 619–23.

Tillander, Herbert. *Diamond Cuts in Historic Jewelry: 1381–1910*. London: Art Books, 1995.

Timoshenko, Stephen. *History of Strength of Materials: With a Brief Account of the History of Theory of Elasticity and Theory of Structures*. London: McGraw-Hill, 1953. Originally published in 1930.

Tinguely, Frédéric. "Introduction." In François Bernier, *Libertin dans l'Inde moghole*, 7–34. Paris: Chandeigne, 2008.

Todericiu, Doru. "Balthasar-Georges Sage (1740–1824), chimiste et minéralogiste français, fondateur de la première École des mines (1783)." *Revue d'histoire des sciences* 37, no. 1 (1984): 29–46.

Tomory, Leslie. "Trade and Industry: An Era of New Chemical Industries." In *A Cultural History of Chemistry in the Eighteenth Century*, edited by Matthew Eddy and Ursula Klein, 137–56. London: Bloomsbury, 2022.

Torlais, Jean. "Chronologie de la vie et des oeuvres de René-Antoine Ferchault de Réaumur." *Revue d'histoire des sciences* 11, no. 1 (1958): 1–12.

Torlais, Jean. *Un esprit encyclopédique en dehors de L'Encyclopédie: Réaumur, d'après des documents inédits.* Paris: Albert Blanchard, 1961. Originally published in 1935.

Torlais, Jean. "La physique expérimentale." In *Enseignement et diffusion des sciences en France au dix-huitième siècle,* 2nd ed., edited by René Taton and Yves Laissus, 619–77. Paris: Hermann, 1986.

Trivellato, Francesca. *The Familiarity of Strangers: The Sephardic Diaspora, Livorno, and Cross-Cultural Trade in the Early Modern Period.* New Haven, CT: Yale University Press, 2009.

Ullmann, Manfred. *Die Natur- und Geheimwissenschaften im Islam.* Leiden: E. J. Brill, 1972.

Valence, Françoise de. "Introduction." In Jean Thévenot, *Les voyages aux Indes orientales,* 7–23. Paris: H. Champion, 2008.

Valette, Guillaume. "La vie de Vauquelin." *Revue d'histoire de la pharmacie* 51, no. 177 (1963): 89–96.

Vanneste, Tijl. *Blood, Sweat and Earth: The Struggle for Control over the World's Diamonds Throughout History.* London: Reaktion Books, 2021.

Vanneste, Tijl. *Global Trade and Commercial Networks: Eighteenth-Century Diamond Merchants.* London: Pickering and Chatto, 2011.

Vatar-Jouannet, François-René-Bénit. *Notice historique sur Cyprien-Prosper Brard, ingénieur civil des mines.* Périgueux: Dupont, 1839.

Vautrin, Guy. *Histoire de la vulgarisation scientifique avant 1900.* Les Ulis, France: EDP Sciences, 2018.

Vœlke-Viscardi, Géraldine. "Les gemmes dans l'*Histoire Naturelle* de Pline l'Ancien: Discours et modes de fonctionnement de l'univers." *Museum Helveticum* 58, no. 2 (2001): 99–122.

Wallis, Patrick, and Catherine Wright. "Evidence, Artisan Experience, and Authority in Early Modern England." In *Ways of Making and Knowing: The Material Culture of Empirical Knowledge,* edited by Pamela Smith, Amy R. W. Meyers, and Harold J. Cook, 138–63. Ann Arbor: University of Michigan Press, 2014.

Warsh, Molly. *American Baroque: Pearls and the Nature of Empire, 1492–1700.* University of North Carolina Press, 2018.

Watson, Katherine D. *Forensic Medicine in Western Society: A History.* London: Routledge, 2010.

Werrett, Simon. *Fireworks: Pyrotechnic Arts and Sciences in European History.* Chicago: University of Chicago Press, 2010.

Werrett, Simon. *Thrifty Science: Making the Most of Materials in the History of Experiment.* Chicago: University of Chicago Press, 2019.

Westfall, Richard. *Never at Rest: A Biography of Isaac Newton.* Cambridge, UK: Cambridge University Press, 1980.

Wiesner-Hanks, Merry E. *Early Modern Europe, 1450–1789.* 3rd ed. Cambridge, UK: Cambridge University Press, 2023. Originally published in 2006.

Wilson, Wendell E. "Fabien Gautier d'Agoty and His *Histoire Naturelle Regne Mineral* (1781)." *Mineralogical Record* 26, no. 4 (1995): 65–76.

Wilson, Wendell E. *The History of Mineral Collecting, 1530–1799.* Tucson, AZ: The Mineralogical Record, 1994.

Wilson, Wendell E. "Nehemiah Grew's Musaeum Regalis Societatis." *Mineralogical Record* 22 (1991): 333–40.

Wootton, David. *The Invention of Science: A New History of the Scientific Revolution.* London: Penguin, 2015.

Wragge-Morley, Alexander. *Aesthetic Science: Representing Nature in the Royal Society of London, 1650–1720.* Chicago: University of Chicago Press, 2020.

Wyckoff, Dorothy. "Introduction." In *Albertus Magnus' Book of Minerals*, translated by Dorothy Wyckoff, xiii–xlii. Oxford: Clarendon Press, 1967.

Yeo, Richard. "Classifying the Sciences." In *History of Science*, vol. 4: *Eighteenth Century Science*, edited by Roy Porter, 241–66. Cambridge, UK: Cambridge University Press, 2003.

Yogev, Gedalia. *Diamonds and Coral: Anglo-Dutch Jews and Eighteenth-Century Trade*. Leicester: Leicester University Press, 1978.

Zylberman, Nicolas. "Boece de Boodt, dernier lapidaire et premier gemmologue." *Revue de l'Association Française de Gemmologie* 177 (2011): 17–22.

Index

Note: Footnotes are included in the index when they contain important information that is not signaled in the main text. For the names of places where gems occur naturally or were thought to do so, see *gem localities*. A page number followed by *f*, *t*, or *a* refers to a figure, table, or appendix, respectively. *Pl.* indicates a color plate. Titles of works cited are not listed in the index, except the *Encyclopédie* of Diderot and d'Alembert.